AMERICAN GEOGRAPHICAL SOCIETY
RESEARCH SERIES NO. 3
W. L. G. JOERG, *Editor*

BATTLEFIELDS OF THE WORLD WAR

Western and Southern Fronts

A Study in Military Geography

BY

DOUGLAS WILSON JOHNSON
Professor of Physiography in Columbia University
Formerly Major, Division of Military Intelligence, U. S. A.

WITH A FOREWORD
BY

GENERAL TASKER H. BLISS
Member of the Inter-Allied Supreme War Council

NEW YORK
OXFORD UNIVERSITY PRESS
AMERICAN BRANCH: 35 WEST 32ND STREET
LONDON, TORONTO, MELBOURNE, AND BOMBAY
1921

THE CONDÉ NAST PRESS
GREENWICH, CONN.

CONTENTS

LIST OF ILLUSTRATIONS

(m = map; bd = block diagram; d = diagram; p = photograph)

Battlefield of Flanders

Cf. also Pl. I (see p. xi)

Battlefield of the Somme

Cf. also Pl. II (see p. xi)

Battlefield of the Marne

Cf. also Pls. III and IX, A (*see p. xi*)

Battlefield of Verdun

BATTLEFIELD OF LORRAINE

BATTLEFIELD OF THE TRENTINO

BATTLEFIELD OF THE BALKANS

PLATES IN SEPARATE POCKET

Figs. 13 and 23 are reproduced with the permission of the Director of Military Intelligence of the War Office, London; Figs. 10, 16, 17, 18, 19, 20, 21, of the Direction des Informations Militaires, Ministère de *la Défense Nationale, Brussels; Figs. 31, 44, 48, 49, 70, 71, 72, 85, 90, 93, 97, 109, 113 and Pls. IX,*A *and IX,*B, *of the Chef de la Section Photographique de l'Armée, Paris; Figs. 127, 128, 129, 130, 131, 132, 133, 138, 139, 141, 143, 144, 145, 146, 147, 148, 149, 150, 154, 155, 158, 164 and Pls. X,*A, *X,*B, *XI,*A, *XI,*B, *of the Direzione del Servizio Fotografico, Comando Supremo, Regio Esercito Italiano, Rome.*

Pls. I to V inclusive are reproduced from the five-color Carte de France, 1:200.000, published by the Service Géographique de l'Armée, Paris. Each plate is made up of the parts of the sheets of this map necessary to cover a given battlefield. The sheets used are: 2 (Dunkerque-Bruges), 4 (Lille), 9 (Amiens), 10 (Mézières), 11 (Longwy), 16 (Paris), 17 (Châlons), 18 (Metz), 19 (Karlsruhe), 25 (Melun), 26 (Troyes), 27 (Nancy), 28 (Strasbourg), 35 (Vesoul), 36 (Zurich).

FOREWORD

Every informed General Staff officer understands, of course, the broad relations between topography and the strategic plans for a campaign or the tactical dispositions for a battle. These were known and applied as well by Alexander, Hannibal, Caesar as by any modern commander. Many of these relations are so simple and self-evident that, in the rudest ages of war, they have been instinctively noted and taken advantage of; and, where not, the resulting disaster may generally be traced to self-confidence or indifference to the plain teachings of Nature. Many a commander has unconsciously taken into his calculations certain basic conditions of geology, just as Molière's grocer spoke prose without knowing it.

But, in war, as between two equally skillful generals, at some all-important time in the game one of them has to play to the other's lead. It may be that then only one of them is favored by all the physical and seasonal conditions of the terrain. The other, most likely, can not defer his action to a more convenient and favorable time or place. His adversary's card is on the table; he must play his own then and there or forfeit the game. All Nature seems to be in alliance with his opponent. It is just then that his success—which may consist in preventing defeat quite as much as in gaining a patent victory—may depend upon a deeper knowledge of Nature and her processes. If the sun, the very stars in their courses, the tides of the sea, the running streams—everything on the surface of the earth and above it has marshaled its influence in favor of his enemy, then, though he may have to surrender topography to that enemy, he himself may have to burrow beneath it all and exert his strength amid the unseen, tumbled, and contorted ruins of old Chaos. Not only there, but on the surface too, he must know how to exert his propaganda on that Nature which seems to have entered into an unholy alliance against him, to play one of her laws or processes against another, to slay her ally with venom-tipped arrows skillfully stolen from her own quiver.

It is difficult to say whether this work of Professor Johnson has been written more for the benefit of the geographer and geologist, or of the military student, especially the student of the operations in the Great War, or of those who like to read charming descriptions in sweetly flowing English, of the physical landscapes in both their gay and gloomy moods—the plains of Flanders, the chalk uplands of Picardy and Artois, the pastures and vineyards of Champagne, the glens and forests of the Vosges, or the mountain valleys and tarns and peaks and precipices of the Alps and the Balkan ranges—which are so clearly portrayed that neither the scientist nor the military student nor the lover of Nature in literature need visit them in order to see and understand them. To all of them it will be a classic, and to none more than to another. But it is to the military student that I especially commend it in order that he may see, among other things, how the art of war has gathered to itself and absorbed not only all trades and arts, all the elder sciences, but has now laid hands on this youngest one of all, this youthful David of human knowledge, to help us to do the one thing that can ennoble our own art and science—to slay the evil giants of wrong and oppression.

TASKER H. BLISS

INTRODUCTION

The Problem Stated

"Do the mountains defend the army, or does the army defend the mountains?" The problem is an old one, and has claimed the attention of military authorities in all countries and in all times. Expressed in broader terms, it is the oft-debated question as to the relative influence of topography upon strategy and tactics under changing conditions of warfare. It is an ever-recurring question, for each "revolution" in methods of combat brings in its train a body of opinion intent on demonstrating that, under the new conditions of fighting, topographic obstacles have lost their significance, strategic gateways no longer exist, and commanding positions no longer "command." Then, as opposing forces share in the new discoveries, or profit in equal measure by new systems, the fundamental importance of topography reasserts itself, and each side maneuvers for an advantageous position on the terrain as one of the prerequisites to victory in battle.

The question is still a live one. The warfare of today employs a variety of inventions and technical devices, each of which may appear to reduce, if not to destroy, the influence of topography upon military operations. What protection is a river channel, when the modern military engineer can throw bridges across it in a few hours, defended by artillery which can reach the enemy many miles beyond the farther bank? What need has the artillery for hill positions, when guns are now commonly concealed in valleys and ravines, firing with marvelous accuracy upon objectives the gunners never see? With sound-ranging and flash-ranging devices to spot enemy batteries, with airplanes and aërial photography to locate these and other objectives and to exercise surveillance over enemy movements, of what significance is a paltry elevation of some few tens or hundreds of feet, dignified in earlier wars as "dominating heights?" So might one multiply, indefinitely, queries the common answer to which would seem to be that in the warfare of the present time the inventions of Man have reduced to insignificance the rôle of Nature.

Strategic Frontiers

If this be true, the matter is one of far-reaching importance, not only to the military leader but also to the statesman. The problem of strategic frontiers and the question of the influence of topography on warfare are inevitably linked the one with the other. If the character of the terrain is of no consequence in modern fighting, much of the argument for strategic frontiers falls to the ground. On the other hand, if a frontier may be strong or weak, impregnable or vulnerable, according to the nature of the terrain which it traverses, then the statesman must weigh this aspect of the territorial problems which come before him for adjustment. Whether he believes that strategic claims are but cloaks covering a multitude of imperialistic sins, or holds the view that strong frontiers may be locks discouraging international burglary and hence aids to maintaining world peace, he must take cognizance of strategic frontiers if they really exist. Others will demand them, and the wise statesman will bring to the consideration of such demands a full knowledge of the strength and weakness of the supporting arguments.

The Value of Terrain

One object of the present volume is to demonstrate the fallacy of the contention that modern methods of warfare have reduced to insignificance the rôle of the terrain as a factor in strategy and tactics. This demonstration might be made by abstract argument: by showing that, despite the enormous improvement in the artillery and other arms of the service, it is still the infantry which must drive back the enemy and conquer the ground on which he stands, and that whatever affects the movement of infantry remains a vital element in the fighting; by pointing out that one of the most effective agents in breaking through wire-defended trench systems, the tank, finds in rivers and marshes more serious obstacles than such topographic barriers ever constituted for lighter and more mobile weapons; by proving that direct and uninterrupted observation of enemy back areas from concealed positions on topographic eminences held constantly in easy communication with the artillery is, for the speedy and accurate control of gun fire, greatly superior to observation from airplanes, because such observation is often interrupted by weather conditions, is limited in some measure as to its efficiency by difficulties of communication, and is reduced in value by the necessity of flying high to avoid enemy bombardments or by the exigencies of aërial combats with enemy planes.

Or the demonstration might rest more heavily upon the expert testimony of those responsible for the gigantic military operations of the World War: upon the opinion of De Castelnau, proved by events to be correct, that the peculiar terrain of the Nancy region would enable limited forces to defeat a numerically superior enemy in a modern battle;[1] upon Haig's dictum that the whole war might be considered a series of struggles for topographic position;[2] upon Ludendorff's statement that by 1916, after the Battles of Verdun and the Somme, "the decisive value of the artillery observation and the consequent necessity of paying great attention to the selection of positions had also become apparent;" upon his testimony, a year later, that "another tactical detail which was emphasized everywhere was the value of ground observation for artillery. Only by that means could the attacking hostile infantry be annihilated, particularly after penetrating our front, or fire be concentrated on decisive points of the battlefield;" and, finally, upon his assertion in 1918, despite the fact that loss of strong defensive positions had by that time weakened his faith in the value of commanding ground in defense, that "in the attack in the war of movement the capture of some high ground brought about the tactical decision. Its possession must therefore be striven for as a matter of principle."[3]

[1] See Chapter X.
[2] Personal communication.
[3] Erich von Ludendorff: Ludendorff's Own Story, August 1914–November, 1918: The Great War from the Siege of Liège to the Signing of the Armistice As Viewed from the Grand Headquarters of the German Army, 2 vols., New York, 1919; references in Vol. 1, p. 324; Vol. 2, pp. 103, 200.

The writer has preferred, however, to let the demonstration rest upon a simple presentation of the record. No one who knows the terrain of the principal battlefields of the World War, and who follows carefully the operations of the contending armies, can doubt that under modern conditions of warfare the terrain is, more than ever before in military history, an important factor in strategy and tactics. This is not to imply that armies should put their whole trust in topographic barriers or commit the error of seeking victory by standing permanently on the defensive upon some strong natural position. The folly of such tactics was repeatedly demonstrated by the Germanic generals of Napoleon's time, and this extreme view of the value of "commanding ground" has been justly condemned by Clausewitz and other writers on the art of war. Nature offers no absolutely impregnable positions. In the words of Foch, "All terrains may be crossed by the enemy unless defended by rifle shots, that is to say, by active and valiant men."[4] It is rather to support the view of Napoleon that strong natural positions are one of the factors which enable an army to maneuver advantageously in the face of a superior or equal enemy; the view of Clausewitz that even the larger army must not despise a source of strength which enables it to hold the enemy at bay with limited forces along one part of its front while it concentrates its major effort elsewhere; and the view of the modern general who finds in the better observation and artillery control offered by a favorable terrain elements which often prove decisive in both defensive and offensive battles.

To the fullest appreciation of the important influence of topography on modern warfare two things are fundamentally essential, two things which unfortunately have not been easily obtainable. The first is a reasonably accurate mental image of each battlefield, a picture of those salient features of the terrain which might be expected to influence the dispositions and movements of armies. The second is a record of the military operations on each battlefield presented, not in the usual terms of advances or retreats to this or that line of villages and towns, but in terms

[4] Ferdinand Foch: Des principes de la guerre, Paris, 1917, p. 29.

of the plateau scarps, ridge crests, valley trenches, marsh barriers, and other topographic features which obviously affected the fighting and its results.

Scope of the Present Work

In the present volume the writer has endeavored to supply these two prerequisites to an understanding of the influence of topography upon the World War. For each important battlefield on the western and southern fronts there is presented, with the aid of maps, perspective diagrams, and photographs taken for the special purpose of showing essential elements of the terrain, such a picture of the stage upon which the drama of war was enacted as is hoped will prove both entertaining and helpful not only to the specialist, but equally to the reader who is neither military expert nor geographer. Throughout the descriptions an effort has been made to unite scientific accuracy of treatment with a phraseology as far as possible devoid of technicalities. If these descriptive chapters prove of service to the future historian of the war who would understand fully the events he chronicles, if they aid the military student in analyzing the operations of each battle and campaign, if they provide the geographer with new illustrations of the influence of topography on one phase of human activities, if they furnish the intelligent traveler with a new form of guide to European battlefields, and if they give to other readers a few hours of pleasure and of profit, the author's labors will be most abundantly rewarded.

For each battlefield described there is a companion chapter recording in proper sequence the major military operations which were carried out upon it. Here for the first time, so far as the author is aware, the reader will find the operations of each battlefield presented as a unit and in terms of the terrain by which the plans of campaign and the movements of armies were most profoundly affected. In these chapters the author has not attempted to write history but to offer a partial interpretation of the significance of certain events in history. Time has not permitted that investigation of original sources essential to an accurate and well-balanced chronicle of the events of the war, and the reader will doubtless discover errors both of omission and commission which would mar the value of the record, did it make any pretense at giving an adequate history of even the major events of the great conflict. It is believed, however, that no such errors will be found to diminish the usefulness of these chapters as a general résumé of the principal battles and campaigns on each sector of the western and southern fronts, as affected by the terrain upon which they occurred.

No treatment of the Rumanian and Russian fronts is included in the present text. The operations on those fronts, in so far as they materially influenced the main struggle, came to an end about the time the author published his "Topography and Strategy in the War." The reader who would complete his survey of the European battlefields with some account of the eastern terrains and the principal operations upon them, may consult that volume. Acknowledgments are due Messrs. Henry Holt and Company, the publishers of "Topography and Strategy in the War," for permission to incorporate in the present work the descriptive matter relating to the northern and central Balkans, as well as certain paragraphs concerning the Italian front and several text figures which appeared in the earlier work.

Basis of the Present Work

From the beginning of the war the author followed closely the military operations on large-scale topographic maps, aided by such special knowledge of the terrain as his previous travels and geographic investigations had given him. The results of these studies appeared from time to time in the publications of the American Geographical Society, and later in the book mentioned above. The cordial reception given these very general essays encouraged the belief, particularly after America entered the struggle, that a more critical and detailed study of the principal battlefields of the World War would give results of value, an idea warmly supported by the American Geographical Society.

An exceptional opportunity was presented when the author,

having previously been ordered by the Secretary of War to report to the Secretary of State for special service abroad, was directed by the latter to proceed to Europe "for the purpose of making special studies in military geography for the use of this Department [of State] in connection with the work being done at the direction of the President by Colonel E. M. House." Preparations were already being made for the Peace Conference which must follow the conclusion of hostilities, and it was considered essential that the American representatives should bring to the discussion of the strategic claims of different countries as full a knowledge as possible of the strategic and tactical value of land-forms under modern conditions of warfare. A part of this knowledge could best be gained by studies on the ground while operations were still in progress, and while the men directing them had all phases of their problems freshly in mind. The American Geographical Society was then housing the staff of the "Inquiry" assembled under Colonel House's direction, and co-operating in its investigations; and it made from its own funds a generous appropriation to assure the author every facility in prosecuting his European studies.

It was under these favorable conditions that the author, accompanied by one of his former students, Lieutenant S. H. Knight, professor of geology in the University of Wyoming, visited the Belgian, British, French, American, Italian, and Balkan fronts. The War Offices at London, Paris, and Rome generously placed at our command large-scale maps and relief models of the battlefields, reports on operations, and other material of value in the prosecution of the work; and also facilitated in every way our studies along the various fronts. As a rule the generals commanding on these fronts not only welcomed us at their messes, where opportunity was afforded to discuss problems of military geography with the most competent leaders in each given sector, but manifested a personal interest in the investigation, often tracing in detail upon maps or models the course of operations, and contributing from their wide experience suggestions of the highest value. A member of the staff familiar with the terrain, its defensive organization, and the operations upon it, was usually assigned to accompany us and to assist in our studies.

Upon the conclusion of hostilities the author was called upon to undertake geographical investigations in preparation for the Peace Conference which were only indirectly related to the study of European battlefields. But at the Conference he had an opportunity, first as Chief of the Division of Boundary Geography on the American delegation and as technical adviser meeting with various commissions, and later as a member of several of the international Territorial Commissions, to participate in extended discussions of the strategic and tactical value of the terrain along certain proposed frontiers.

If the author has thus enjoyed some unusual advantages in the prosecution of the studies upon which the present volume is based, it is but fair to add that other circumstances limited the extent to which these advantages could be utilized. The enemy was still in possession of large areas which are of necessity included in any adequate survey of the principal battlefields, and it has been possible to visit but a limited portion of his former holdings since the armistice. The time available for the battlefield studies was seriously curtailed by the necessity of carrying on other investigations for the "Inquiry," the results of which were needed in preparation for the Peace Conference. Before a beginning on this volume could be made, the assignment to service at the Peace Conference postponed the work for many months, a postponement which was prolonged by further service as an adviser on geographical questions to the Department of State after the author's return from Paris. Not until February, 1920, could the considerable task of preparing this volume be taken up, and then only in connection with regular university duties and under special conditions which made absolutely necessary the completion of the manuscript by early autumn. Thus have circumstances beyond the author's control limited unduly the time at his disposal, while lack of access to a large library during half the working period proved a further handicap. If the reader feels that inadequate consideration is given to certain works on the great war which have recently appeared, and discovers other deficiencies, of which the author is only too conscious, it is hoped he will agree that the omissions do not materially affect the main purpose of the volume, and that he will show such indulgence as the circumstances cited may seem to deserve.

Acknowledgments

It is evident that the conditions under which the author visited the battle fronts and prosecuted his studies must place him in an embarrassing position in acknowledging his indebtedness to the many who have assisted in this work. The number of those who rendered valuable aid is so large that merely to name them would be impracticable. It must suffice to express here my profound gratitude to all of them, and to record my sense of obligation to each and every one who in government bureaus or on the battlefields placed his materials, his expert knowledge, and his invaluable assistance at my service. A special measure of appreciation is due to Marshal Joffre, General de Castelnau, General Gouraud, and General Bourgeois of the French Army; Field Marshal Sir Douglas Haig, General Rawlinson, and General Biddulph of the British Army; General Diaz, General Badoglio, and General Ferrero of the Italian Army; and General Henrys and General Milne of the French and British forces in the Army of the Orient, for critical analyses of certain military operations on different terrains, for valuable suggestions and criticisms of the problem under investigation, for providing unusual facilities for prosecuting the work, or for personal discussions and explanations in the field. To all of these the author desires to acknowledge his sense of obligation, without in the slightest degree making them responsible for any statements of fact or any interpretations which appear in the following pages. To Colonel Delfino De Ambrosis, military geographer on the Italian General Staff, for invaluable assistance and advice at the War Office at Rome, at the Military Geographical Institute at Florence, and along the Italian front; to the Italian geographer, Colonel Filippo De Filippi, and his gallant brother, Captain De Filippi of the

Royal Italian Navy, who gave his life to save others when his destroyer was sunk in the Adriatic, for manifold services at London, Paris, Rome, and in Albania; to Lieutenant Colonel T. Edgeworth David, Chief Geologist of the British Expeditionary Force, for important data on the geological conditions encountered on the Somme front and in Flanders; to Captain Alan G. Ogilvie, geographer with the British forces in the Balkans, for valuable criticisms of the text describing the Balkan front; and to my assistant and companion in the field, Lieutenant S. H. Knight, for efficient aid at all times, my cordial thanks are gladly given. It is a special pleasure to express my gratitude to my former chiefs, General Marlborough Churchill and Colonel R. H. Van Deman, for their unfailing and generous assistance throughout my service under them, given both directly and through their representatives at London, Paris, and Rome.

In the descriptions of the terrain the author has profited much from the published works of French students, especially the series of admirable geographic monographs which includes Raoul Blanchard's "La Flandre," Albert Demangeon's "La Picardie," Émile Chantriot's "La Champagne," and the smaller work of Bertrand Auerbach entitled "Le plateau lorrain." Frequent citations of these works will be found in the text, but it is only fair to state that in endeavoring to paint for the reader a picture of each battlefield the author has, consciously and doubtless also unconsciously, drawn upon the works in question to an extent which cannot fully be indicated by specific citations. His own mental image of the battlefields is due to the excellent descriptions of his French colleagues, as well as to his personal studies on the ground; and if the attempt to transmit this image to the reader has been successful, no small share of the credit must go to the distinguished Frenchmen who are well-known authorities on the geography of their country. For detailed illustrations of the influence of different topographic elements upon certain phases of the fighting, Arthur Conan Doyle's volumes on the British campaigns in France and Flanders have repeatedly been laid under tribute, because they present in a telling manner the reaction of individual units to the difficulties which opposed them.

For those photographic illustrations which are not from the author's own negatives, acknowledgments are due to the Photographic Services of the French, British, Italian, and Belgian armies, as indicated under each reproduction in the text. The block diagrams, with one exception, were prepared under the author's general direction by three of his former students: Professor F. K. Morris, of Pei Yang University, Tientsin, China; Professor A. K. Lobeck, of the University of Wisconsin; and Professor S. H. Knight of the University of Wyoming. As the author has modified these drawings to meet special needs discovered during the preparation of the text, it is but just for him to credit their excellence to his three colleagues, without making them responsible for the precise form in which the drawings now appear. In drawing the generalized sketch maps of the battlefields the author has adopted the simple and effective method employed by Professor W. M. Davis in his "Handbook of Northern France," and in some cases has utilized portions of his maps, modified to meet the needs of the present volume.

To Miss Ellen Churchill Semple thanks are due for permission to use certain data assembled by her relating to the passes and routes of the Alps. The Library of Columbia University has extended exceptional courtesies both in the matter of placing large numbers of books, maps, and other documents at the author's disposal while he was absent from the city, and in providing expert assistance in tracing records and verifying data essential to the work. Special acknowledgments are made to Miss I. G. Mudge, Reference Librarian, for invaluable assistance throughout the preparation of the manuscript. A very heavy debt of gratitude is owed my colleagues in the Department of Geology at Columbia University, whose generous sympathy and fraternal assistance have alone made it possible to complete the work in the allotted time.

To the American Geographical Society, and to its Director, Dr. Isaiah Bowman, it is a pleasure to record the author's hearty appreciation of the warm support they have given his undertaking from its very inception. Not only has the financial support been prompt and generous, but in such matters as the providing of needed secretarial assistance, the preparation of maps and diagrams, and the meeting of other needs which arose in the course of the work, the author's every request has met an immediate and cordial response from the Society and its officers. Many improvements in the text have resulted from the valuable suggestions of the Society's Research Editor, Mr. W. L. G. Joerg, to whose untiring co-operation the author is indebted for much aid which lay far beyond the usual limits of editorial assistance.

Finally, it is the author's privilege to express his thanks to General Tasker H. Bliss, who generously undertook to read the proof sheets of this volume, and who brought to the task an unequaled breadth of knowledge based on his long and distinguished military career, his membership on the Inter-Allied Supreme War Council and on the Armistice Commission, and his service as one of the five American Commissioners to Negotiate Peace. A similar friendly service was performed by General Francis Vinton Greene, whose intimate knowledge of military history and well-known ability as an author and critic made his suggestions and criticisms extremely helpful. To both General Bliss and General Greene the author would acknowledge a very great debt of gratitude. Many improvements in the text have resulted from their generous assistance; but on the author alone rests the responsibility for statements of fact and expressions of views which the text sets forth.

DOUGLAS WILSON JOHNSON

Columbia University
December 31, 1920

CHAPTER I

THE BATTLEFIELD OF FLANDERS: THE WET CLAY PLAIN BATTLEFIELD

One afternoon in September, 1918, a British staff automobile left corps headquarters "somewhere in Flanders," sped eastward over good roads for a few miles, then plowed into the ruts and mud holes of a newly recaptured portion of His Majesty King Albert's dominions. It was typical Flanders weather, for a drizzling rain was falling and low clouds or banks of fog drifted over the plain. The machine skidded into a slimy shell hole and half overturned. Three officers stepped out, belted their raincoats tighter about them, and slopped on through the sea of mud, into the wall of mist.

A quarter of an hour later you might have seen them toiling up the slope of a low hill along a sandy path winding picturesquely through a wood of conifers, each heavy step weighted down with a mass of clay which clung tenaciously to boot and spur and picked up sand, leaves, and twigs from the narrow trail. Muddy and bedraggled, the three crept in between the sand bag walls of a shelter near the crest and waited. The chill wind drove the lowering clouds about them, and through the gray curtain there came from the north, east, and south the rumble and crash of a great battle. Then, as if raised by magic hands, the fog curtain slowly lifted and parted. A flood of golden sunshine burst through, lighting up a vast green-carpeted plain on which rivers and ponds glittered like silver spangles. Stretching in a vast crescent across the stage thus revealed to the waiting observers was a line of flashing tongues of flame, a semicircle of steel and fire which, from the sea on the northwest to the uplands of Artois far away to the southward, was slowly blasting the

NOTE. For Chapters I and II the reader should constantly consult the detailed map of the battlefield in the pocket (Pl. I) and the block diagrams (Figs. 14 and 36).

Kaiser's hordes back into the Fatherland. It was the last great Battle of Flanders in full swing, and from the summit of Mont Kemmel our officers were with their glasses sweeping the vast field

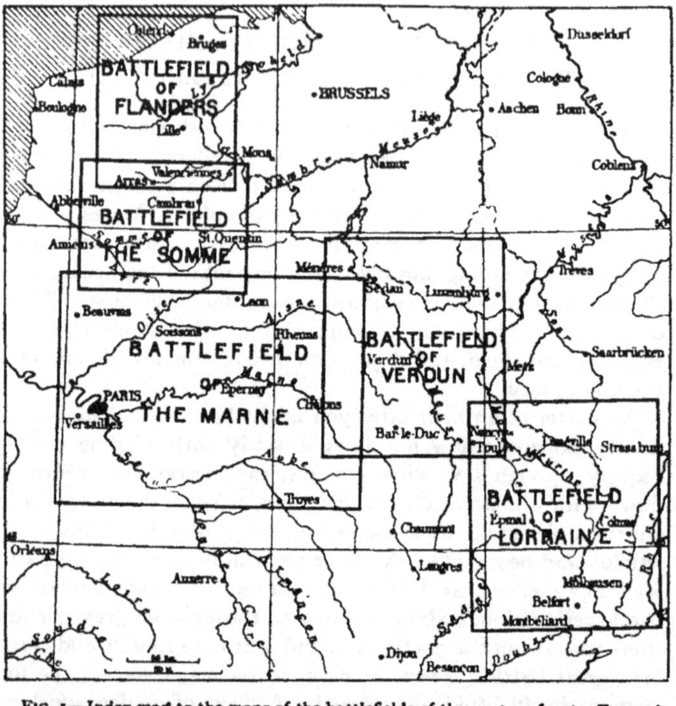

FIG. 1—Index map to the maps of the battlefields of the western front. For each rectangular area shown here there will be found in the text a generalized sketch map (Figs. 2, 24, 56, 79, 98), and in the pocket at the back of the volume a detailed large-scale map (Pls. I–V), both of which should be consulted while reading the two chapters devoted to each battlefield.

from end to end. For the low sandy hill, scarcely to be discovered when one stands among the trees and orchards of the plain, is none other than the famous "Mount" (Fig. 3), stained red with the blood of many thousands and standing in the very center of a battlefield which numbers its victims by the hundreds of thousands. Its summit rises barely 500 feet above the sea and only 350 feet over the adjacent plain; yet it is one of the dominating points of Flanders. In all the vast plain bounded on the northwest by the sea, on the south by the chalk uplands of the Somme battlefield and the foothills of the Ardennes, and extending eastward beyond Brussels and Antwerp, there are only two other points higher; and both of these are in the same range of hills as Mont Kemmel. Look westward and one may see them, if the trees are not too much in the way: Mont des Cats (Fig. 4), some 550 feet high, with the lesser heights of Mont Rouge, Mont Noir, and others lying between; and farther on, at the western end of the ridge, the king of the Flemish mountains, Mont Cassel, boasting all of 560 feet.

FIG. 2—Generalized sketch map of the Battlefield of Flanders. Ruled areas are uplands; the dotted area is a coal field. For the topographic details and place names referred to in the text see Pl. I and Figs. 14 and 36.

But one does ill to scoff at these tiny mounds, which over-enthusiastic patriots have compared with the Alps. Each summit commands a view of which a real mountain would have no cause to be ashamed: in the foreground a charming landscape of green meadows bordered by rows of trees and dotted with red-tiled farmhouses; here and there a shady grove or picturesque Flemish village; far away to the northwest the green of trees and meadows, bounded by a thin band of white, the sand dunes along the coast. The eye sweeps the vast plain (Fig. 23) to the north, east, and south and is arrested only when it discovers to the southwest, barely emerging from the blue mist on the horizon, a low elevation which can be nothing else than the historic Vimy Ridge.

The view is broad, not because the hill is high but because the plain is so extremely flat. One is tempted to say "absolutely flat," and the exaggeration would be pardonable. Very few places

FIG. 3—Mont Kemmel, a strong point near the apex of the natural "Ypres bastion." Although of moderate elevation only, it offered commanding observation over a vast expanse of the plain; and hence its possession was bitterly contested in important battles. When the photograph was taken the Germans held the eastern (right-hand) portion of the hill, the British the main crest.

in Europe can show a surface so nearly level. Roads, railways, and canals cross the country in every direction, and usually in straight lines, for seldom is there even a modest elevation to be avoided. The whole extent of Flanders provides but a single railway tunnel, and a canal may run for many miles without a lock. The rivers seem to flow *on* the plain instead of cutting valleys *in* it, and hills are so rare that even the most insignificant attracts more attention than many alpine crags. Houses and trees are usually the most imposing features in the landscape. Surely if ever a region was fully entitled to the term plain, Flanders is that region.

Of the battlefields of the World War the rolling surface of the Somme region just to the south can best be compared with the Flanders plain. Yet how few the elements of correspondence, how many the points of contrast! Two different worlds seem to come together where the chalk of the Somme region meets the clay of Flanders. One is a rolling upland, the other a level lowland. The even-flowing, clear streams of the chalk country do

FIG. 4—Mont des Cats, in the middle distance, rising slightly above the general level of the plain of Flanders and serving as a natural observatory of the highest importance.

not in the least suggest the unruly rivers whose turbid waters repeatedly flood the fields and homes of the Flemish peasant. The boundless horizon of the treeless Somme plain is poorly counterfeited in the usual view on the plain of Flanders, where apparent vastness is nullified by the trees which limit one's vision. Nothing could be farther apart than the aridity of the chalk surface and the humidity of the water-soaked clay plain. The dreary stretches of open fields on the chalk, devoid of human habitation except where thickly clustered dwellings form a village, find no counterpart in the green, tree-bordered meadows and gardens among which are scattered the homes of the widely disseminated Flemish population (cf. Figs. 11 and 12). Surely the imagination can picture no two geographic regions belonging in the same class which are more dissimilar than the plain of Artois and Picardy and the plain of Flanders.

The real dissimilarities are so pronounced that they lead to belief in others which do not exist. Ask the peasant who knows both regions, whether it rains more in Artois or Flanders, and he will tell you the latter receives much more rain. Yet there is only one place in all Flanders where the rainfall is not less than the *lowest* rainfall on the Artois upland.[1] The smaller rainfall on the flat clay plain cannot quickly escape, and produces a far greater impression on man than does the heavier rainfall, which soon disappears into the fissured chalk. Again, ask the peasant whether the wet and dry seasons are the same in Artois and in Flanders, and he may answer that there is no dry season in Flanders; that it rains often and much, the year round. But the Flemish rivers rise in the summer and autumn, and a study of statistics demonstrates that, while the sky of this country is much of the time overcast and some rain falls on an average nearly every other day throughout the year, the greatest precipitation occurs in the summer and autumn months. The peasant remarks the many days of rain but does not measure the quantity; and the soldier is apt to be like him. He is more affected by the length of time he shivers in wet clothing and stands in wet trenches than by the amount of water that falls on him. Therefore he will tell you that the low belt of plain nearest the coast is the rainiest part of Flanders, when in reality it is the zone of least precipitation; and will yearn for the "less humid country" to the south, where nearly twice as much rain falls! The physical conditions of the Flanders battlefield responsible for such effects will soon claim our attention.

Strategic Importance of the Flanders Battlefield

To one standing on the Mont des Cats and looking out across the low plain of Flanders eastward toward Brussels, Antwerp, and the gateway of Liège, southward toward Arras and the

[1] Raoul Blanchard: La Flandre: Étude géographique de la plaine flamande en France, Belgique et Hollande, Paris, 1906, p. 27.

gateway to Paris, westward toward the sea and the Channel ports, the strategic significance of the region cannot be wholly lost. It forms part of that great belt of plain across which one may travel by rail from the Pyrenees to northeastern Russia without passing through a single tunnel and without rising 600 feet above the level of the sea. Across it lies the only path by which armies may advance into France without encountering formidable mountain or plateau barriers. But it is more than a pathway; it is one of the important meeting places of northern and southern Europe. English, Dutch, Danes, Scandinavians, and Germans from the north, Venetians, Genoese, Spaniards, and Frenchmen from the south, have throughout the centuries met on the Flanders plain in commercial intercourse or armed conflict. Whether one prefers the view that "Flanders occupies perhaps the most superb international situation which exists in Europe," or the more pessimistic conception that "when God had made this good Flanders he put it between all in order that it might be devoured by one after another," he cannot doubt the very great significance of its location. On its level surface have been fought out some of the greatest struggles of history: a few of them slowly and unimpressively, like the conflict between the Latin and Teutonic tongues; many of them in the rush and furor of battle, when the warriors of many nations clashed in arms and some of the world's famous generals made or lost their reputations in the mud and marshes of the plain. This battlefield of the World War has always been the battleground of Europe.

In 1914 the Flanders plain offered the German General Staff something more than the smoothest pathway between mountains and sea along which to launch its enveloping movement designed to crush the French armies in the space of a few weeks. The plain was provided with that abundant network of roads, railways, and canals which is the natural product of a dense population inhabiting a region of very little relief. Nowhere else could the Germans find such admirable facilities for transporting and supplying a great army. An intelligent agriculture had made the loamy parts of the plain highly productive, while the region was also noted for its rich pastures and its cattle. Here, then, was the food supply necessary for the support of unexampled concentrations of men. Once in undisputed possession of the plain, the Teuton armies could continue southward through the low gateway (p. 92) between the Ardennes Mountains and the upland of Artois, to outflank the French armies and capture Paris; or could strike farther west to gain the Channel ports, threaten British communications with the Continent, and bring to the very doors of England the menace of imminent invasion. The complete extinction of a country which had dared to antagonize the German colossus would be accomplished when all the Flanders plain was occupied, and a valuable lesson thereby given to other countries still neutral; while the political advantages to be derived from possession of Flanders at the end of the war would be enormous. Along the southern part of the Flanders plain lay the rich coal fields of northern France. About them had grown up densely populated industrial centers whose products were vital to the economic and military power of the Republic. To seize this region would strengthen Germany's fighting machine and weaken the opposition it would have to overcome. In German opinion the blow might well realize the aspiration of the Teutonic militarists, as expressed by Bernhardi, "to crush France so completely that she can never again come across our path."

But if the Flanders plain offered Germany high inducements as a route for the invasion of France, it imposed obstacles peculiarly formidable. Far more important than any topographic barrier was the neutrality of Belgium, the violation of which by Germany would call into action moral and material forces capable of exerting a decisive influence upon the issue of the conflict. Even the topographic obstacles were far more serious than they may have appeared; and as soon as Allied man power should be assembled for their defense they would prove impassable. Notwithstanding the fact that both the Central Powers and the Allies repeatedly launched major offensives on the Flanders plain, not one of them was successful after the first German onrush in 1914 before the natural barriers were adequately defended, until the Allied advance late in 1918 after the German armies had already been defeated elsewhere and compelled to begin their withdrawal to shorter lines. On the Somme plain, along the Chemin des Dames, farther south on the Marne plateau, and elsewhere, great drives succeeded; but on the Flanders plain all failed because of physical obstacles which deserve our careful consideration.

It is to the detailed topography of the Flanders plain that we must now turn our attention. Beginning thus with the lowest, flattest, and most monotonous of all the battlefields of the World War, we shall pass next to the slightly higher and more dissected plain of the Somme, then in orderly sequence to the trenched plateau of the Marne, the still higher plateaus of the Verdun country with their intervening lowland belts, the loftier domes and ridges of the Vosges, and finally to the towering peaks and precipices of the Alps and the complex mazes of the Balkan ranges.

Surface Features of the Flanders Battlefield

Look over the Flanders plain from the summit of Mont des Cats and you will see no evidence of system in the topography. Rivers and canals seem to run at will in every direction across the level surface, while the few low hills are scattered irregularly here and there. Now take a detailed topographic map of the region and search there for any sign of symmetry in the surface of the country which might escape the more limited vision of the unaided eye. Again all seems hopeless confusion. Rivers change to canals and back again to rivers, and wander aimlessly toward the ocean or directly away from it. Roads and railways make a confused network without suggesting any definite pattern imposed by the form of the land. Here and there an isolated hill or low ridge rises suddenly out of the vast expanse of plain. The ridge may trend east-west or north-south. And who shall say whether the Mont Cassel-Mont Kemmel ridge should be continued eastward to connect with the scattered hills beyond the valley of the Lys, or whether it bends sharply northward to form an integral part of the Messines-Passchendaele swelling? As for the forest patches, what could be more "hit-or-miss" than their distribution?

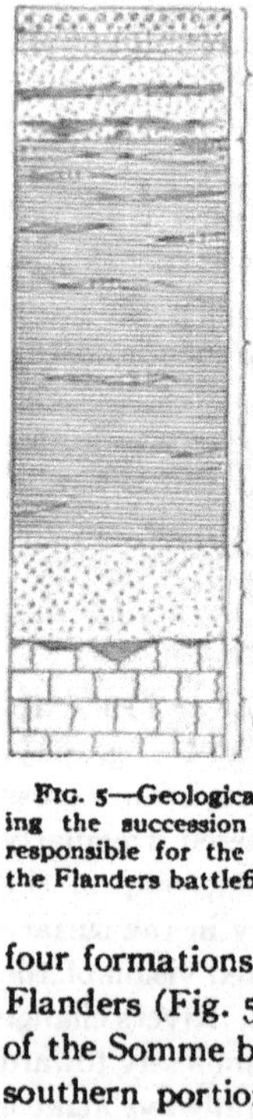

FIG. 5—Geological column showing the succession of formations responsible for the topography of the Flanders battlefield.

Nevertheless there is order in the apparent chaos. To see and appreciate it we need only know the very simple geological structure of the region. The rocks are the skeleton upon which Nature fashions the varied forms which we behold upon the earth's surface; and, just as the artist must begin by studying the human skeleton if he would end by painting aright the beauty of the human form, so must a knowledge of rock structure precede any understanding of the beauties of natural landscapes. Fortunately in this case the task is an easy one.

If we omit certain beds, of importance only beyond the limits of the region which particularly interests us, we need consider but four formations as responsible for the form of the Battlefield of Flanders (Fig. 5). At the base is the chalk, really characteristic of the Somme battlefield to the south and only appearing in the southern portion of the Flanders field to give a transition zone possessing some features common to both regions. Patches of clay of variable thickness lie in irregular depressions in its surface. Next above comes a series of argillaceous sands, sometimes partially consolidated to form sandstones, from 130 to 170 feet in total thickness. For sake of convenience we will call this "the lower sand formation." Overlying it is a great deposit of clay, in places nearly 600 feet thick, containing lenses of sand scattered throughout its mass. It is the formation which more than all others is responsible for the typical Flemish landscape. Let us call it simply "the clay formation." In its upper portion the clay formation gets more and more sandy until it merges with the overlying series of beds, "the upper sand formation." This consists for the most part of loose sands with layers of clay frequently interspersed, but toward the top the sands are partially consolidated into more resistant sandstone, while one layer is of conglomerate. The total thickness of the upper sand formation is not great in the central part of the region under discussion but reaches several hundred feet farther north. Geologists have subdivided these several formations into a much larger number of beds and given each a special name; but for our purposes the simpler division is sufficient.

Well records show that all four formations dip gradually downward toward the north or northeast (Fig. 6). Now it is the rule that, when a series of inclined beds is beveled across by erosion to form a plain, that plain is "belted," i. e. has parallel zones differing in surface form or character of soil, or both, according as different formations are exposed. Thus parts of our own Atlantic coastal plain have a cotton belt, a pine belt, and a rice belt, named from the products of the soil formed on three different rock types successively appearing at the surface. The Flanders plain is no exception to the rule, as will appear from Figures 6 and 7. Each of the four formations described above produces a characteristic surface zone, except that the chalk, disappearing and reappearing as a result of changes in the inclination of the beds, gives a mixed or transition zone toward the southeast, but does not show at all in the plain southwest of Hazebrouck (Fig. 7). The lower sand formation is represented by a narrow southern sand belt, the thick clay formation by a broad clay belt, and the upper sand formation by a northern sand belt.

Turn again to the map (Pl. I) and, remembering that sandy soils are apt to be poor and hence not so completely stripped of their forests, note the distribution of woodlands. It will now appear that there is a southern zone where woods are more numerous than usual, extending from near St. Omer to the vicinity of Orchies, southeast of Lille. These woods are either on the southern sand belt or near the southern edge of the clay belt where the clay is thinnest; except that about Orchies itself they are in the transition belt. A geological map of the Orchies region would show that there is here an oval basin in the chalk, with the clay formation in its center and the lower sand formation outcropping all around the sides, between the chalk and clay, as shown

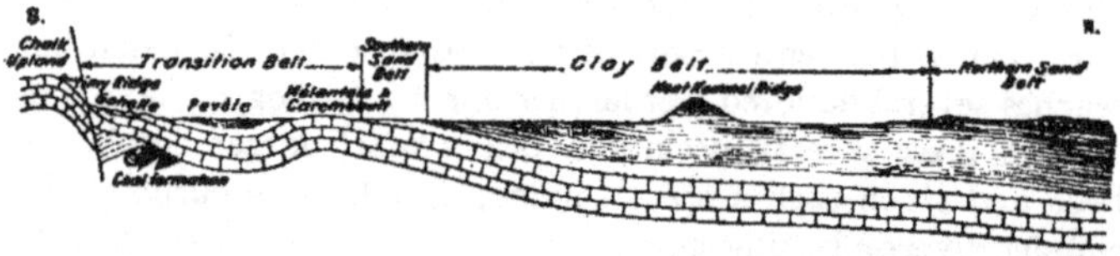

FIG. 6—Generalized north-south cross-section across the Battlefield of Flanders, showing the "belted" plain resulting from the northward inclination of the geological formations.

near the right side of Figure 7. The forests really describe an oval surrounding Orchies, following the pattern of the lower sand formation; but the eastern end of the oval lies beyond the limits of our map. North of Roulers there is a second zone of more abundant forested areas, corresponding to the northern sand belt. Between the two zones stretches a less forested region, in the midst of which the sand-capped ridge of Mont Cassel-Mont Kemmel is partly wooded. Assuredly there are other patches of forest the presence of which is to be explained on various grounds; but already there begins to appear some semblance of system in the surface features of the Flanders plain.

If the different formations of the plain have varying resistances to erosion, the dipping beds ought to produce, in addition to soil belts, corresponding belts of varying relief. Thus the dipping beds of the coastal plain of England give the parallel sloping uplands, or "cuestas", of the Chiltern Hills and the Cotswold Hills, separated by the Oxford lowland; and the inclined layers of the

Paris Basin form a remarkable series of parallel cuestas which exercised an important influence on the grand strategy of the war.[2] In both of these cases the rocks making the cuesta uplands were very resistant as compared with the weaker beds eroded into lowlands. When, as in the Flanders plain, there are no really resistant beds, but all are weak and worn down to a low surface, we can expect to find only a faint development of uplands, corresponding to such faint differences of resistance to erosion as may exist between unconsolidated clays and unconsolidated sands. When we study the map with this moderate expectation in mind, we are not disappointed. The Mont Cassel-Mont Kemmel ridge, capped by the upper sand formation in which there are some layers partially consolidated, appears to be the remnant of an east-west trending upland or cuesta which had its continuation in hills extending on eastward beyond the limits of the battlefield (Fig. 2).

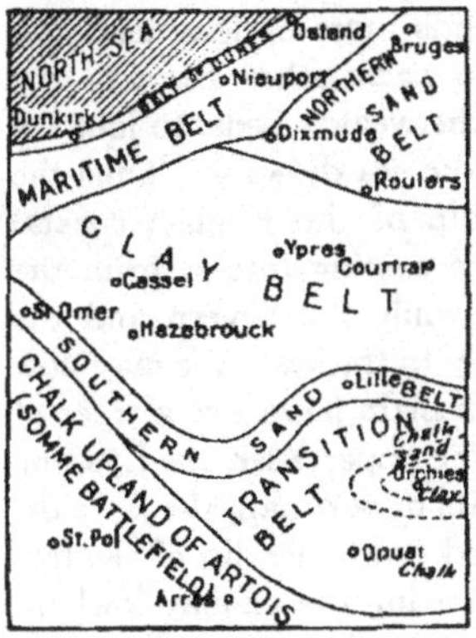

FIG. 7—Diagrammatic sketch map of the topographic belts of the Flanders battlefield.

From near Dixmude a second line of low hills—much lower, it is true, and more irregular in form, but having generally a steeper slope toward the south and a gently sloping upland toward the north—suggests by its typical asymmetrical form and by its position the remnants of a cuesta trending parallel to the Mont Cassel-Mont Kemmel ridge. Just north of it is a third cuesta-like elevation, also very low and inconspicuous but not wholly without significance. Both of these northern cuestas (Fig. 2) are located in the upper sand formation (Figs. 5 and 6). On the basis of this interpretation the Messines-Passchendaele ridge is part of the gentle backslope of the Mont Cassel-Mont

[2] D. W. Johnson: Topography and Strategy in the War, New York, 1917, pp. 1–49. See also Chapters V–X, below.

Kemmel cuesta, left uneroded between the headwaters of certain branches of the Lys River on the east and of the Yser River on the west (Fig. 2).

Where sloping uplands of the cuesta type are developed, the river system ordinarily shows a certain systematic relation to the cuestas and lowlands. Some rivers, often the principal ones, flow down the initial slope of the plain, which is down the dip of the beds, crossing cuestas and lowlands more or less nearly at right angles; others flow parallel to each other through the lowlands which they have eroded on the weaker beds, to join the streams of the first type or to enter the sea directly. Thus the Thames flows southeast down the dip of the English coastal plain beds, joined by smaller streams coming into it from the Oxford and other parallel lowlands, while the Severn and the Cam flow through the lowlands directly to the sea. We may now discover that the rivers of the Flemish plain have also a trace of this same systematic arrangement. The upper Yser, the Lys, the Deule past Lille, and, beyond the limits of our map, the Scheldt, Dendre, and others flow northeastward down the dip of the formations. The branch of the Yser following the marshy lowland from near Lichtervelde westward past Handzaeme to Dixmude lies in the lowland between the first and second cuestas; the stream which now forms the Bruges-Ghent canal follows the lowland between the second and third cuestas described above, and other short streams or canals show a similar orientation; while the lower Aa flows from St. Omer northwestward directly to the sea. If one shades heavilyon a map those parts of the rivers and canals trending from southwest to northeast down the dip of the rocks, or northwest-southeast (often nearly east-west) at right angles to the direction of dip, he will be surprised to note what a large proportion of the drainage system falls into one or the other of these two classes. The fact that the principal streams trend strongly toward the northeast, while the streams in the lowlands are usually not at right angles to them but more nearly east-west in their trend, has been explained by the fact that the uppermost beds, on which the principal rivers first took their course to the sea, had more of a northeastward slant than the lower beds, in which the branch rivers later cut their lowlands. However that may be, we need only remember that there is a system of northeast-flowing rivers and canals in Flanders, crossed by another system trending more or less nearly east-west, however much they may be obscured by other watercourses traversing the level plain in all possible directions.

After the four principal formations of the region had been beveled by erosion to form the four topographic belts already described, the northwestern part of the plain was depressed below the sea and covered with 60 to 70 feet of marine sands. Then the sea withdrew, and the submerged area changed to a fresh-water morass in which peat was formed in thicknesses varying from a few inches to as much as ten feet or more. During the fourth and fifth centuries of the Christian era the sea again covered this maritime zone, burying the peat under five or ten feet of sand and clay. As these deposits gradually built up to the surface of ordinary high tide, and as the formation of the dune ridge along the coast and the building of artificial dikes helped to keep out the sea, we now have a fifth belt some ten miles broad, cutting across the other belts at an oblique angle and often called the "maritime plain" (Fig. 7).

Over the surface of Flanders there is spread a thin coating of loam which usually conceals from view the older formations previously described. This loam is sandy toward the bottom, more clayey in its upper part. Where it mantles the clay formation of the extensive clay belt it is as a whole much more argillaceous than where it covers the sand belts. Apparently the underlying formation influences the character of the thin overlying loam. Certainly it is the fundamental beds which determine the topographic belts of Flanders, for the superficial mantle of loam affects the topography but little. Only on the chalk of the transition belt do we find the loam fundamentally different from the basal formation. There it plays an independent rôle of some importance, as we shall see when we study its typical development in the Battlefield of the Somme. In Flanders the loam

Fig. 8—Typical view of the Gohelle portion of the transition belt of the southern Flanders plain, showing mining villages and mounds of waste from the coal mines. The view was taken from the crest of Vimy Ridge.

reflects the form and character of the underlying beds so well that one is apt to lose sight of it in describing the features of the country.

THE TRANSITION BELT

Let us, having in mind the general picture of the Flanders battlefield which we have just gained, consider next those features of the different topographic belts most likely to affect military operations. The transition belt lies lower than the Somme battlefield to the south but higher than the typical Flanders plain farther north. From the base of Vimy Ridge the rolling chalk surface, partly coated with loam, slopes gradually downward toward the northeast, forming the faintly inclined plane which the natives call the Gohelle (Fig. 6). It is in the deeper rocks, under this part of the chalk, that are found the coal deposits mined at Lens and neighboring localities.

A typical view of the Gohelle may be had from the summit of Vimy Ridge (Fig. 8). The nearly treeless undulations of the chalk appear barren and dreary in contrast with the tree-bordered gardens and meadows of the clay plain beyond. An occasional woodland patch and lines of trees along the roads relieve the monotony of open fields. Villages are numerous, but they are usually the ugly *corons*, long rows of red-roofed miners' houses, built together and all of the same plain style of architecture. Uglier still are the huge black dumps of waste from the mines, often wrongly called slag piles, which may rise a hundred feet or more above the surface of the plain and form the most prominent feature in the landscape. Their "command" of the adjacent country gave them a high military value. For, while less flat than the clay plain farther north, the Gohelle is nevertheless a surface of such faint relief that from an elevation of a hundred feet the observation is really commanding. From our vantage point on Vimy Ridge not only are Givenchy-en-Gohelle, Vimy, Oppy, and other near-by villages in full view, but the great mining center of Lens, Drocourt where began the Drocourt-Quéant switch of the Hindenburg Line, Loos of bloody memory, and even the more distant Vermelles and the towers of Wingles are easily distinguished, together with a score of other villages and towns which help to make up this important mining district. Topographic barriers are few; but the underground workings of the mines provide secure shelter for vast quantities of military stores; the *corons* and other villages and towns are well adapted to serving as fortified strong points in a network of trenches; cellars and caverns in the chalk offer concealment and protection from bombardment to large bodies of troops; the mine dumps give valuable observation; while the network of roads, tramways, and railways necessary to serve a mining region furnish abundant means of transport for military purposes. Except for its lower relief and the presence of the coal mines and their accessory features, the military aspects of the Gohelle resemble those of the Somme battlefield farther south. We will reserve fuller discussion of the effect of the chalk upon military operations until we consider its typical development in that field.

Among the few topographic barriers in the chalk of the transition belt the marshy valleys of the Scarpe and Sensée Rivers deserve special attention. The Sensée might be taken as the boundary between the transition belt of the Flanders plain and the typical chalk plain of the Somme; and both it and the Scarpe have their sources in the latter region. But their military history has perhaps been more particularly bound up with that of the Battlefield of Flanders. Both rivers have long been regarded by French military authorities as important links in the northern defensive system of their country. The broad marshy valley of the Scarpe (Fig. 9) prolonged eastward, from near Arras, the natural barriers of Vimy Ridge and the rectilinear valleys of the Canche and Authie. Today the stream is canalized for much of its course, and the marshes partly drained; but inundations can still turn the valley into a serious military obstacle. At its junction with the Scheldt the river helps to make a natural quadrilateral, formerly of much military importance, enclosing the great Forest of Raismes, bounded on three sides by marshy river valleys capable of being inundated at will and further strengthened by military engineering works at three of the corners. The marshes of the Sensée are broader than those of the Scarpe, and much of the area "is at all times of the year a large bog or quagmire and, following heavy rains, a veritable lake almost a kilometer in breadth, at some points filling the river basin. On each bank peat bogs prevent access to the stream. The Sensée

FIG. 9—The marshes of the Scarpe, from earliest days one of the important natural defensive barriers of northern France. Compare this humid, verdure-clad valley with the dry chalk upland shown in Figs. 29 and 43. (From a photograph by A. Demangeon.)

Canal and the stream itself make the valley still more impassable. . . . In case of war, locks permit raising the water surface 2 meters higher." The junction of the marshy Sensée with the marshy Scheldt at Bouchain (Fig. 42) has always been a point of much strategic value. The northern side of the famous quadrilateral known as Caesar's Camp (p. 125) was formed by the Sensée marshes, and Bouchain defended one of its corners.

When the French under Marshal Villars were striving in 1710–1711 to prolong the struggle against the Allies, led by the brilliant Marlborough, they took refuge behind the Scarpe and Sensée barriers. Villars inundated the Scarpe to Biache eight miles east of Arras by damming it, whence the line crossed to the Sensée marshes by a canalized depression which was likewise flooded. Thus were these natural obstacles organized into a part of the famous *ne plus ultra* line which taxed the military genius of Marlborough to the utmost. The floods of the Sensée could be crossed only on two causeways, one at Arleux* and the other at Aubigny, both of which were protected by fortifications. So formidable was the barrier that it could not be taken by direct assault, and Marlborough had recourse to a most remarkable and complicated stratagem which resulted in giving him possession of the crossings.[3] Marlborough's despatches are filled with references to the "morasses" of the Sensée, Scarpe, and Scheldt which now prevented him from attacking the French, now protected him from enemy assaults.

In 1794 the French and their Allied enemies were again facing each other across the barrier of the Sensée marshes. These are but scattered instances in a long history in which the Scarpe and Sensée played an important rôle, the later chapters of which saw the French Army of the North retreating behind the line of the Scarpe after being defeated by the Prussians near Amiens in 1870, and the Germans shielding themselves from Allied attacks by flooding the Sensée marshes during the war of 1914–1918.

The chalk of the Gohelle dips under the oval basin containing the lower sand and the clay formations (Fig. 7). Within this oval, rudely outlined by forest patches following the lower sand outcrops (p. 12), one finds the typical landscapes of the low clay plain characteristic of most of the Flanders battlefield. It is a bit of real Flanders in the midst of the chalk, and has been given a special name, the Pevèle. The regional name is sometimes attached to a village name, after the French custom, as in the case of Mons-en-Pevèle. Beyond the basin the chalk again bulges up to the surface, forming the region called Mélantois just south of Lille and

* Unless otherwise stated, places mentioned in Chapters I and II may readily be located on Pl. I in the pocket, on or near the river or other topographic feature with which the names are associated in the text.

[3] J. W. Fortescue: A History of the British Army, 7 vols. to date, London, 1899–1912; reference in Vol. 1, pp. 540–547.

Carembault farther southwest (Fig. 6). Both these regions repeat the topographic features characteristic of the chalk plain of Gohelle, except that the elements contributed by the coal mines are largely lacking.

THE SOUTHERN SAND BELT

As the lower sand formation outcrops only as a narrow band and is there characterized by no particularly resistant or non-resistant layers, it does not exercise a striking effect upon the landscape. Forest patches are, however, a little more frequent on or near it, and the bare hills of the chalk begin to give place to a more pleasing landscape of meadows and gardens interspersed with trees. The sand formation has, furthermore, real importance as a source of water supply. Rainfall penetrating the porous sand descends to great depths, held in by the clay covering of the chalk below and by the great clay formation of Flanders above. Wells driven through the clay formation to pierce the lower sands give some of the most abundant flows of water in all Flanders. This is an item of considerable military importance, for we shall discover that, curiously enough, in the water-soaked plain of Flanders it is a difficult matter to secure sufficient uncontaminated water to supply the large demands of a great army.

THE CLAY BELT

It is on the low, flat clay plain that the topography of Flanders finds its typical expression. The fact that the clay is fine-grained and not yet consolidated into rock makes it a more ready prey to erosion than the chalk. Hence it has been worn down more rapidly and to a much lower level than the upland of Artois, even lower than the chalk plain of Gohelle in the transition belt. As a whole the clay plain rises but slightly above the level of the sea. The rivers are very little lower than the interstream areas, and pronounced valleys like those in the chalk plain farther south are quite unknown. It has well been said that the whole of the plain is one great valley. If a river happens to cut against a low elevation in the clay so as to give a wall a few yards high, it is a topographic feature worthy of special remark.

On so flat a surface the rainfall finds it difficult to flow away. Nor can it escape readily underground, for the clay is one of the most impervious formations imaginable. Unable to sink downward or to flow laterally, the water remains stagnant over large areas, forming ponds and marshes, or rises until it slowly creeps, halting and hesitating as to what course to take, toward one of the sluggish rivers which wander with apparent aimlessness over the level land. The ground is saturated with moisture, whether it be the clay itself outcropping at the surface or the thin deposit of clayey loam resting as a mantle upon it. Either one gives a sticky, slippery mud which is the abomination of Flanders. Even when it dries, as it does when some days of sunshine interrupt for longer than usual the succession of cloud, mist, and rain, the clay is an enemy of man; for the peasant cultivating the hardened soil must pour water on the plow and get men to help force the share into the ground as it moves painfully forward. The hardened mass cracks open in all directions, letting the next rain descend into the fissures to make the sides of the clay blocks so slippery that the divided mass glides on itself, thus producing a landslide where there is slope enough to permit its descent to obstruct some railway, road, or canal.

Flanders Mud

Flanders mud demands more than mere passing notice. In the early days, before metaled roads and railways had made their appearance and before the system of canals was so extensive as now, the mud made large parts of the country difficult of access—some parts absolutely inaccessible. This condition endured through the seventeenth and eighteenth centuries, for

> in 1635 Bailleul could barely procure the necessaries of life during winter, and that thanks to navigation upon the *becque*, the roads being too bad. The intendant Barentin states in 1699 that the rich castellany of Cassel could not, on account of the lack of passable roads, distribute the provisions it produced nor assist neighboring castellanies suffering want. A project for the defense of the country after 1735 declares that between Armentières and Cassel the main highways are impracticable almost the entire year. . . . The prefect Dieudonné in 1804 refers

sorrowfully to the bad condition of the roads "which have, in the nature of the soil, a perpetual cause of deterioration, the low, wet, and often peaty soil having no solidity." The state of the roads was still worse in the plain of the Lys. The character of the ground, even more clayey and wet than around Cassel and Ypres, rendered them impracticable in all seasons. In 1761 the town of St. Venant complained of being "inaccessible through the bad condition of the roads;" and in 1766 the sheriffs explain that in wet periods it is the mud which prevents approach to the town, while in dry periods it is the ruts which are so frightful "that it is impossible to pass either on horseback or in carriage without risk of smashing everything." For several years the old Roman road from Cassel to Arras remained impassable between Estaires and La Bassée at a spot called 'le Trou Gallot', "where opens an abyss which must cost the lives of all who pass that way." The only means for those on foot to traverse the country in winter (for travel by carriage was scarcely to be considered) was to jump step by step, on the blocks of Béthune sandstone [from the southern sand belt] which the foresight of the authorities had caused to be placed along the side of each road and which were called *pierres de marchepied*, or stepping stones; shod with shoes heavily metaled to prevent slipping upon the stones and disappearing in the mire up to the waist and armed with long poles to aid in jumping from one block to another, the people of the country became so used to this mode of traveling that accidents were rare.[4]

Today the main highways are paved and many of the lesser roads improved. But an army cannot restrict its movements to a limited number of roads; and, when it must fight in any given region, modern methods of warfare compel operations on almost every square foot of the terrain. Thus the mud of Flanders remains an element of the first importance with which military leaders perforce must reckon. When Philip Augustus let his army become trapped in the morass southwest of Ypres in 1197, he but shared the experience of many before and after him, from the days before the Roman conquest to the days when the German General Staff and the Allied commanders saw one great offensive after another stop short in the mud. It would be difficult to enumerate all the ways in which the seemingly unfathomable mud of Flanders affected the fighting powers of the opposing armies in the fateful years 1914–1918. The damage done by

[4] Blanchard, La Flandre, pp. 446–447.

artillery fire was greatly reduced when the shells exploded in a sticky clay. Shell holes filled with water which could not drain away, turning the battlefield into an almost impassable morass which blocked the advance for which the bombardment was supposed to be a preparation. Munitions and other supplies could not be hurried forward in time to support an advance properly, and reinforcements of troops floundered in the mire behind the lines while their hard-pressed comrades were forced to relinquish captured positions for want of prompt assistance. Heavy guns could not leave the metaled roads, were delayed by the congested traffic confined to these narrow paths through the wet plain, and could not be distributed at will when they reached the firing line. Even the light field artillery used in supporting an attack had to be equipped with material for making the shell-torn surface passable before it could advance, and one artillery officer is reported to have said: "I am carrying forward my guns and ammunition, the material for making my road as I go along, and the material for fortifying my new position. . . I am half expecting orders to bring along an acre of ground with me, too." [5] Assaulting columns found it difficult to scramble out of the slippery trenches and were mowed down by enemy fire as they advanced slowly through a tenacious clay into which they sank more than ankle deep. Rifles became so clogged that they could not be fired; and, when they were wrapped in cloth to keep the mechanism clean, were not ready for instant use. The wounded lay half buried in the mud, and many were suffocated. Even the well and strong were caught in fatal mud traps, for detailed official reports on the operations in Flanders contain not infrequently such statements as "part of Company bogged in communication trench south of St. Eloi; two men smothered" and "three men suffocated in mud near Voormezeele." In a British assault on the low clay mound near St. Eloi in April, 1916, the attackers had to lie flat and distribute their weight evenly in order to prevent sinking into the mire. As it was, a number of the men were engulfed and suffocated. In many parts of the plain trenches

[5] Frank Fox: The Battles of the Ridges, London, 1918, p. 95.

slumped in so fast that new positions could not be consolidated, and important points captured at a heavy cost in lives had to be relinquished because the survivors could not protect themselves in the soft clay before heavy counterattacks were delivered. Even well-established trenches required constant repair.

The effect of ever-present, everlasting mud on the morale of an army is a factor difficult to evaluate but certainly not to be ignored. The mere labor of keeping rifles, guns, shells, and other equipment dry and clean is alone a heavy task. The cleaning of gun carriages, automobiles, trucks, horses, and other means of transport adds to the burden. Superhuman efforts would be necessary to keep men and equipment up to that high standard of cleanliness which has earned for the British troops the enviable if somewhat unpoetic name of "the spit-and-polish army." In Flanders the soldier's best efforts left him discouraged. Forever busy cleaning the sticky deposit from himself and everything about him, he forever found himself and his equipment caked with the mire. Cold, wet, tired, and disgusted, the unhappy fighter in Flanders would crawl into his straw-floored dugout, leaving his clay-coated shoes at the entrance, and lie shivering as he cursed the eternal mud which was by far his worst enemy. The author has visited the fighting fronts from the sand dunes of the Belgian coast to the entrenched camp of Saloniki and observed during the conflict the conditions under which men fought from the polders below sea level to the glacier-clad heights of the Alps. He has no hesitation in saying that, of all the combatants, those who fought on the plain of Flanders endured the most terrible physical conditions.

The Rivers of the Clay Plain

Mud is not the only enemy of an army condemned to operate on the clay plain. Although the rainfall is moderate and notably less than that on the arid chalk uplands of Artois, we have seen that it is held at the surface by the impervious clay and cannot escape rapidly from so flat a region. This excess of water makes trouble in numberless ways. Where it flows into the sluggish rivers they are flooded, for with their faint gradients they cannot rapidly discharge the large total volumes thus received. When they rise, their flood waters spread far and wide, because the country is so low and flat. Almost all the streams of the plain, from the smallest rivulet to larger rivers like the Lys, have one or more floods every year, especially in the late summer and fall and in the early winter. Half a dozen floods in the same stream in a single year are not unusual. In 1882 the Yser overflowed its banks fourteen times. The floods of the Lys are the most dangerous of all in the Flemish plain, and the river has not inaptly been called "the scourge of Flanders." The rising waters not only submerge its broad, shallow valley and flood the city of Ghent, but force the Scheldt to back up and overflow its banks, fill the Bruges-Ghent canal to overflowing, and interfere with the régime of other canals connecting with its valley.

Elaborate works have been undertaken to control the rivers. Many have been straightened and canalized, their rising waters are held in by dikes, their great breadth has often been reduced to a narrow channel, and the extensive marshes bordering their courses have been partly reclaimed. Yet even today disastrous floods are not unknown, and the marshes are still numerous. Where marsh and forest combine, as at Ploegsteert Wood in wet weather, the obstacle is peculiarly formidable and can only be threaded on wooden footbridges, or *passerelles*, raised above the morass. The river channels are barriers which armies may pass only with the aid of bridges, while the valley floors can be flooded by opening the dikes and the difficulty of passage thus greatly augmented. Almost every phase of the defensive value of rivers, marshes, and inundations discussed by Von Clausewitz in his classic work "On War" finds exemplification on the Battlefield of Flanders.[6]

The tactical and strategic value of the rivers of Flanders cannot be doubted by one who reads even a few pages of the long military history of this region. Froissart's "Chroniques" contain many references to the line of the Lys, among them the detailed

[6] Carl von Clausewitz: On War, London, 1911, Vol. 2, pp. 263–295.

and quaint description of how in the latter part of the fourteenth century Philip van Artevelde commanded the destruction of all bridges over the river on a front of fifty miles or more, in order that it should be "not in the power of the King of France nor of his Frenchmen, that they should pass the river of the Lis." The advisers of the French King held the barrier in profound respect, and when asked, "This river of the Lis, is it so evil to pass that one cannot find passage save at the well-known crossings?" one of them replied, "Sire, yes; there is no ford, and the stream flows upon marshes which one cannot traverse." Some advised ascending the river until the French army could pass around its source, others counseled turning northeast into another part of Flanders where the advance would be blocked by no such obstacle. Bolder spirits urged an attack on the river barrier, assuring the others that God would aid them to make a crossing; and their advice prevailed. But so difficult did the task prove that even those who urged the attack repented that they had not taken the long route round the river sources or marched in some other direction across an easier terrain. When success was finally achieved, it was by means of a surprise crossing effected secretly with small boats at a poorly guarded point.[7] In the succeeding pages of the "Chroniques" one reads the same complaints of marsh and mud, the same references to soldiers standing knee-deep in the mire, which became so familiar in the despatches of 1914–1918. If we turn to the wars of the French Revolution, we find the Allied armies holding the line of the Lys early in 1794, then fleeing in disorder when the French capture a section of the barrier. Next it is the French who make the line of the Lys their position, advancing from and returning to it with the exigencies of the campaign. In the disastrous Allied offensive in May, known as the Battle of Tourcoing, Clerfayt's corps, one-fourth of the whole army, was held up by the French standing behind the Lys at Wervicq and so counted for nothing in the first day's operations. Two other columns were held up at the Marque, "a stream impassable except by bridges, owing to soft bottom and

[7] Jean Froissart: Chroniques, edit. de Lettenhove, Brussels, 1870, Vol. 10, pp. 106–126.

swampy banks." In the preceding year the line of the Yser at Bambecque was held so effectively by the Allies that ten times their number of Frenchmen were unable to dislodge them by direct attacks; while the line of the Yser-Ypres canal from Nieuport to Ypres, which was to figure so largely in the World War of 1914–1918, served as the main defensive position of the Allied armies' right wing in 1794.[8]

It is unnecessary to multiply instances in which the rivers of Flanders have in the past served as natural lines of defense. In the recent war the Lys again played an important defensive rôle, the Yser blocked some of the greatest offensives launched by the Germans, while other streams, including so small an example as Kemmel Brook, contributed in no small degree to the results of operations in the clay plain. Rivers and canals (Fig. 10) combined with the mud of the plain to oppose a serious obstacle to one of the most effective of modern offensive weapons. In most parts of the clay plain it was practically impossible to use tanks; in others, especially in more sandy and less humid areas, they were utilized, sometimes by equipping them with fascines of wood which could be dropped in front as the tanks advanced. But, notwithstanding this device to improve the terrain over which they moved, the work of the tanks on the low plain was never so effective as on the chalk uplands farther south. As for cavalry, military writers have from earliest times pictured the difficulties occasioned that arm of the service by the endless network of small drainage ditches, the marshes, the larger canals, and the rivers. Artillery and even infantry find the drainage ditches an obstacle which forces them to keep to the roads or to the larger dikes of rivers and principal canals.

Submerged Trenches and Dugouts

The presence of an excess of water on or close below the surface proved a serious embarrassment in the operations of trench warfare. Where water stood in ponds or marshes in faint depressions of the low plain, trenches were impracticable. Elsewhere even shallow trenches might encounter the zone of permanent

[8] Fortescue, Vol. 4, Part I, Chs. 5 and 10.

saturation close below the surface, and so be permanently inundated. Even where the land was slightly higher, trenches in the clay caught rainfall which could not drain off through the impervious material. Hence the terrible sufferings of men compelled to stand in cold water or liquid mud ankle-deep, knee-deep, sometimes even waist-deep. What this means to an army can

FIG. 10—A canal used as a defense line and crossed by a temporary footbridge. (Belgian official photograph.)

only be fully understood by one who has experienced it; but some idea of the truth can be gleaned from reading the reports of officers of all grades, now become part of the British War Office archives. As one report modestly observes, "The difficulties of this part of the country are worthy of note. The trenches are very wet, and the water is up to the men's knees in most places." Such phrases as "men knee-deep in water," "trenches full of liquid mud 2 to 3 feet deep," "trenches full of water 3 feet deep in places," "trenches untenable owing to flooding," "ground so wet only able to dig down 2 feet," occur in endless repetition.

One officer reports his men as "in pitiable condition coming out of trenches; wet through, caked with stinking mud from head to foot, and perishing with cold." The state of the men's feet became unbearable, and much space is devoted to casualties from this cause. Eight hundred men unable to walk were left behind in

FIG. 11—Scattered population on the low plain of Flanders, where abundance of water close to the surface enables every farmer to have his well and his home in the midst of his fields. (From Lille sheet of 1:80,000 topographic map of France.) Compare with Fig. 12.

a single village. The terse report of another officer speaks volumes: "Trenches full of liquid mud. Smelt horribly. Full of dead Frenchmen too bad to touch. Men quite nauseated." Can one marvel that the physical conditions of the Flanders battlefield tried the souls of men as they were tried nowhere else on the long fighting front?

If deep trenches were hardly practicable, it goes without saying that still deeper dugouts and all that elaborate system of subterranean fortifications which honeycombed the chalk of Artois and Picardy were impossible in the low clay plain. Small, shallow dugouts were excavated in places; but shelters built on the surface of bags of earth and other material were essential over broad areas, while railway embankments, dikes, and other

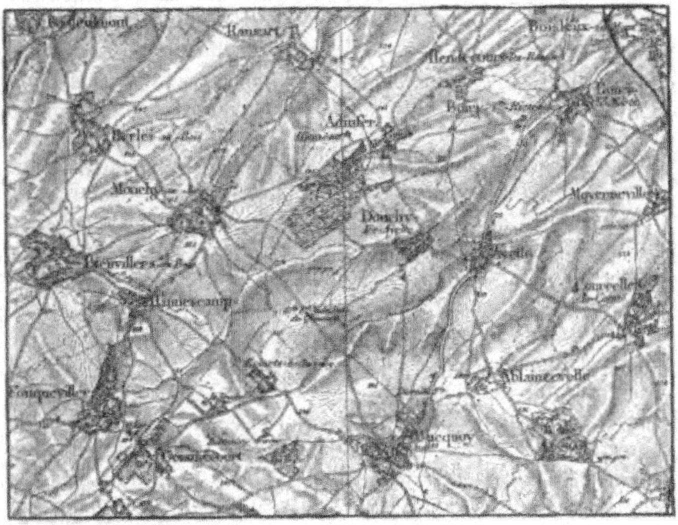

FIG. 12—Compact villages grouped about wells in the Somme region, separated by uninhabited open country. The great depth to water in the chalk formation makes wells few and costly. (From Amiens sheet of 1:80,000 topographic map of France.) Compare with Fig. 11.

artificial structures were much utilized (Fig. 17). Even the "trenches" were often mere breastworks of sandbags, and attacking parties carried with them the quantity of bags necessary to consolidate and hold a new position.

Contaminated Water Supplies

It might be supposed that with such excesses of water on every hand, the fighting forces would at least be spared the trouble of securing proper water supplies. There is, of course, no lack of quantity, and it is easily accessible. The sandy lower portion of the loam mantle covering the plain is full of water which cannot sink deeper because of the impervious clay immediately beneath. One has only to sink a shallow well some few feet through the more clayey upper half of the loam to reach this water horizon. Every peasant can afford to have his own well; and, since travel over the muddy plain is difficult, he digs his well and builds his home conveniently in the midst of his fields. This is in striking contrast with the conditions on the chalk plain of the Somme battlefield, where all the peasants of a district will cluster their homes close about a few deep wells dug at the common expense. A map of the Flanders plain (Fig. 11) contrasts strongly with a corresponding map of the chalk region, with its compact villages and uninhabited open country (Fig. 12). In the defensive systems organized by the opposing armies in Flanders there was little to suggest that network of isolated fortified towns and villages which figured so largely in the Somme campaigns.

But if water was close at hand it was also easily contaminated. Surface wash, carrying with it impurities from every possible source of filth, came to rest in the lower part of the loam, where it was held by the clay below. The muddy rivers and canals, discolored and poisoned by refuse from the manufacturing cities and towns along their courses, not only were unfit for use themselves but often contaminated the shallow water zone of the loam into which the peasants sank their wells. The loam horizon was, furthermore, too limited to supply the huge quantities of water demanded by countless great manufacturing establishments. Hence the water-soaked Flanders plain became a region of numerous deep and costly artesian wells.

When vast armies camped upon the plain in 1914 the water problem became aggravated, and especially so after drainage washed the decaying bodies of the battlefield and passed into the water-bearing horizons. Even the deeper waters were often poisoned by surface supplies which penetrated down the sand formations or down the sandy lenses in the clay formation. Hence the development of sanitary supplies of water in the enormous quantities required by large armies created a difficult problem. Nothing less than a full knowledge of the underground structure of the battlefield could enable an army to deal successfully with the vital matter of determining which layers in the different formations would carry uncontaminated water and at what depths they would be encountered in different parts of the field.

Vegetation of the Clay Plain

Another consequence of the humidity of the clay plain is the vigorous growth of vegetation which covers it with a mantle of green. There is not now, nor does there ever appear to have been, any such vast forest as once covered the rolling chalk plain of the Somme region. In its original state a succession of grassy meadows and swamps studded with trees and interspersed with more continuous woodland patches, the vegetation has since been modified by man so that today rectangular patches of meadow, field, and garden cover most of the land, the scattered trees are less numerous and are largely confined to the borders of roads, canals, and rivers, or to the margins of meadows, fields, and gardens; while the patches of veritable woodland are much more restricted than formerly. Wherever one may be on the clay plain, he has the impression of standing in the midst of a large and rather intensively cultivated clearing in a forest. When he advances to where the forest seemed unbroken, the woodland dissolves into scattered groups and rows of trees bordering cultivated areas or transport lines, and the scene is the same as before. Distant views are rare, and one has no such sense of the vastness of the plain as he gets on the barren but much more uneven upland of Artois. It is only when he mounts to the summit of some hill and views the flat country from above (Fig. 13) that he sees over and beyond the trees to more distant parts of the plain. Hence the incalculable military value of even faint elevations on so level a surface. Histories of military campaigns in Flanders

Fig. 13—Looking southeast from Mont des Cats over the ruins of Bailleul and the Flanders plain. Part of a British official panorama of the plain showing some of the points important in controlling artillery fire.

contain frequent reference to the difficulty of observing enemy movements on account of the trees.

Aside from the wooded areas on the upper sand formation, found capping the scattered hills and ridges left isolated by erosion here and there over the plain, the principal forest patches are those rooted in the clay formation where it lacks the loam covering. These surface exposures of the clay are called *clyttes*, and, as they furnish a cold, wet, sticky, inhospitable soil, poorly adapted to cultivation, the peasants often leave them covered with trees. The Houthulst Forest north of Ypres is a good example of a type of woodland in which forest fighting is at its worst, because to the other difficulties usually presented by a fortified wood are added the horrors of a water-soaked, tenacious soil giving a morass in which troops cannot entrench.

Hills on the Clay Plain

The hills of the clay plain are perhaps its most important topographic feature from the military point of view. A general can well afford to sacrifice thousands of men in order to conquer and hold a small hummock a hundred feet or less in height; for an observer on that hill can save tens of thousands of his comrades by directing artillery fire against enemy positions, batteries, troop concentrations, railways, and roads, and other objectives easily located from his point of vantage but absolutely hidden behind a wall of trees to one who stands on the plain itself.

An insignificant mound rising only 30 feet above the plain and called "the Bluff" was the scene of terrific fighting in the early part of 1916 because its value for artillery observation made it the key to a section of the British line. To secure control of this one point the enemy excavated tunnels and exploded five mines under it and then launched a successful infantry attack. From the captured elevation the Germans could enfilade other British positions, and two weeks of British counterattacks failed to regain the lost mound. So valuable was the tiny hillock that the British now made more elaborate preparations for its recapture. A frontal attack was necessary, since one side of the hill was protected by a canal barrier, the other by a marsh. The story of the struggle, of the final successful British assault, and of the later unsuccessful German counterattacks is clearly pictured by Conan Doyle in the third volume of his detailed account of "The British Campaign in France and Flanders," while Sir Douglas Haig states that between January 16, 1916, and June 7, 1917, a total of twenty-seven mines were exploded at this point alone. Anyone who doubts the significance of a 30-foot hill on a flat plain should read the pages of unofficial and official reports on these operations and try to measure the labor expended and the blood spilt in contests for possession of the Bluff.

A still more sanguinary conflict raged for six long weeks in the spring of 1916 for possession of "the Mound" near St. Eloi, a small bump of clay about 25 feet high. The British positions on the Mound were violently attacked in the middle of March, following a heavy bombardment and the explosion of a mine under the hillock. The attack was successful, and the defenders had to evacuate a considerable stretch of their line, dominated as it was by the Mound. In unsuccessful counterattacks the British, swept by a hail of projectiles from the higher point, in a short time lost over 900 men, including 40 officers. Thereupon they excavated five mines under the Mound, placing as much as 30,000 pounds of ammonal in a single one. These were exploded the last of March, and a furious combat lasting several days resulted in a British victory. Then began a series of German counterattacks continuing for some weeks, in the course of which the mine craters, filled with pools of blood and mud in which the men fought waist deep with rifle and machine guns clogged with dirt, changed hands repeatedly. Not until the end of April was the position again definitely in British possession.

To gain Hill 60, a low ridge fifty feet high in the Zillebeek region from which the Germans dominated a broad stretch of the plain in front while their own back areas were concealed, the British on April 17 discharged six subterranean mines and began a contest which raged with fury for weeks. Within four days the British casualties rose to more than 3,000 men and 100 officers on a front hardly longer than two or three city blocks, and the blood-drenched hillock had changed hands several times. Still the battle raged. The Germans delivered their first bombardment with poison gas shells, and in the first days of May discharged several poison gas waves which alone cost the British another thousand men and drove them from the position. In the fight for Mt. Sorel and Observatory Hill in June of the same year the Canadians sacrificed 7,000 men.[9]

The Ypres Bastion

If an insignificant hill may be of such vital importance, who shall set a value on such heights as the east-west ridge from Mont Kemmel to Mont Cassel, rising several hundred feet above the plain? Dominating the lowland from a central point, these "Monts de Flandre," as they are called, should form the key position of Flanders. It is perhaps not too much to accord them the dignity of such a title. In Roman times Mont Cassel was crowned with fortifications, from which radiated a system of Roman roads to bind the surrounding plain to the central stronghold. The eleventh, fourteenth, and seventeenth centuries saw fierce battles rage about it. When Philip of Valois overthrew the Flemish rebels holding Mont Cassel in 1328, all Flanders submitted. While the French held southern Flanders against the combined forces of the Allies in the wars of the French Revolution, they made of Mont Cassel an entrenched camp, from which military operations were carried out in all directions. So throughout history one might trace the important rôle played by this ridge of hills and particularly by its culminating point, Mont Cassel, which was in 1914 to serve as the observatory from which Foch would direct the first battles of Flanders and from which he could, in fact, on a clear day see the flashes of the guns from the dunes at Nieuport to the chalk upland at Vimy Ridge.

But the real strength of the hill position is only apparent when taken in connection with that remnant of the backslope of the

[9] A. Conan Doyle: The British Campaign in France and Flanders: 1916, London, 1917.

former cuesta, or asymmetrical upland, which trends northward from near Mont Kemmel to the vicinity of Passchendaele (Fig. 2). This Messines-Passchendaele ridge is lower than the Mont Cassel-Mont Kemmel remnant of the southern crest of the cuesta, and, being part of the gentle back slope of the former upland, it gets progressively lower toward the north, where it is in a sense continued by fragments of the second cuesta near Staden so as to give a line of heights bending northwest nearly to Dixmude. In front of this northern sector, and strengthening it, lies the marshy Handzaeme lowland, a natural moat before the upland barrier. The Messines-Passchendaele crest is broader than the higher ridge, affording space for large bodies of troops; and it commands a wide stretch of the flat plain to the east and west. Especially toward the east, the direction of chief danger, the crest dominates the vast plain like a watch-tower on a castle wall. In the angle between the east-west and north-south ridges lay Ypres, defended by these higher lands on the east and south, and on the west by several parallel branches of the Yser River and the Yser Canal. About Ypres there is, furthermore, a girdle of forested areas, partly on the arenaceous soils of the upper sand formation capping the higher portions of the ridges, partly on the compact clay exposures, or *clyttes*, which here protrude through the loam mantle in unusual numbers and which, we have seen, are so inhospitable to agriculture that they remain wooded. Altogether the Ypres region is a military stronghold of the highest importance, and this importance is fully attested by the number of pages in history which record bloody struggles for its defense or conquest. Siege and assault have followed each other in a long succession of which the great Battles of Ypres of the World War were merely the most recent chapters.

The Mont Cassel-Mont Kemmel ridge and the Messines-Passchendaele ridge meet in an apex near Mont Kemmel, forming a "bastion" pointing southeast, the flanks of which present a most formidable natural obstacle to an enemy advancing from either the south or the east. Neither to the north nor to the south of it could hostile troops press very far westward without incurring heavy risks, so long as the obstacle remained unshaken to threaten their flank and rear. On the south especially would a

FIG. 14—Block diagram of the Flanders and Somme battlefields, showing salient features of the topography.

westward advance of necessity be limited, for the Vimy Ridge side of the Arras bastion (p. 189) and the Mont Cassel-Mont Kemmel wall of the Ypres bastion formed the two jaws of a giant trap which might at any time prove fatal to an army pushing too far over the plain between them (Figs. 14 and 36). The Mont Kemmel-Mont Cassel ridge is continued westward by the hills east and west of Watten until it meets the northwestern prolongation of Vimy Ridge west of St. Omer. Between the two bastions there is thus a triangular re-entrant, or "curtain," of low plain hemmed in by higher land. Into this curtain an enemy could safely venture only if both bastions were conquered and danger from the higher land removed. We shall not be surprised, therefore, to find much of the fighting in the Flanders plain centering on the attempted destruction of the Ypres bastion, "the key of Flanders," nor to find the Germans, after pushing a salient across the plain between the Ypres and Arras bastions, voluntarily withdrawing from the dangerous position when repeated attempts to destroy the two jaws of the trap had failed.

The hills are apt to be drier than the adjacent portions of the plain, either because they are capped by the upper sand formation or because they rise higher above the permanent groundwater level and shed rainfall more readily from their sloping sides. Holders of the hills thus possess more comfortable positions, as well as better observation. But they are exposed to a new danger, for in the better drained ground subterranean mine warfare can be prosecuted on a scale impossible in the water-soaked lowland. It is true that tunneling operations must encounter dangers from underground water in the hills, but where layers of sand alternate with layers of clay, engineers guided by geological advice may escape much of the trouble suffered by those who remain ignorant of the detailed structure of the terrain. Even in rainy Flanders there is a "dry" season, which causes the groundwater level to sink lower than usual; but, owing to a lag in its movement, it continues to sink for some time after the wet season has begun, and the rise does not cease until after the beginning of the dry season. Failure of the Germans to realize this fact resulted in the drowning of some of their tunnels driven just above the water level as determined when the heavier rains ceased. The chief geologist of the British forces, familiar with the subterranean movements of the water table, saved his army all trouble from this source. The high importance of mining operations in Flanders, and the necessity of a knowledge of underground structure in directing them, were fully demonstrated in the Battle of Messines Ridge, which began with the greatest mining operations in the history of warfare.

THE NORTHERN SAND BELT

It is only by way of contrast with the clay belt that the northern zone may be called sandy. The land is so low and flat that water stands always on or near the surface just as in the clay country. The sands themselves are usually argillaceous, and layers of true clay alternate with them. So also the loam covering contains a large clay content, even though more sandy than farther south. Hence the water-soaked soil is muddy, its humidity favors a vigorous growth of vegetation, and one who passes from the clay belt to the sand belt is not impressed with any sudden or marked change in the character of the country. Seldom is there even a trace of that appearance of aridity which one usually associates with a sandy region.

We have already seen (p. 12) that forested areas are more numerous in the sand belts and that there are remnants of two faint cuestas (three if we count a very short intervening one) indicated by the steeper southern and more gentle northern slopes of two slightly elevated zones. In a belt where the "elevations" are matters of a few feet only and where the "steeper slopes" are scarcely noticeable as such, the topography would be called "flat" by any one who did not give it a very careful examination. Nothing but the fact that on a flat battlefield elevations of half a dozen feet may be significant makes it worth while to emphasize such faint topographic elements.

It is in the northern sand belt that contamination of deeper waters is most likely to occur. This is due to the fact that the clay layers interspersed with the sands are not continuous over broad areas. Impure waters carried downward through the sand to the first clay bed move along its surface until its margin is reached, then descend to the next layer, and so on. The fact that a well is deep is therefore no guarantee that its water is fit for use. "At Bruges wells 40 meters deep are as dangerous as those which barely penetrate below the surface."[10]

In general, however, there is little in the northern sand belt to differentiate it from that part of the Flanders battlefield just to the south. Certainly the points of resemblance are far more striking than the differences. It is therefore unnecessary to repeat the previous descriptions of features which find their typical expression in the clay plain and which are only modified in a moderate degree in the northern sand belt.

THE MARITIME BELT

Low, flat, and wet as is the rest of Flanders, the maritime belt is lower, flatter, and wetter. Only an occasional area rises above the level of high tide, and vast stretches lie well below tide level. The dunes and the dikes alone prevent the belt from being almost completely submerged when the tide is high, at which times some areas would have 10 feet of water over them. The flat plain of the clay belt is undulating, almost hilly, when compared with the remarkable flatness of the low land bordering the sea. Trees are rare, houses few, and villages still less numerous; hence wide vistas open to the view. Yet the eye sweeps the broad expanse of fields and pastures in vain for an elevation sufficiently high to be detected by the unaided vision. Only careful measurement would reveal the fact that occasionally the surface swells gently a very few feet above the average level. Here one has an impression of vastness which is lacking in the tree-studded plain of the interior.

On so low and so level a surface the problem of excess humidity is necessarily present in its most difficult form. With the whole belt menaced by invasion from the sea on one side, by flooding from the rivers which flow into the lowland on the other side, by deluges of rain from a leaden sky above, and by eruption of brackish water from the earth beneath; and with no chance for all these waters to escape with the aid of man's labor, it must be evi-

[10] Blanchard, La Flandre, p. 56.

dent that only a severe and never-ending struggle can keep the land fit for human habitation. The line of dunes must be guarded and strengthened, and breaches in them blocked by artificial dikes. When during the great siege of Ostend in 1601 the Dutch removed the dunes northeast of the town in order to prevent the Spaniards from attacking it on that side, the sea poured

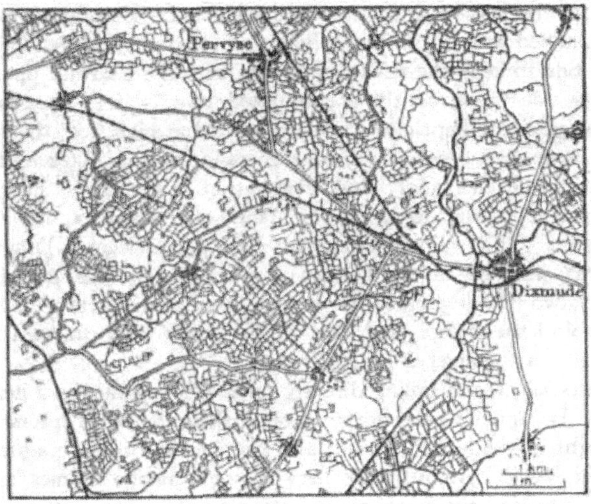

Fig. 15—Drainage ditches and canals in the maritime belt of the Flanders battlefield.

through the gap and flooded many square miles of the maritime plain. During the War of the Spanish Succession, when the French under Marshal Saxe were besieging the English base at Ostend, the commander-in-chief of the British forces in vain urged his Dutch and Austrian allies to open the dikes and flood the country as a measure of defense. In this case failure to profit by the natural topographic advantages of the region led to the fall of Ostend. In 1793 the French under Souham, besieged in Dunkirk by the Allied armies, opened the sluices and deluged all the former marsh region east of the fortress, thereby contributing largely to the ultimate success of the defense. These are but isolated instances of the many occasions on which inundation of the low plain has been resorted to as a means of defense in war.

The river waters must not be permitted to enter the plain but must be carried across it between dikes raised high enough to confine the waters until they reach the sea. Here gates must be installed and faithful watchmen see that they are closed at high tide to keep the sea from flowing into the land but opened at low tide to permit the accumulated river waters to escape. If the gates are kept closed, or are opened at high tide, the land may be flooded, especially if sluice gates in the river dikes are also opened. On the other hand, evacuation of the canals may be resorted to in order to deprive an enemy of transport at a critical moment. This was demonstrated during the siege of Dunkirk in 1793, when the defeated Allies on retiring had to abandon their thirty-two heavy guns to the French because the latter, "having control of the sluices, had shut off the water from the canal, so that it was no longer of use for transport."[11]

Disposal of rain falling directly on the flat surface is a heavy task. It cannot sink into the water-soaked ground; evaporation is slight in the humid region; and the neighboring rivers are at higher levels. An intricate network of drainage ditches must be dug to lead the water to collecting points, whence it may be raised by windmills or steam pumps to the diked rivers. In the lowest part of the plain the complex of ditches and canals (Fig. 15) is particularly striking. Even detailed maps show but a fraction of the total number of ditches, for they are sometimes spaced only 30 or 40 yards apart. They break the country up into tiny morsels of ground bounded by water and oppose to an army a formidable obstacle. The uselessness of cavalry on such a terrain has often been commented upon in past campaigns; the artillery finds itself embarrassed at every turn; while that heavy modern arm of the service, the tanks, can scarcely negotiate the

[11] Fortescue, Vol. 4, Part I, p. 132.

FIG. 16—*Passerelle* across the flood waters at the border of the submerged zone along the Yser. Many of the advanced positions could be reached only by means of these narrow footbridges. (Belgian official photograph.)

FIG. 17—Sandbag shelters and defenses commanding the flooded valley of the Yser. Note the wooden footpath over the muddy surface and the fact that the water-soaked soil compels resort to exposed sandbag shelters in place of concealed underground dugouts. (Belgian official photograph.)

larger channels. Even infantry finds such terrain difficult to cross, as will readily appear from perusing accounts of the sieges of Dunkirk. This position is protected on the east by a former marsh, the Moëres, or Great Moor, occupying the lowest part of the maritime plain, which was drained by a system of rectangular ditches in 1624, but partially flooded during the sieges.

FIG. 18—The water-soaked plain of Flanders between Ypres and Dixmude, showing the morass of mud, shell craters, and fallen tree branches produced by artillery fire on a tree-covered part of the low surface. (Belgian official photograph.)

One naturally expects in so low and humid a region to find unsanitary conditions and sickness among the difficulties to be combated by large military forces. Such great marshes as the Moëres near Dunkirk did indeed curse all the surrounding plain with malarial fevers, and even today the evil is not completely eliminated. The *clairs*, or openings, left by the extraction of great quantities of peat when that labor was the chief industry of the plain, contain large bodies of stagnant water, and not all the marshes have been drained. The water near the surface is contaminated with all sorts of impurities; that deeper down is brackish. "All the wells are bad, or ought to be so considered."[12]

[12] Blanchard, La Flandre, p. 292.

The brackish waters, which penetrate through the sands below the maritime plain under pressure from the higher sea, not only destroy the value of wells but may even push to the surface, especially near the dunes, and kill trees and gardens. During the sieges of Dunkirk the Allied armies found the region "wholly destitute of drinking water, that in the canals being brackish, and that found in the wells unpalatable;" and in 1793 it was the combination of "incessant fighting, a swampy encampment, bad drinking water, and fever" which brought the besieging armies to the verge of disaster.[18]

That man should have so long maintained the struggle against his liquid foes is due in part to the fertility of the loam which covers the maritime plain. The low-lying reclaimed polders have the richest soil in all Flanders, and despite its clayey consistency it rewarded the toiler with abundant harvests. Its natural richness is soon reduced, but wise methods of agriculture have maintained the productiveness of the plain. Fields and pastures together are able to support a population which, if not so dense, is more prosperous than that of neighboring belts. The scarcity of trees is explained not through any defect of the soil; nor, as in the case of the Somme plain, by the fact that the fertility of the soil caused all the land to be cleared and cultivated. It seems due rather to the greater exposure of the plain to the winds, which, sweeping in from the open ocean, bend and deform such trees as persist in growing on the level expanse.

As in the rest of Flanders, so in the maritime plain the clay is a most important element to be reckoned with. The layer of peat which underlies almost the entire region, but shows at the surface over limited areas, is usually covered with the argillaceous loam. Here also, therefore, a dense, sticky, slippery mud is characteristic of the water-soaked land and opposes to military operations all those embarrassments described on previous pages. The awful mud of the Yser is the strongest memory which one carries away from that part of the battle front. Even the *passerelles* above the flood waters (Fig. 16) were coated with the

[18] Fortescue, Vol. 4, Part I, pp. 124, 132.

slimy deposit tracked there by countless feet, and one slipped and slid, continually in imminent peril of gliding into the waters. Undrained marshes still exist, and mud, water, and marsh, especially when torn into a frightful morass by shell fire, produced a surface horrible to contemplate (Fig. 18).

Deep trenches and dugouts were of necessity little known on the lower parts of this front. Those which existed were often filled with water and liquid mud, and life in them was a nightmare. Breastworks and shelters of sandbags were brought into use (Fig. 17), and every faint knoll rising above the general level became a fortified island in the sea of mud. Railway embankments and dikes were much-prized strips of higher and drier land, often used as principal lines of defense (Fig. 17). But it was the flood waters, released from canals and diked rivers and permitted to flow in from the ocean, which formed the ultimate line of resistance to an enemy attack. It is true that even an empty canal, with slimy sides and mire in its bottom, was no easy obstacle to cross (Fig. 19.) Yet this was as nothing compared with the vast stretch of waters which filled the so-called valley of the Yser when the gates at Nieuport were opened and the whole country from the coast to Dixmude, formerly an estuary of the sea, reverted to its original state. The line of the Yser, repeatedly utilized as a defensive barrier in the wars of the past in Flanders, was to play a grand rôle in the World War.

THE DUNE BELT

Between the low maritime plain and the sea stretches the great barrier of sand, from one to several miles broad, capped by dunes which keep out the ocean waters. The dunes are from 30 to 100 feet high and, while sheltering more trees than the level plain, show large spaces of barren white or yellowish sand, conspicuous even at great distances (Fig. 20). The sandy soil is much drier than that found on the plain; but the water lies close below the surface, as the Duke of York discovered when he found it impossible, during his siege of Dunkirk, to protect his flanks by trenches because they encountered water at a depth of two feet.[14] In form the individual dunes are extremely irregular but often arranged in two or more bands parallel to the sea and enclosing broad depressions or basins called *pannes*, occasionally containing ponds of fresh water. The higher parts of the dunes permit the digging of dry trenches, which must, however, be specially walled to prevent the slumping of the loose sand (Fig.

FIG. 19—Defensive position sited along a canal draining the low Flanders plain. Even when the water in the canal is low, its muddy bottom and marshy borders make it a moat of some tactical value. (Belgian official photograph.)

21). Shells bursting in dune sands do less damage than in almost any other type of terrain.

Because the dunes are higher and less humid than the plain, they contain an unusually dense population. They are out of the reach of floods and at all seasons of the year serve as a highway for traffic along the coast in preference to the low and muddy plain. Both the flat strand of the beach between the dunes and the sea, and the depressions between the principal dune ridges, parallel

[14] Fortescue, Vol. 4, Part I, p. 128.

to the sea, have been used as routes of travel and for the advance of armies. The dunes contain the least objectionable waters for drinking purposes, as the rain falling on them and sinking into the loose sand forms a water table highest in the center and declining toward the sea on the outside and toward the polders on the inside. Wells a few feet deep get a limited supply fit for household uses.

FIG. 20—Barbed wire defenses to prevent an enemy advance along the dry highway formed by the strip of dunes along the coast. Because of the water-soaked character of the maritime plain, the dunes have always been a favorite route for advancing armies. (Belgian official photograph.)

Higher, drier, better provided with drinking water, and affording some shelter from an enemy, the dunes are far better adapted to military operations than is the adjacent plain with its ditches and canals, its water and its mud. Before the modern day of military operations on a grand scale one could truly say: "When an army ventures into the maritime plain, it is not on the vast open fields that it delivers battle; it is upon the sands that it encounters the adversary. The rare combats which have taken place in the plain have had for their theater the sands: the Battle of the Dunes of Dunkirk, the Battle of the Dunes of Nieuport. It is by the dunes that the Duke of Guise lays hand upon Calais; by the dunes that Condé, Turenne take Dunkirk, that the Duke of York approaches it in 1793, that the Archduke Albert persists in the attack on Ostend."[15] In the great Battle of the Dunes of Dunkirk, in 1658, the Spanish took up a strong position in the sand

Fig. 21—Extension of the defense lines from the coastal dunes across the sandy beach into the sea, to block the natural pathway formed by the strand. Like the strip of dunes, the strand has in successive wars served as an easy line of advance for hostile troops. (Belgian official photograph.)

hills, with their right flank protected by the sea, their left by a canal in the plain, and the key to their position held by four regiments on a particularly high mass of dunes. So formidable was the position and so well protected its flanks that the commander of the English troops for a time despaired of forcing it. Success was achieved only when the dominating point on the highest dune was stormed at a heavy sacrifice.

The line of the dunes is occasionally broken through by rivers

[15] Blanchard, La Flandre, p. 226.

or canals, the mouths of which have been transformed into harbors with the aid of artificial locks, basins, and other engineering works. Such are the harbors of Calais, Dunkirk, Nieuport, Ostend, and Zeebrugge. All are exposed to the danger of silting up; but the outwash from rivers and canals, sometimes aided by tidewater accumulated in special basins for the purpose, usually serves to keep them open. Any obstacle placed in the entrance by checking the outflowing currents will speedily cause the blocking of the port, a fact of which the British took advantage in their brilliant exploit in closing the German submarine base at Zeebrugge. In front of the harbors the bottom of the sea is ridged with submarine banks of sand parallel to the shore which break the force of storm waves, while between the banks and the coast are channels, often 30 feet or more deep, where vessels may lie in comparative security until a favorable moment for entering port. Opposite Nieuport these banks reach their maximum development and so afford protection to naval craft which might desire to protect the flank of an army standing behind the line of the Yser from possible attack by enemy forces advancing along the dunes.

We have now sketched in outline the salient features of each topographic belt in the Flanders battlefield. In all of them we have discovered that the points of resemblance are more striking than the points of difference. There is thus a real geographic unity in the Flanders plain, which the several phases of its topography do not materially affect. Everywhere the plain is low, everywhere it is comparatively flat, everywhere except on the limited hill and dune areas it is a sea of sticky mud in wet weather. In all parts are found sluggish rivers and canals which may be used in time of war as defense lines by opening the sluice gates and inundating the broad, shallow valleys. Few parts do not have the terrain cut up by countless ditches filled with water and the ground permanently saturated below a very shallow depth. Everywhere the water-supply problem is a serious one in spite of the extreme humidity of the country. The Battlefield of Flanders has its own consistent and striking character and is like no other battlefield of the World War.

CHAPTER II

MILITARY OPERATIONS ON THE BATTLEFIELD OF FLANDERS

About the middle of October, 1914, the Allied forces, following the Battle of the Marne and the check at the Aisne, were prolonging their great flanking maneuver, sometimes called "the race for the sea," northward from the Somme region into the low country south of Ypres. At the same time the Belgian army, retreating from Antwerp, was falling back toward the line of the Yser. Thus from opposite directions were in progress the movements which would soon crystallize the battle front across the Flanders plain.

On October 11 British cavalry entered the southern edge of the Battlefield of Flanders, crossing the southern sand belt and driving the enemy cavalry from the Forest of Nieppe south of Hazebrouck, one of the woodland areas characteristic of the southern margin of the clay belt and the sand belt. Pushing on across the plain to the Mont Cassel-Mont Kemmel ridge, it drove an outpost of the German army from the Mont des Cats (Fig. 4) after a stiff fight and captured the high ground farther east. Thus was the securing of the Ypres bastion one of the first objective of Allied strategy in this field. With a view to a further advance eastward Sir John French "ordered General Allenby, on the 15th, to reconnoitre the line of the River Lys, and endeavour to secure the passages on the opposite bank." But the barrier was so strongly held by the Germans to the northeast of a point near Armentières that "the Cavalry corps was unable to secure passages or to establish a permanent footing on the eastern bank of the river."[1]

The infantry and artillery on entering the plain swung eastward to support the left wing of the French, which was already

[1] Sir John French: Despatches: Mons; The Marne; The Aisne; Flanders, London, 1914, pp. 119, 128.

engaged with the Germans southwest of Lille. Immediately Sir John French found himself embarrassed by the nature of the terrain, and in his despatch covering the campaign he early notes that "the ground throughout this country is remarkably flat, rendering effective artillery support very difficult."[1] Meanwhile both British and French made fruitless attempts to dislodge the Germans from their strong position at La Bassée, where they had fortified themselves on a slightly higher bit of dry land surrounded by low, flat, marshy ground partly underlain by peat bog. "This position of La Bassée has throughout the battle defied all attempts at capture either by the French or the British," reports the Commander-in-Chief of the British forces. Elsewhere the Allies pushed on, much handicapped by the ditches and dikes and unable to reach Lille.

Meanwhile General Rawlinson's forces, which had been aiding the Belgians in their retreat from Antwerp, fell back to the Ypres bastion, taking up a position on the Messines-Passchendaele ridge and facing east in the neighborhood of Gheluvelt, while French reserves were massed inside the bastion, in and west of the town of Ypres. Still anxious to obtain a footing on the east bank of the Lys the Commander-in-Chief directed that Rawlinson's forces should advance eastward from the ridge to the river at Menin and force a passage over the barrier there. But Rawlinson feared to abandon his advantageous position on the ridge, in view of the fact that overwhelming enemy forces were threatening him from the east and northeast. He accordingly made representations to this effect and after a short advance returned to his lines on the eastern side of the Ypres bastion, while French cavalry on his left drove back advanced parties of the enemy beyond the Forest of Houthulst in order to secure the northwestern continuation of the ridge toward Dixmude. They were soon pressed back from these northern hills to the line of the Yser River, but most of the Ypres bastion was still in Allied hands, while the enemy held the important line of the Lys just to the southeast, along which he was assembling very heavy forces.

[1] Sir John French, Despatches, p. 121.

During these movements in the south and center of the Flanders plain, the Belgian army, supported by French forces, was falling back from Antwerp. King Albert, "judging that no other line offered as great advantages, decided to establish the army on the Yser and to place this line in a state of defense."[8] On October 15 the Belgians and French took up the new position along the west bank of the Yser from Zuydschoote, five miles north of Ypres, past Dixmude to the sea at Nieuport, leaving outposts on the eastern bank. The Allied front was now based on the river-canal barrier from the sea to a point south of Dixmude, thence on the ridge positions from near Passchendaele southward beyond Gheluvelt, whence it cut across the plain to follow southwest up the western side of the Lys nearly to Armentières. From here it ran across the plain past La Bassée to the edge of the chalk upland of Artois. In the north this position was a strong one, but it was held by wholly inadequate forces in the face of an enemy already greatly superior in numbers and constantly growing stronger.

The Battle of the Yser

The moment had now arrived when the German armies, confronted by a continuous line of enemies from Switzerland to the sea, must give up any hope of outmaneuvering the Allied line by a turning movement around its left end and launch a direct attack against some part of the front. The sector selected was the plain of Flanders, and the two immediate objectives were to force the Yser barrier and smash the Ypres bastion. If these two formidable natural obstacles could be conquered, the Channel ports might be seized, direct communication between England and the Continent endangered, and the left wing of the Allies rolled up or forced to fall back on the Artois upland. In the latter case the Allied line would form a dangerous strategic salient with its apex near Arras, which, broken at any point by a German offensive, would compel an Allied withdrawal to the line of the Somme.

[8] Military Operations of Belgium in Defence of the Country and To Uphold Her Neutrality: Report Compiled by the Commander-in-Chief of the Belgian Army for the Period July 31st to December 31st, 1914, London, 1915, p. 65.

Thus the German line would be greatly shortened and the fertile fields of Picardy added to the German holdings. The way for an advance on Paris would once more lie open.

The first attack was directed against the line of the Yser from Dixmude to the sea at Nieuport, and both Belgians and French

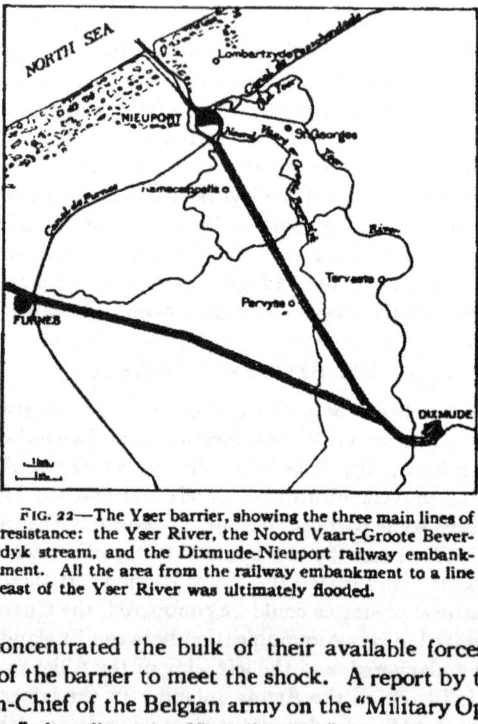

FIG. 22—The Yser barrier, showing the three main lines of resistance: the Yser River, the Noord Vaart-Groote Beverdyk stream, and the Dixmude-Nieuport railway embankment. All the area from the railway embankment to a line east of the Yser River was ultimately flooded.

quickly concentrated the bulk of their available forces behind this part of the barrier to meet the shock. A report by the Commander-in-Chief of the Belgian army on the "Military Operations of Belgium" describes the Yser line as "an excellent defensive position" and "tactically a strong one." The left flank rested on the sea, which was in friendly hands and where the submerged banks off Nieuport offered fair shelter to a fleet engaged in supporting land operations by its fire. The river itself was about 65 feet wide, diked on both sides, and passable only by bridge or boat. Fortunately the western bank commanded the eastern one, being about 6 feet higher. Roughly parallel to the river and just west of it runs another stream about 30 feet wide, called the Groote Beverdyk in its upper portion, the Noord Vaart farther down near where it joins the Yser at Nieuport (Fig. 22). This stream forms a good second line of resistance. Just west of it is a third line, the embankment of the Dixmude-Nieuport railway, raised from 3 to 6 feet above the wet plain. All this region, a gulf only a few centuries ago as we have already seen, was low, marshy, and muddy, intersected with a network of ditches filled with water, largely devoid of cover for attacking forces, and subject to inundation at any time (Fig. 15). The principal point of weakness in the line was the large meander of the river near Tervaete, half way between Dixmude and the sea, called "the Tervaete bend." As this meander is convex toward the north, an enemy on the northern side can concentrate fire from three directions upon the defenders within, render it untenable, and then effect a crossing.[4]

At its two ends the line was buttressed by strong positions. Nieuport is a center upon which six canals or streams converge, all of them passing east or north of the town, while the dunes offer cover for troops holding a bridgehead north of the watercourses. This bridgehead was of vital importance, for at Nieuport were located the gates which would admit the sea to inundate the Yser valley, should this become necessary as a means of defense. The security of this important buttress was further guaranteed by Allied control of the sea. Dixmude, at the other end of the line, was defended by strong positions in its suburbs, protected by the partly submerged marshy bottom of the Handzaeme lowland on the east, by the Handzaeme canal, and by a railway embankment on the east and south. Altogether the line was one of formidable strength.

Never before had an enemy attempted an invasion by this route. Ludendorff chides the Allied commanders for their

[4] Military Operations of Belgium, p. 72.

stupidity in seeking to attain great strategic ends which tactical conditions rendered impossible of achievement. The best example which the war affords of such unwisdom is the attempt of the German high command to gain the grandiose strategic ends outlined above by an assault on the impregnable natural barrier of the Yser. Such a blunder can be excused, if at all, only on the ground that the German leaders believed their overwhelming numerical superiority was more than a match for any natural obstacle, however strong, when defended by nothing more than a wretched remnant of the little Belgian army and a few French forces hastily rushed to their assistance.

On October 16 the Battle of the Yser began with a reconnaisance in force against Dixmude. By the 18th Nieuport and parts of the river front between the two buttresses were feeling the enemy pressure; a British flotilla, supported by French ships, was shelling the Germans attacking along the historic highway, and "dryway," of the dunes; and the battle was becoming general. One determined assault after another, launched against the Nieuport bridgehead, was beaten off with heavy losses to the enemy. Violent attacks against the Dixmude bridgehead, some of them delivered in the darkness of night, nearly achieved success at one moment, but the lost positions were recovered. A furious bombardment deluged the whole front and the Belgian back areas with shells, while heavy assaults on the Tervaete bend, fully recognized by the Germans as the weakest point along the barrier, menaced the Belgian center. By October 21 seven German divisions were flinging their whole weight furiously against the line of the Yser. That night, under cover of darkness, a crossing was effected on a temporary bridge at the dangerous Tervaete bend, and, despite the efforts of the Belgians to execute the order to hold the chord drawn across the base of the meander "at all costs by clinging to every inch of the ground," the condition was so grave there on October 23 that the Belgian command appealed to the French for support, at the same time again ordering the Belgian center "to hold out to the last extremity." Fortunately, every attempt made by the Germans to cross the barrier elsewhere had proved a costly failure. On the other hand, an Allied offensive by troops debouching from behind the barrier had to be abandoned in order to help the imperiled Belgian center.

The German commanders now concentrated all their efforts against the weak center at Tervaete bend and drove the defenders from their position along the chord. The latter thereupon fell back behind the line of the Groote Beverdyk. This exposed the right flank of those defending the river farther west, and they fell back behind the lower course of the same stream, where it is called the Noord Vaart (Fig. 22). Only the Dixmude end of the main river position still held, and the enemy now made a supreme effort to capture that part of the line. Fifteen assaults in close succession were hurled back by the gallant Belgians and the heroic French marine fusiliers, and still the enemy returned to the onslaught with dogged perseverance.[5]

The defenders, hopelessly outnumbered and reduced to a state of extreme exhaustion, lacking reserves, and faced with a shortage of munitions, now required all the aid that the natural advantages of their position could give them. "The key of the position was not at Dixmude, nor at Pervyse, nor at Ramscappelle, nor at Ypres; but in the pocket of the chief hydrographic officer who guarded the locks at Nieuport."[6]

A council of war on October 25 decided to flood all the Yser region east of the railway embankment forming the third line of resistance. Accordingly all aqueducts through the embankment were sealed up, and the sluices at Nieuport were opened when the tide was high in order to let in the sea, and closed when the tide was low to prevent the accumulating river, rain, and marine waters from escaping. It was none too early, for on the 26th the Germans in irresistible numbers drove the Allies from behind the Beverdyk, compelling them to seek protection behind the railway embankment. But the terrible strain of continual assaults through mire and water, over rivers, canals, and ditches,

[5] Military Operations of Belgium, p. 81.

[6] Charles Le Goffic: Dixmude, *Rev. des Deux Mondes*, Vol. 26, 1915, pp. 169–194, 370–403; reference on p. 390.

was beginning to tell on the invaders. As they paused for breath, the new ally of the Allies made its appearance. About the 28th the Germans discovered that water was rising in the ditches and beginning to flow over the flat land. The peril was imminent, and on the 29th they began a new series of furious attacks, hoping to break through before it should be forever too late. Their only hope lay in capturing the Nieuport buttress, containing the engineering works by which alone the floods could be controlled. But this strong position, supported by the Allied flotilla off the coast, defied every assault. Just to the east the gray-clad invaders, struggling through the mire and the rising waters, "hideous, drenched to the waist, and smeared with mud to the tops of their heads," stormed the railway embankment and entered Ramscappelle. They maintained themselves in the village for a brief space but were soon thrown back. Elsewhere the line held.

The waters continued to rise. Soon from the Dixmude-Nieuport railway embankment the Allied defenders looked out across a flood two or three miles wide, covered with débris and swollen German corpses (Fig. 17), which effectively blocked the Kaiser's frantic drive toward the Channel ports. The Teutonic hordes were faced by a liquid trap which yawned to engulf them. Beneath the muddy waters were hidden ditches and canals into which men would suddenly plunge over their heads, and bottomless mud which would hold them fast in the flood. The level surface of the liquid could offer no shelter from the fire of sharpshooter and machine gunner, and those who fell wounded must drown where they dropped. Numerical superiority and Prussian discipline were alike helpless before such a barrier. The Germans retired northward beyond the flood, abandoning quantities of arms and ammunition. And while they might later seize the ruins of Dixmude north of the river, the barrier itself was absolutely impregnable. The Battle of the Yser was over.

The First Battle of Ypres

While the Battle of the Yser was still in progress the Germans were preparing the blow designed to shatter the Ypres bastion.

The Allied command, judging that the best defense would be an offensive movement eastward and northeastward from the bastion, and seeing in the situation along the Yser an added reason for attacking the German forces opposed to them, on October 21 ordered an advance northeastward against the northern part of the Messines-Passchendaele ridge, still held by the Germans. After taking the ridge the offensive would be developed as far as possible in the direction of Roulers and Thourout, thereby flanking the Germans attacking the line of the Yser, with Bruges and Ghent as more remote objectives. A little ground was gained, but the stiff German resistance, the growing menace of increasing German numbers, and the necessity of awaiting Allied reinforcements made further attacks against the northern end of the ridge inadvisable. The troops were then ordered to strengthen their positions along the ridge east and southeast of Ypres from Zonnebeke to Messines and to hold fast until a new French army hurrying northward should enter into the line.

On October 22 and 23 the Germans attacked with great determination, gaining some ground but suffering heavy losses. The Allies counterattacked with equal energy and tried on the 23rd to continue their offensive, but were held up at several points, their center in particular being unable to cross the small brook flowing from near Passchendaele down the west side of the ridge. For nearly a week the Allied forces endeavored without much success to drive the enemy from the northern part of the ridge about Passchendaele.

On the 29th the Germans, exhorted by their leaders to strike "the decisive blow," returned to the assault with redoubled energy. They assembled enormous forces opposite the three-mile sector of the ridge between Gheluvelt and Hollebeke and for two days hurled them against the barrier with almost irresistible fury. On either side the main attack was supported by other troops. At a terrible cost they won the crest in the Gheluvelt region and at several points farther south. The moment was critical in the extreme. But additional French forces were thrown into the fight; the anxious leaders were encouraged by the imperturbable optimism of the genius now directing the battle from the highest point in Flanders as later he would direct the whole gigantic struggle from the highest position of military authority; and the men responded magnificently to the demands of their leaders. The enemy was hurled back down the eastern slope, and the Kaiser, now on the front, awaited in vain the moment for entering Ypres. Furious counterattacks again gave the Germans a foothold on the crest, but the bulk of the territory previously held by the Allies remained in Allied possession. On November 5 the Kaiser returned to Germany.

The First Battle of Ypres was in reality a *bataille de rencontre*. Two offensives, hurled against each other, both came to grief. If the Germans were unable to secure full possession of that part of the Messines-Passchendaele ridge southeast of Ypres, the Allies failed equally to gain the portion northeast of the town. The Ypres bastion was not securely in the hands of either combatant, and a resumption of the struggle for mastery of this critically important terrain could not long be delayed. For the Allies the tactical situation was anything but satisfactory, because their line described an awkward salient about Ypres, permitting a concentration of enemy fire upon that junction point of all important roads, railways, and canals in the vicinity. German possession of part of the ridge seriously complicated the situation and called for an opening out of the salient by pushing the enemy northeastward and southeastward beyond the heights and into the plain below. For the Germans there was no hope of victory in Flanders, now that passage of the Yser barrier was definitely excluded, until their hold on the Ypres bastion had been extended into conquest of the entire obstacle.

Both combatants prepared an immediate renewal of the struggle. A continuous and violent bombardment by the German artillery and the massing of new forces behind their front made their intentions clear. On the Allied side orders were given November 6 for an offensive which should clear the enemy from his footholds on the bastion northeast and southeast of Ypres and open out the salient by advancing its two sides. But every attempt at progress encountered violent resistance. On November 9 the attacks and counterattacks became more violent along the whole line from Dixmude to beyond Messines, and on the 10th the Germans launched their supreme assault. A division of the Guard, secretly brought to Flanders and advised by the Kaiser that he counted on them to succeed where their comrades had failed, was hurled against the Allied front.

Along the Yser Canal from Dixmude southward, now flooded by the waters backing up from the locks at Nieuport, a subsidiary attack gave to the enemy that part of Dixmude east of the barrier and secured him a precarious foothold on the west bank at the Maison du Passeur, halfway between Dixmude and Ypres. Elsewhere the barrier held firm. But the grand assault was delivered along the Ypres bastion farther southeast. It made progress, but only at terrible cost. "The regiments of the Guard . . . had been, north of Gheluvelt and between Zonnebeke and Passchendaele, so badly used up that they were gasping. Elsewhere certain regiments . . . had been, in the region of Poelcappelle, almost completely annihilated. 'On the 10th,' wrote one of the soldiers, 'we launched an assault in which almost the entire batallion was wiped out. In my company, in one hour, all fell except one officer and fifty men.'"[7] Even at such a price the gains were very moderate. The Messines end of the Messines-Passchendaele ridge was made securely German for the time, while the northern, or Passchendaele, end they still firmly held. In the center it had proved impossible to dislodge the Allies from their hold on the crest about Gheluvelt. Capture of a small part of the difficult terrain had exhausted the enemy's offensive power.

This second phase of the battle, sometimes called the "Second Battle of Ypres," left the fate of the bastion still undecided. The Allies held the southern wall from Mont Kemmel westward and a central section of the eastern wall; while the Germans were

[7] Louis Madelin: La Bataille des Flandres, *Rev. des Deux Mondes*, Vol. 40, 1917, pp. 241–276, 506–539; reference on p. 530.

firmly established along the two ends of the eastern wall. Neither offensive had succeeded. Ypres was exposed to direct enemy observation from the Messines-Passchendaele ridge and suffered from accurate artillery fire. The salient, instead of being widened, had been rendered more cramped and dangerous than ever. On the other hand, the German effort to crush the obstacle had failed, the left wing of the Allies was still secure, and the Channel ports seemed farther away than ever. Neither side could succeed in any large operations in Flanders until the Ypres bastion was wholly conquered. Certainly the struggle for mastery was merely adjourned.

The Second Battle of Ypres

During the winter of 1914–1915 there were many local combats around the Ypres bastion, in the course of which the Allies gained a little terrain in the direction of Passchendaele and elsewhere but lost more of the ridge crest near Gheluvelt and east of Mont Kemmel. It was on the morning of April 23, 1915, that the French and Canadian troops holding the naturally weak segment of the front connecting the Yser River-Canal barrier near Lizerne* with the Messines-Passchendaele ridge barrier near Zonnebeke, saw a greenish-yellow cloud rolling toward them across the flat plain. In order to flank the Allies from their last hold on the eastern wall of the Ypres bastion, the Germans were resorting to a new and barbarous offensive weapon, poison gas. Completely taken by surprise, strangled by an enemy they could not combat, the Allies retreated in disorder. On the northwest the Germans succeeded in crossing the canal in the confusion and established bridgeheads on the west bank at Lizerne and neighboring points. The northern side of the salient was pushed a couple of miles nearer to Ypres, and the crest of the ridge cleared of Allied troops. Nothing but the most heroic action on the part of the defenders of the bastion prevented a complete break in their front and the loss of the entire stronghold. Fortunately the rush was stopped in front of Ypres; long and bitter fighting ejected the Germans from the west bank of the canal, and the

* Five miles north of Ypres. Not shown on Pl. I.

Yser barrier remained intact. But the east wall of the bastion was in German hands, and the defenders of Ypres were subjected to all the tortures of an inferno as accurately controlled artillery fire poured upon them from every part of the salient.

Local Combats

In the southern part of the Flanders plain La Bassée had continued to prove a strong point which held firm despite repeated local fluctuations of some magnitude north and south of it. To the north the British in March introduced, at Neuve Chapelle, the system of massed artillery fire on a limited front; but, despite its success in destroying the Germans' front trenches, only a small advance could be driven across the level plain where the attackers were fully exposed to a deadly fire. The British lost over 10,000 men on a narrow front in a few hours. During the first battle of Vimy Ridge in May the British again lost 8,000 men on the same terrain. South of La Bassée the British co-operated in the second battle of Vimy Ridge (September) by undertaking an advance on Lens across the more rolling plain of the Gohelle, in the transition belt. Loos was captured, and the low but important Hill 70, dominating Lens on the north, was seized but could not be held. The German positions in the chalk were strong, the British organization for supporting and relieving attacking troops on the badly exposed terrain was very defective, and the losses in a few hours were so heavy that the operation was regarded as a disaster. Not until command of the plain had been secured by full Allied control of Vimy Ridge could operations on the plain below be carried on to advantage. The battles for Vimy Ridge and the Arras bastion are treated in a later chapter.

Throughout the rest of 1915 and all of 1916 the battle line in Flanders remained practically stationary. There were local struggles at various points on the front as each side endeavored to secure some topographic advantage, some point of better observation, in preparation for larger operations in the future. The war on the surface was now accompanied by the war of subterranean mines on a large scale. So valuable were hill positions on the level plain, and so difficult to take by surface operations alone, that extensive tunnels were run under the hills and the summits, in some cases, literally blown off. In the confusion the hill might be seized by the attacking forces, the lines of the former holders deluged with artillery fire accurately directed from the point of vantage, enemy movements in the plain so well observed as to make surprise counterattacks difficult or impossible, while the new possessors of the elevation would enjoy comparative immunity from direct observation in their back areas. It was during the first half of 1916 that the bloody struggles for individual hills, mentioned on earlier pages (pp. 35–37), took place.

Many of these local operations were hampered or defeated by the water and mud of the Flanders plain. Sir Douglas Haig's despatches covering this period abound with references to the extraordinary difficulties. During the fighting for the Bluff "heavy rain turned the ground into a quagmire so that progress was difficult for the attacking force." At St. Eloi "the work of consolidating our new position . . . . proved extremely difficult, owing to the wet soil, heavy shelling, and mine explosions; though pumps were brought up and efforts at draining were instituted, the result achieved was comparatively small. By dint of much heavy work the brigade holding these trenches . . . succeeded in reducing the water in the trenches by two feet by the morning of the 5th. This state of affairs could not, even so, be regarded as satisfactory; and during the 5th the enemy's bombardment increased in intensity, and the new trenches practically ceased to exist."[8] As one soldier is reported to have said, when told to "consolidate" his position: "It is impossible to consolidate porridge." Men were swallowed up in the mire and suffocated, while rifles became so caked with mud that firing was impossible. Bayonets and bombs alone could be used in some of the assaults.[9] The fields became bogs and the trenches canals. Little progress could be made in such a terrain.

[8] Sir Douglas Haig's Despatches, London, 1919, pp. 6, 8.
[9] A. Conan Doyle: The British Campaign in France and Flanders: 1916, London, 1917.

The Third Battle of Ypres (Battle of Messines Ridge)

In the meantime the British were at one place carrying on mining operations on a scale never before attempted in warfare. The Ypres bastion was again to become the center of a violent struggle, in which the British would seek to dislodge the enemy from its eastern wall, the Messines-Passchendaele ridge.

The positions held by us in Ypres salient since May, 1915, were far from satisfactory [writes Sir Douglas Haig]. They were completely overlooked by the enemy. . . . They were certain to be costly to maintain against a serious attack, in which the enemy would enjoy all the advantages in observation and in the placing of his artillery. Our positions would be much improved by the capture of the Messines-Wytschaete Ridge*, and of the high ground which extends thence north-eastwards for some seven miles and then trends north through Broodseinde and Passchendaele. . . . The village of Messines, situated on the southern spur of the ridge, commands a wide view of the valley of the Lys, and enfiladed the British lines to the south. Northwest of Messines the village of Wytschaete, situated . . . on the highest part of the ridge, from its height of about 260 feet commands even more completely the town of Ypres and the whole of the old British positions in the Ypres salient. . . . The natural advantages of the pos.tion were exceptional, and during more than two years of occupation the enemy had devoted the greatest skill and industry to developing them to the utmost.[10]

Ludendorff emphasizes the importance of the ridge to the Germans, referring particularly to the value of direct observation of enemy positions from the ground and of having his own back areas shielded from the view of the enemy.[11]

But experience had shown that any operation against the formidable obstacles of the ridge must, in order to be successful, be conducted on an elaborate scale, with large forces, and only after the most careful preliminary preparation. Accordingly early in 1916, nearly a year and a half before the blow was launched, the British began the building of the network of roads

* Southern part of the Messines-Passchendaele ridge.

[10] Haig, pp. 82, 105.

[11] Erich von Ludendorff: Ludendorff's Own Story, August, 1914–November, 1918: The Great War from the Siege of Liège to the Signing of the Armistice, As Viewed from the Grand Headquarters of the German Army, 2 vols., New York, 1919; reference in Vol. 2, p. 101.

and railways necessary to move troops, guns, munitions, and other supplies in enormous quantities with great speed, and the excavating of a series of tunnels under the ridge in order to blow the Germans out of their strong position by a series of mine explosions. The work of preparation was peculiarly difficult on account of the nature of the terrain. On the flat plain there was nothing to conceal the new roads and railways undergoing construction, and enemy observers on the ridge crest directed the accurate registering of these for artillery fire when they should come into use. Yet without an abundance of good roads as well as railways the attackers would find their task impossible in a region of mud and marsh intersected by countless streams, canals, and ditches. There was therefore no possibility of a surprise attack. The enemy would know what was coming, and prepare for it. Only in the case of the mines could a wholly unexpected element be injected into the attack, and there was doubt as to whether the Germans might not even be forewarned of this danger. Sir Douglas Haig speaks of the countermining carried on by the Germans and says they realized the peril,[12] but Ludendorff claims that mining operations had largely ceased, that no sound of underground work on the part of the enemy could be heard, and implies that the explosions were totally unexpected.[13] In any event, so saturated with water were the underground formations and so much trouble did the Germans themselves encounter during their tunneling, that they probably had no conception of the stupendous scale of the British operations, even if they suspected mining. The difficulties were indeed enormous; but they were surmounted through the scientific skill of the chief geologist of the British Expeditionary Force, Lieutenant Colonel T. Edgeworth David, who made a careful study of the alternate layers of sand and clay in the ridge and of the variations of the water levels in each, with the result that tunneling on a grand scale was successfully carried out. After fifteen months of titanic labor twenty-four giant mines were

[12] Haig, p. 104.
[13] Ludendorff, Vol. 2, p. 31.

placed under the ridge, charged with over one million pounds of high explosives.

In the water-soaked plain one of the difficult problems to be solved was the supply of sanitary water to the attacking forces. Advantage was taken of natural lakes and of catch pits constructed on Mont Kemmel, from which water was piped to the front. Sterilizing barges were established on the Lys, and materials assembled for extending the pipe lines up the ridge as the attack progressed. With the aid of pack animals and carrying parties good water reached the troops within twenty to forty minutes after the capture of new positions, while a week after the assault began water was being piped to the crest of the ridge at the rate of between 450,000 and 600,000 gallons daily.[14]

Full appreciation of the formidable character of the ridge obstacle was evidenced not alone by the grand scale on which preparations for the attack were made but also by the care with which the troops were instructed as to the form of the terrain on which they were to operate. Relief topographic models, showing every detail of ridge, spur, ravine, valley, wood, as well as all artificial features, were prepared and carefully studied. The different operations of the attack were rehearsed on the models; and when officers and men knew well both the parts they were to play and the stage on which they were to play them, all was ready for the opening act.

On June 7, 1917, at ten minutes past three in the early dawn, Lloyd George, sitting at his telephone in London, heard the simultaneous explosion of nineteen mines under Messines Ridge. The battle was on. Nineteen volcanoes vomited fire, mud, and men, the country shook with a great earthquake, craters as much as 140 yards in diameter opened in the crest of the ridge and engulfed the defenders. In the awful stillness which succeeded the subterranean upheaval observers at a distance stood appalled, and so unearthly was the scene before their eyes that strong men suffered the impression of having awakened in the hereafter to behold a glimpse of the inferno. Then began the victorious

[14] Haig, p. 103.

advance. Terrified and disorganized, smothered by a terrific bombardment, and drenched with liquid fire pouring on them from airplanes in the sky, the Germans were no match for the British forces which swept over them with the precision of clockwork. Ludendorff admits that "the moral effect of the explosions was simply staggering." In a few hours Messines Ridge for a distance of some seven miles, from its southern extremity to the vicinity of Gheluvelt, was once more in Allied hands.

The consequences of the victory were clearly expressed by two British officers who, according to one account of the battle, stood on the captured crest, the first looking back into the plain which had been a literal hell for him and his comrades during all the long months that Germany had held the ridge, the second looking eastward to where the Germans in the plain below would now become the chief sufferers. "My God," cried the first, "it's a wonder they let us live there at all!" "It's great to look eastward!" said the other.

The Fourth Battle of Ypres

With the southern end of the Messines-Passchendaele ridge in Allied hands, preparations could be pushed for the attack on the northern end. Again the preparations on the flat plain were fully exposed to German observation from the ridge north of Gheluvelt, and from minor elevations, like Pilckem ridge,* within the salient; while the water-soaked clay permitted no such deep dugouts and subterranean cellars and caverns as could be used in the chalk upland of the Somme plain to shelter troops assembling for attack. Sir Douglas Haig complained of both these difficulties of the terrain. "On no previous occasion," he writes, "not excepting the attack on the Messines-Wytschaete Ridge, had the whole of the ground from which we had to attack been so completely exposed to the enemy's observation. Even after the enemy had been driven from the Messines-Wytschaete Ridge, he still possessed excellent direct observation over the salient from the east and southeast, as well as from the Pilckem ridge to the

* Three miles north of Ypres. Spelled Pilken on Pl. I.

north. Nothing existed at Ypres to correspond with the vast caves and cellars which proved of such value in the days prior to the Arras battle, and the provision of shelter for the troops presented a very serious problem."

Preparations were completed by the last of July, and on the morning of the 31st the fourth great battle for possession of the Ypres bastion, often called the "Third Battle of Ypres" because the Messines operation was given a special name, opened on a front of fifteen miles. The main attack was directed northeastward toward the ridge, from that part of the front lying between the Gheluvelt region and the Yser barrier. Fortunately the Germans for some unknown reason, possibly from fear of new mine explosions, had withdrawn some distance from the east side of the Yser Canal barrier north of Ypres, thus permitting the Allied forces holding that part of the line to debouch from behind the obstacle and assist in driving the enemy northeast. The British Commander-in-Chief points out that "this operation greatly facilitated the task of the Allied troops on this part of the battle front, to whose attack the Yser Canal had previously presented a formidable obstacle."

As the attack progressed, the greatest opposition was encountered east of Ypres, where the enemy enjoyed the great advantages of the main ridge position and where forest patches on the sandy formation had been organized into strong points of resistance. Tanks were employed, but encountered great difficulty from the clay soil. Their employment in the preceding battle had not been a great success, partly because they could not advance rapidly enough to be of service. The Allied hold on the ridge crest was extended a little, but only at the cost of heroic exertions. Then intervened in full force the worst enemy of an offensive. Rain began falling that night and continued for four days without intermission, turning the battlefield into a quagmire.

The low-lying, clayey soil, torn by shells and sodden with rain, turned to a succession of vast muddy pools. The valleys of the choked and overflowing streams were speedily transformed into long stretches of bog, impassable except by a few well-defined tracks, which became marks for the enemy's artillery. To leave these tracks was to risk death by drowning, and in the course of the subsequent fighting on several occasions both men and pack animals were lost in this way. In these conditions operations of any magnitude became impossible, and the resumption of our offensive was necessarily postponed until a period of fine weather should allow the ground to recover. . . . This unavoidable delay in the development of our offensive was of the greatest service to the enemy.[18]

In the middle of August the battle was resumed and some ground gained despite great difficulties encountered in the character of the terrain. But again the rain fell, the ground became impassable, and the offensive was reduced to a series of small local operations. It was not until September 20 that major operations were once more feasible. Heavy attacks on an eight-mile front on that day gained additional ground on the ridge and widened the salient farther north. The battle raged for several days, and on the 26th another grand assault on the ridge captured the famous Polygon Wood stronghold and Zonnebeke. Here the struggle raged with the utmost fury, and the ground gained was only held at a heavy cost. On October 4 the British battering ram once more launched its powerful blows against that part of the ridge east of Zonnebeke and despite serious obstacles gained an additional section of the crest. On the plain to the northwest a little mound called 19 Metre Hill caused the attackers considerable embarrassment.

Rains were now frequent, and the ground was in terrible condition. Nevertheless, another assault was attempted October 9, and progress was made under most painful conditions. Fighting in the mire continued, and on the 12th the indomitable British again plowed forward. But they were attempting the impossible. "The valleys of the streams which run westward from the main ridge were found to be impassable. It was therefore determined not to persist in the attack, and the advance towards our more distant objectives was cancelled." Natural barriers were bringing the great drive to a halt. "It was the difficulty of movement," reported the British Commander-in-Chief, "far more

[18] Haig, p. 116.

than hostile resistance, which continued to limit our progress, and now made it doubtful whether the capture of the remainder of the ridge before winter finally set in was possible."

However, the struggle was not given up without further efforts to gain all of the vital crest position. Local battles of the greatest intensity were waged almost daily, and, every time the sun or a favorable wind dried the ground even a little, new assaults on a larger scale were attempted. The men plowed on, "in spite of immense difficulties from marsh and floods in the more low-lying ground," and of vigorous resistance from the enemy posted in excellent positions on the spurs and in the undulations of the main ridge. Ground gained was sometimes yielded because in the advance the rifles of the infantry became so choked with mud that it was impossible to use them when the enemy counterattacked. This was what happened at Gheluvelt. "We went with our rifles and Lewis guns bound up with flannel," wrote one officer according to Conan Doyle, "so as to keep the mud out, and with special cleaning apparatus in our pockets; but you can't clean a rifle when your own hands are covered an inch thick." Nor can one keep the flannel wrappings on guns and at the same time use them in pushing an attack. Ludendorff paints the picture from the German side: "The horror of the shell hole area of Verdun was surpassed. It was no longer life at all. It was mere unspeakable suffering. And through this world of mud the attackers dragged themselves, slowly but steadily, and in dense masses. Caught in the advanced zone of our hail of fire they often collapsed, and the lonely man in the shell hole breathed again. Rifle and machine gun jammed with the mud. Man fought against man, and only too often the mass was successful."[16]

Several successive assaults on the ridge at Passchendaele gave the Allies possession of that village on November 6, but the forces of Nature finally triumphed over the forces of man. The lowland had long been one vast morass. Marshy *beeks*, or brooks, such as the Steenbeek, the Brombeek, the Watervlietbeek, and

[16] Ludendorff, Vol. 2, p. 105.

others, were flooded marshes transformed into an awful mire by heavy shelling. In the seemingly bottomless mud the men could not outmaneuver the concrete "pill boxes" and, fully exposed to their murderous fire, fell by the thousands. On the ridge, where the ground was a little less difficult so far as concerns its physical condition, the form of the surface afforded compensating advantages to the defenders. Artillery support of attacks was difficult because guns stuck in the mud and because shells bursting in the plastic material did less than the normal damage. "Throughout the major part of the Ypres Battle, and especially in its latter stages, the condition of the ground made the use of tanks difficult or impossible." As the mud became worse the task of keeping weapons clean and serviceable became more and more difficult and was an important element in bringing the offensive to a close. The long battle, which had raged for three months and a half with a persistence and a fury theretofore equaled only by the 1916 Battle of the Somme (p. 144), was ended by the intervention of insuperable natural obstacles before its objective had been fully attained. Much had been won at great cost. Most of the Ypres bastion was in Allied hands; the line of the Yser, now flooded from Nieuport to Ypres, impregnable to frontal attacks and solidly buttressed on the sea at one end, was at last reasonably secure from the danger of a turning movement at the other end. But the northern end of the Messines-Passchendaele ridge, the eastern wall of the Ypres bastion, still remained in enemy hands. The whole bastion would, in the opinion of the British Commander-in-Chief, have been captured in a few weeks had not the character of the terrain offered greater difficulties than did the military power of the German armies. Looking back on the whole history of the operations and giving full credit to the fighting ability of the enemy, he could say: "Despite the magnitude of his efforts, it was the immense natural difficulties, accentuated manifold by the abnormally wet weather, rather than the enemy's resistance, which limited our progress and prevented the complete capture of the ridge." [17]

[17] Haig, p. 133.

The Battle of the Lys

While the fourth battle of Ypres was was still in progress, Canadian troops in a brilliant local operation on August 15 seized and held Hill 70, dominating Lens on the north. This low elevation rose enough above the general level of the plain of the Gohelle in the transition belt to give good observation over a wide range of country. With it in German hands the Allies had suffered severely from accurate artillery fire; now that it was in their own hands they could direct a well-controlled fire upon the defenses of Lens. Other local operations occupied the armies on the Battlefield of Flanders during the remainder of the year 1917, but it was not until April 9, 1918, that the plain was to witness another battle on the grander scale.

The great German advance across the plain of the Somme, begun March 21 as the first move in a desperate effort to achieve victory before American strength should be fully developed, was wearing itself out, and Ludendorff was ready to launch a new attack. He had rejected the Flanders plain as the scene of the first offensive operation because in the south the Lys barrier, against which the main force of the blow would strike, was at that season too formidable an obstacle. The possibility of a direct attack against the still more formidable barriers of the Yser floods and the Ypres bastion was not even considered. But in April exceptionally favorable weather had dried the muds and marshes of the Lys valley to some extent, and it seemed feasible to launch an offensive across that part of the plain.[18]

Ludendorff fully realized that the operation could not achieve any real success unless the two strong natural bastions defending the plain on the north and south were captured. He had just assaulted the Arras bastion in vain (p. 188), but would try again by a flank attack from the north as soon as he had pushed westward across the plain some distance. "To have the high ground in our possession," he writes, "was bound to be decisive in any fighting in the plain of the Lys." He would likewise assault the Ypres bastion from the south as progress was made in the plain.

[18] Ludendorff, Vol. 2, pp. 220, 238.

FIG. 23—The Flanders plain, looking southward from Mont des Cats, showing commanding observation obtained from a very slight elevation. The arrows indicate points important in controlling artillery fire. (British official photograph.)

It is possible, indeed, that the bastion, being the key to the whole situation in Flanders, was the main objective of the operations, and that the Battle of the Lys was as a whole merely another Battle of Ypres.

Early on the morning of April 9, under cover of a dense mist or fog, the Germans struck on a ten-mile front from the strong point on slightly higher and drier land in the La Bassée-Givenchy region to a point just southwest of Armentières. Portuguese troops holding several miles of this front were routed, the attackers poured through the gap, and the defenders fell back at once on the natural barriers of the Lys River and its tributary, the Lawe. Bridges over the streams were unfortunately not everywhere completely destroyed, and the pursuers were able to cross at several points. Nevertheless the battle continued furiously along the rivers, and parties of Germans reaching the west bank were repeatedly thrown back. Ludendorff complained that "toward Estaires we did not penetrate far enough, but stuck fast in the Lys position; toward the Lawe, also, our progress was but slight." Other features of the terrain were giving him trouble: "The ground was still soft in places, and this made it very difficult to bridge the enemy's trench system . . . The detachments of tanks we had employed had proved a hindrance. It took a long time to get up guns and ammunition;"[19] while lack of observation on the flat plain made proper artillery support difficult. Gradually, however, the outflanking of their defenses by the German advance across the southern edge of the plain and at the crossings which they had been able to retain compelled the British to abandon the line of the Lys, and the Germans pushed deeper into the pocket. The defenders found it difficult to follow closely the enemy's movements and intentions in a flat country where numerous trees cut off the view, especially in the middle of the plain, far from the rimming heights which alone could offer good direct observation; while their field of fire was frequently cut short by willows and other growth along the watercourses.

Fortunately, Nature had set a limit to the depth of the advance

[19] Ludendorff, Vol. 2, pp. 240, 245.

possible. As the pocket was deepened and widened it was at every point coming closer to the rimming heights which narrow the plain westward until the bounding highlands meet near St. Omer. Soon the advancing enemy would begin to fight with direct observation favoring the defenders (Fig. 23), and then affairs would wear a totally new complexion. Already, indeed, small outlying elevations in the plain were beginning to tell their story. There were fierce combats at Mont de Lille, Windmill Hill, and Mont de Merris, three low hills on an east-west line from 10 to 15 miles southwest of Ypres. These checked the assaults and destroyed the vigor of the offensive. The bottom of the pocket was approaching the eastern edge of the great Forest of Nieppe south of Hazebrouck.

It was high time that the two bastions threatening the Germans on the north and south should be conquered, for the pocket was already so deep that their rear and flanks were seriously menaced. To continue westward into the trap would be folly unless its jaws could be broken. The attack on the Ypres bastion had begun April 10 but was making slow progress. On the 18th a violent assault directed at the Arras bastion was launched on a front extending from Givenchy westward. Here the bastion was protected by the outlying barrier formed by the Aire-La Bassée Canal. At Hinges, northwest of Béthune, the canal has a northward bend around a hill which commands long stretches of the barrier to the east and west. If the Germans could seize this hill they might breach the preliminary defense and attack the main obstacle. Eighteen gray-clad battalions leaped to the assault, but the line held. A well-directed British fire mowed down the enemy troops where they attempted to cross bridges over the canal, and the strong point at Givenchy resisted every effort at capture. Along the whole barrier the attack met a bloody repulse. The southern bastion was unshakable.

The Fifth Battle of Ypres

On the north the preliminary attacks were more successful. As already stated, the Ypres bastion was assaulted April 10, the

Germans sweeping forward on a wide front from Ploegsteert Wood north of Armentières to Hollebeke. Filtering along the valleys of the Douve River and other streams heading against the ridge, under cover of a mist which made defense difficult, parties of the enemy outflanked the British positions at Ploegsteert Wood and Messines and secured a foothold on the southern end of the barrier. The lower crest at Neuve Église, parallel to, and an outer defense of, the main east-west Mont Kemmel-Mont des Cats ridge, was bitterly contested for two days; and when the British were forced from it they had to yield at the same time other portions of their lines dominated from it. The southern wall of the bastion was being gravely menaced.

Day after day the struggle raged with unabating fury, the enemy increasing his hold on the southern end of the eastern wall of the bastion inch by inch until both Messines and Wytschaete were in his possession. Then on April 17 the attack was concentrated on the adjacent Mont Kemmel (Fig. 3), forming the eastern end of the southern wall. Two heavy assaults were repulsed on that day. French troops now took over the defense, while the Germans prepared to renew their attempts to capture the commanding position. A week later the slopes of Kemmel were flowing blood as German Alpine troops, skilled in hill fighting, were storming the position, supported by a grand assault against the whole eastern end of the ridge from north of Bailleul to west of Wytschaete, and by "a frightful charge of airplanes like tempestuous cavalry" which rained fire from the sky.[20] With prodigal disregard of the cost in lives, Von Arnim flung his masses forward again and again until their very numbers smothered the defenders and Mont Kemmel passed into German hands.

The situation was now serious indeed. The apex of the bastion, including the ends of both the southern and eastern walls where they joined, was held by the enemy. From Mont Kemmel the Germans overlooked the entire system of defense and all communication lines behind the eastern wall, where the British still

[20] Louis Gillet: La bataille des Monts de Flandre, *Rev. des Deux Mondes*, Vol. 51, 1919, pp. 640–670; reference on p. 663.

held their advanced positions along the crest, won at such a terrible price the preceding year. Their position was too perilous to endure. Already the British Commander-in-Chief had reduced the risks by withdrawing from Passchendaele behind the marshy valley of the Steenbeek, thus yielding the more northern portion of his ridge holdings and decreasing the number of troops in the apex of the Ypres salient. But so valuable to the enemy was the dominating height of Mont Kemmel that surrender of the entire eastern wall became a painful necessity. With heavy hearts the British retired from the Messines-Passchendaele ridge April 26 and 27, and drew their lines tighter about Ypres.

It remained for the Germans to take the rest of the southern wall of the bastion. With the Mont Kemmel-Mont Cassel ridge wholly theirs, the Allies would be compelled, as Ludendorff points out, to abandon the impregnable Yser barrier. All Belgium would be conquered, the Channel ports gained, and the Allies forced back on the Artois upland with their front in the form of a dangerous salient which they could hardly hope to hold.

Without delay the Germans pressed forward to complete the great task. French and British counterattacks to recapture Mont Kemmel were first beaten off, the new defenders of the hill using the marshy Kemmel Brook to excellent advantage. The French were completely held up at this small but difficult obstacle. British troops crossed waist deep in the water but could not effect a permanent lodging on the slopes beyond. Now the Germans in their turn rushed to the assault. But the little valley could not be passed. Four times they swept down the smooth slopes to drive the Allied forces from behind the barrier, and four times a murderous fire mowed them down as with an invisible scythe. Only a heavy blow on a large scale could shatter the defense, and Von Arnim's army was incapable of striking such a blow. Too large a proportion of it lay dead on the sloping walls of the bastion, and the survivors were too exhausted for any more supreme efforts. If the bastion had not been strong enough to withstand entirely the blows of the German battering ram, it had at least smashed the machine beyond repair. Ludendorff might dream of renewing the offensive in this region and completing his difficult undertaking; but the dream would never come true. The Ypres bastion, its walls broken and bloodstained, had saved the whole left wing of the Allied armies from incalculable disaster.

Withdrawal from the Lys Salient

It was Ludendorff's intention, after his offensives on the Aisne and Marne had weakened the Allies' power of resistance and had led them to reduce their forces in Flanders, to strike again on the Flanders front and attempt the complete reduction of the Ypres bastion, the key to the Allies' whole position on the low plain.[21] But Foch's great counteroffensive blasted all such hopes. Eight days after Foch launched his attack on the Marne plateau (p. 312) the German leaders knew their Flanders offensive was doomed. The German armies must economize men by shortening their front. About July 26 they began the evacuation of the vast accumulations of munitions and stores from the dangerous salient which they had made in the hope of conquering the Ypres bastion and in which they had maintained themselves only at a heavy cost in casualties inflicted by Allied artillery fire concentrated from all sides and accurately directed from the rimming heights. Early in August local withdrawals of troops in the salient began, and at the end of the month the retirement on a grand scale was under way. German rear guards made excellent use of the topographic features of the plain, a few men protected by natural obstacles holding the Allies at bay from point to point so that the retreat was able to proceed in an orderly fashion. Along the Neuve Église ridge and at Hill 63 the resistance was particularly hard to overcome. But the enemy was not trying to stay in the salient, he was trying to get out; and for this reason topographic advantages were utilized for the moment only, being yielded up whenever the pressure of the pursuers became uncomfortably strong. Early in September the Ger-

[21] Ludendorff, Vol. 2, pp. 253, 278.

mans were out of the trap, Mont Kemmel was surrendered, and the battle line ran across the plain from Givenchy to the apex of the Ypres bastion near Messines. The Germans now held the eastern wall of the bastion, the Allies all of the southern wall.

The Sixth Battle of Ypres

On the 28th day of September, 1918, four British divisions sprang out of their trenches at the foot of the Messines-Passchendaele ridge east of Ypres, without preliminary bombardment, and in a few hours swept the German defenders from the crest and down into the plain to the east. To the south the crest of the ridge about Messines and Wytschaete was attained, while Belgian troops farther north cleared Houthulst Forest and reached the northern continuation of the ridge west of Staden. The King of the Belgians was leading an Allied army of Belgians, French, and British out of the Ypres bastion to final victory.

Opposed to them was a weakened and demoralized German army. Less than five divisions were found defending the vital ridge crest east of Ypres. Their quality was far below that of the German army of former days. For more than two months an unbroken series of disasters had been sapping their confidence and lowering their morale, while the disintegrating influence of a skillful propaganda was beginning to destroy their discipline. No topographic barrier, however strong, could be held by such an army. In a single day the British won at slight cost what they had fought long bloody months to attain a year before.

A few days later King Albert's forces held all the Ypres bastion. On the north the Belgians had passed beyond the ridge from Dixmude to Staden and were facing the marshy lowland of Handzaeme. Farther south the British were far out over the plain to the east and approaching the lower Lys. Between them the French were beyond the ridge and advancing on Roulers. On October 2 the Germans began a further retirement on the southern Flanders front, from the edge of the chalk upland south of Lens to the Lys at Armentières.

The Retreat from Flanders

Henceforth it was an unbroken story of retreat, sometimes voluntary as to the given sector because of disaster elsewhere, sometimes forced at that point by a new Allied offensive; always pressed heavily by the armies of His Majesty the King of the Belgians, which now included two divisions of American troops. Throughout the retreat it was the series of more or less parallel rivers, whose northeastward courses were determined by the initial slope of the plain, which the Germans utilized most extensively as temporary lines of defense. By the middle of October the Germans were back behind the Lys barrier on a twenty-five mile front, from near Armentières to northeast of Courtrai. Farther north the retreat was in rapid progress; Ostend and Bruges were evacuated, and in a few days the northern sand belt west of the Eecloo canal was free of an enemy who had not paused to take much advantage of such topographic defense lines as existed. South of the Lys the retiring Germans stood for a time along the Deule, offering strong resistance behind its marshes and canal. A week later it was the Scheldt from Valenciennes to Avelghem which protected the defeated hordes of the Kaiser on a front of 40 to 50 miles.

The Scheldt was part of the famous "Hermann Stellung," which near Avelghem crossed to the Lys, followed it northward to the Eecloo canal, then followed the canal to the Dutch border (Fig. 55). South of Valenciennes the Hermann Line followed up the Selle River as we shall see in a later chapter (p. 211). This great defensive position, based almost entirely on natural obstacles, was reconnoitered by the Germans, and the work of strengthening it begun as soon as it became evident that a retreat from Flanders was unavoidable. The entire Hermann Stellung, across the Flanders plain and into the chalk country farther south, was now occupied by the Germans, who were greatly relieved to have a practically unbroken barrier of marshy valley, river, and canal, 150 miles long, between them and their enemies. When the Hermann Line was finally shattered along the Lys, the Scheldt, and the Selle, the Germans fled eastward toward the Antwerp-Meuse Line, their last important defensive system west of the Rhine. The war in Flanders was over.

CHAPTER III

THE BATTLEFIELD OF THE SOMME: THE DRY CHALK PLAIN BATTLEFIELD

In the early summer of 1914 a dusty traveler on the highway from Amiens to St. Quentin paused in the open expanse of the rolling plain to look about him. North, east, south, or west, everywhere the same monotonous landscape met his gaze, the same dreary waste of country stretching away in gentle undulations to the level horizon. The main highway descended a gentle slope at his feet and in a straight line crossed a shallow depression hardly deserving the name of valley, to mount the gentle slope beyond, disappearing and reappearing as successive low hills were crossed, like a narrow white ribbon laid down upon the uneven surface with geometrical precision. Near by a minor road curved gracefully around a projecting hill spur and disappeared into a shallow ravine soon lost in the maze of undulations. Far to the north a double row of poplars, silhouetted against the sky line, showed where a more distant highway took its course across the plain, while two dark patches of trees, tiny remnants of once greater forests, rose faintly above the southeastern horizon. An occasional isolated cluster of peasant homes with green shade trees spreading above thatched or tiled roofs, the tall chimney of a sugar refinery, and the smoke of a distant locomotive were the only signs of life in the sleepy landscape. Elsewhere treeless low hills and treeless shallow valleys succeeded each other in endless procession as far as the eye could reach. Such in 1914 was the stage upon which was soon to be played one of the most tragic acts in the World War drama.

Could our traveler have visited the plain of the Somme in

NOTE—For Chapters III and IV the reader should constantly consult the detailed map of the battlefield in the pocket (Pl. II) and the block diagrams (Figs. 14, 36, and 64).

the days before the hand of man changed the features which Nature gave it, he would have found it almost entirely covered by vast forests. Where now he could sweep with his eye a range of country almost equal to that visible on the rolling Great Plains of western America, then he would perforce have

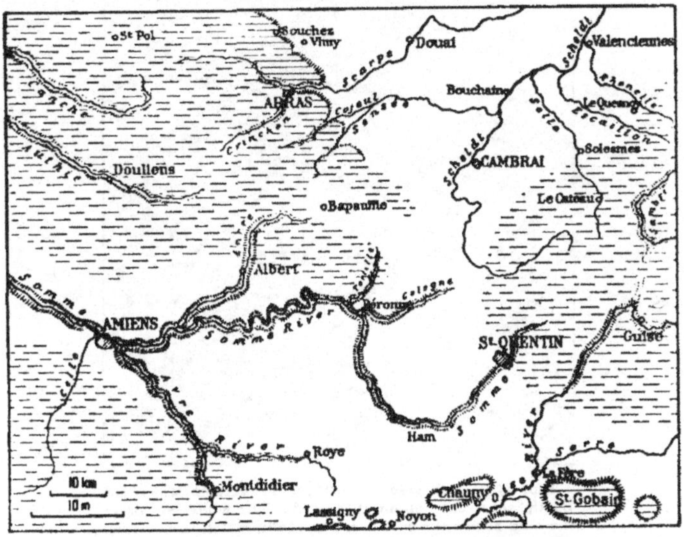

FIG. 24—Generalized sketch map of the Battlefield of the Somme. Ruled areas are higher than adjacent unruled areas, and especially so where the ruling is continuous. For topographic details and place names referred to in the text, see Pl. II and Figs. 14, 36, and 64.

had to thread his way through a woodland so dense that it concealed even what was close at hand. But very early in history the destruction of the forests was begun in order to make way for a more productive agriculture. When the Roman legions first entered this part of Gaul they found large clearings already effected and proceeded to extend their limits. With the coming of the Benedictines in the seventh century and the growth of their agricultural communities the destruction of the forests was carried on more vigorously than ever. In the tenth and eleventh centuries there began another period of rapid deforestation, coinciding with the new period of colonization and agricultural activity by the monks which followed the era of agricultural stagnation due to the Norman invasions.

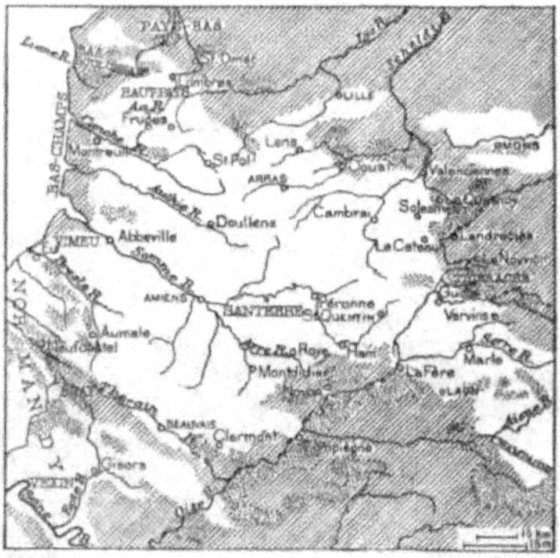

Fig. 25—Chalk area of northern France, on which the Battlefield of the Somme is located. The white portion of the map shows the extent of the chalk, which is, however, often partially concealed by overlying deposits of loam, loess, and clay. Ruled areas represent other rocks. Note that the forests (dotted areas) have been almost entirely cut away from the chalk area, because of the value of the land for agriculture. (Modified after Demangeon.)

The rulers of state and of church, the great lords and the proprietors of vast domains, eager to increase their wealth by turning forest land into productive fields, encouraged their subjects and serfs by rich rewards to cut down the trees and entrusted the exploitation of the cleared lands to the energetic occupants of the monasteries. By the opening of the eighteenth century the forests were reduced to scattered remnants separated by broader areas of cultivated fields. Still the process of destroying the woodlands continued, especially when in the nineteenth century the number of small farmers seeking new lands to cultivate greatly increased. In the Department of the Somme there were cut down in the four decades between the years 1792 and 1833 between 7,000 and 8,000 acres of forest. But it was the introduction of sugar beet culture which finally gave to the woodlands their *coup de grâce*. In the feverish desire to gain every square inch of land for the new source of wealth, infertile tracts which could support nothing but trees were robbed of their only possible produce. Such is the history which transformed the vast forests of an earlier day into that dreary expanse of open, rolling plain (Fig. 25) over which swept the waves of German invasion in 1914.

During the war what was merely dreary became the acme of desolation. Four times the contending armies carried their work of destruction forward or backward over the plain. The productive fields were trampled under foot, seamed with an endless network of trenches, and pitted with the shell craters of countless bombardments. Shapeless heaps of stone, brick, mortar, and tiles reveal the site of former villages; the sugar refineries and other evidences of human industry are gaunt ruins: stumps show where the double rows of poplars lined the roads and where occasional orchards added their wealth to the scattered communities; while the remaining patches of woodland, some the product of reforestation on infertile tracts, are only marked by shattered trunks and splintered branches of the few skeleton trees which still rise from the ruin about them.

Strategic Position of the Somme Battlefield

A glance at a map of northern France shows that the Somme is only one of a series of rivers which flow in remarkably straight courses northwestward to the sea (Fig. 26). The Seine, Béthune, Bresle, Authie, and Canche, as well as the Avre branch of the Somme and a large number of smaller streams, have valleys as rigidly geometrical and as distinctly parallel as if Nature had laid down their courses with the same giant ruler. Even the Thérain, which flows in the opposite direction southeast to the Oise, has a valley parallel to those just mentioned. The secret seems to be that the rocks of the plain are faintly folded, the axes of the shallow folds trending northwest-southeast. Although so faint as usually to be unnoticeable to the eye, these folds were sufficient to guide the running waters and so to determine the location of parallel valleys. The rocks were sometimes broken as well as bent, and the northwest-southeast breaks, or *faults*, likewise determined the course of stream erosion, thus emphasizing the parallel topographic features dependent upon rock structure.

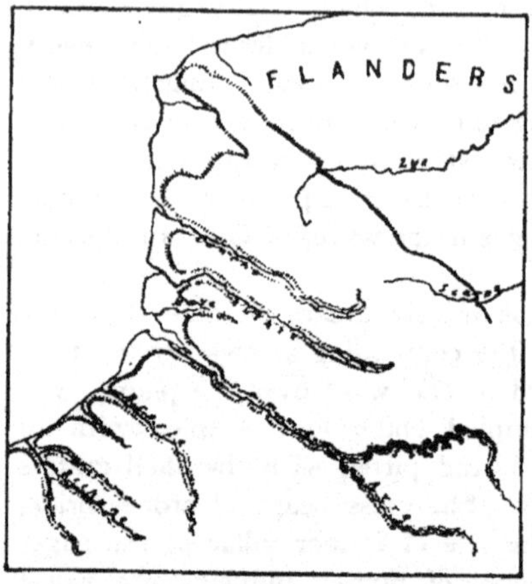

FIG. 26—Parallel valleys of the Somme and neighboring rivers, due to parallel folds and faults in the underlying rocks, and constituting natural defense lines of great historic importance.

Geological structure, by determining surface form, exercises a profound influence upon the military history of a region. It is evident that the valleys just described must form a succession of natural defense lines against an enemy advancing north or south across the chalk uplands of Artois and Picardy. Indeed, it has well been said that these valleys give to northern France a distinctly military character. History offers ample confirmation of this conclusion. During the War of the Spanish Succession the French armies took their final stand in the campaign of 1711 behind the Canche, Marshal Villars establishing his famous *ne plus ultra* lines on the southern side of this natural trench from its mouth to its source, whence they continued eastward behind the marshes of the Scarpe and Sensée. The Duke of Marlborough found the French position along the Canche practically impregnable and directed his strategy against a portion of the line farther east. The Authie next south repeatedly served as an important line of defense, and the fortified town of Doullens, which guarded the eastern end of the valley barrier, so often suffered the horrors of fire and sword that certain etymologists would derive its name from *vallum dolens*, the vale of sorrow. Even the small valley of the Maye takes a significant rank in military geography, for it was along the northern slope of this depression, where it makes a re-entrant into a small side ravine, that the English army deployed to receive the shock of the French attack at the historic battle of Crécy. The soldiers of France, advancing from the south, were forced to cross the natural trench under fire from the English archers. These were placed in rows one above the other on the terraced northwestern wall of the ravine, which was almost impossible of ascent by the French cavalry. Military critics have ascribed the overwhelming victory of the English not merely to the indiscipline of the French forces but in considerable part also to the magnificent natural position chosen by King Edward for the stand of his army.

The River Somme has always been one of the most important military barriers of northern France. It served as a formidable line of defense for the Roman Empire during the time of its decadence, and the valley walls are still dotted with traces of Roman defensive works for a distance of more than fifty miles. In the wars of France against the English, against the Burgundians, and against the Imperialists first one army and then another hurled itself against this natural moat with its steeply sloping walls and its floor of river and marsh. When Edward III invaded France and moved north toward Flanders, it was on the line of the Somme that the French under King Philip first tried to stop him; and Froissart[1] records in detail

[1] Jean Froissart: Chroniques (edit. de Lettenhove), Vol. 5, pp. 1–22, Brussels, 1870.

the repeated failures of the English soldiers to cross the obstacle, over which Edward assured his disheartened followers that "God, the Mother of God, and St. George have provided a passage, I know not where." When, seventy years later, Henry V repeated the invasion and turned north toward Calais, it was again the line of the Somme that the French forces elected to defend, and with such effect that the English army marched along the southern bank some sixty or seventy miles, from near the sea to a point between Péronne and Ham, before finding a crossing over the marshy barrier. In each case it was only through the treachery of a French peasant that a little-known passage, insufficiently guarded, was discovered to the English; in the first instance over shallows at low tide near the sea, in the second by a pathway through the marshes to a ford across the river. And just as Edward's passage of the Somme was the prelude to the English victory of Crécy, so Henry's passage of the same obstacle was the prelude to the English victory of Agincourt.

In 1536 Henry of Nassau moved first on St. Quentin, then on Péronne, with the object of forcing the Somme barrier, which blocked the way to Paris. Philip II besieged St. Quentin in 1557 with the same design of breaching the line of the Somme, but the barrier held him at bay long enough for the King of France to reconstitute his army on the Oise and prevent the advance on his capital. Near the end of the sixteenth century the Spaniards were fighting to force a passage at Amiens. The names of St. Valéry and Le Crotoy near the mouth of the Somme, of Abbeville, Amiens, Corbie, Péronne, Ham, and St. Quentin, places which guard the more important strategic points along the barrier, awaken the memories of many a siege and battle. The first-named suffered sixteen different sieges in four centuries, while Le Crotoy was taken, retaken, and burned twenty times within the same period. So also for Abbeville, Amiens, and the rest of the list one might catalogue abundant proofs that the line of the Somme looms large in military history. Even in the Franco-Prussian War, after the military power of France was practically crushed and Paris was invested, the French Army of the North assembled behind the Somme barrier and disputed with the Germans for its possession until the capture of Amiens, Ham, and Péronne by the invaders sealed its fate. Thenceforth the Germans utilized it as a protective barrier against the remnants of the French army which still held the field farther north, strongly guarding every passage across the river and its marshes while their own ranks were being reinforced. The rectilinear lower portion of the Somme trench is continued southeastward by the marshy valley of the Avre. At its eastern end the Somme-Avre line is protected by the town of Roye (Fig. 24); the importance of which may be gauged from the fact that it sustained no less than thirteen sieges between the tenth and seventeenth centuries.

Enough has been said to demonstrate the enormous military significance of the parallel valleys of Picardy and Artois. It should be noted that these valley trenches are most impressive in their lower courses, in part because the rivers cut deeper and have wider flood plains as they approach the sea, and in part also because the upland rises gradually to the northwest, especially in Artois. Add to this the further facts that the rising upland is more thoroughly dissected into rugged hills as the sea is approached and that it terminates toward the northeast in the formidable barrier of the Vimy Ridge escarpment (p. 101), and one can readily appreciate why invading armies might in general seek to avoid a terrain so difficult to cross, and to choose for their operations the more gently rolling plains farther east, where the shorter valleys have disappeared and where the Somme alone presents a barrier of primary importance. Now it is precisely this latter region that is included in the modern Battlefield of the Somme.

The significance of the position occupied by the Somme battlefield becomes clearer when one notes that to the eastward the land rises to merge in the rough country of the Ardennes Mountains. Between the difficult terrain to the west in Artois and Picardy, already described, and the difficult terrain of the

Ardennes Mountains and foothills to the east there is a low saddle or gap. This is in effect a natural gateway connecting the Paris region on the south with the low country of Flanders on the north; and, since the region of the Somme includes this gateway (Fig. 27) with parts of its northern and southern approaches, one is not surprised to hear it spoken of by the French as the "Seuil de (threshold of) Vermandois." It was inevitable that this threshold or gateway should become the theater of a historic struggle between the French on the one hand and, on the other, the powers controlling or seeking to control Flanders. The great battles of the Somme during the World War were but episodes in the latest chapter of that long, historic struggle.

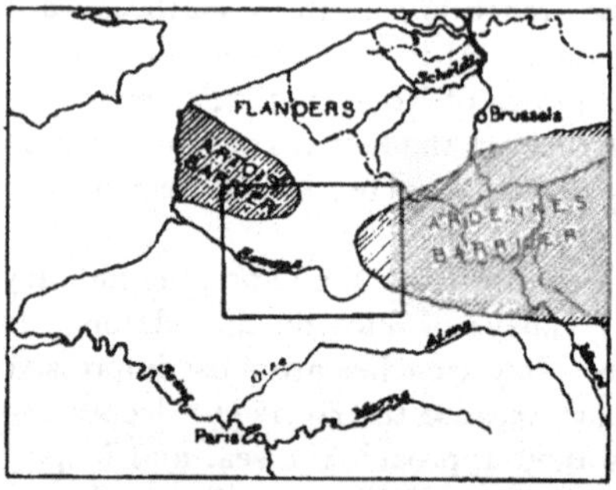

Fig. 27—Location of the Somme battlefield at the natural gateway (between the Artois and Ardennes barriers) connecting Flanders with the Paris region.

But it is not merely in its military aspects that the strategic position of the Somme area becomes apparent. In its economic and cultural relations it has always been little influenced by the lands to the east and west, but profoundly affected by the products and peoples of French Flanders, Belgium, Holland, and even England to the north, and similarly by those of the Paris region to the south; for it was the gateway through which peoples and products moved from one of these regions to the other, and both areas eagerly looked to it for the rich harvests gathered from its fertile plain. In manufactures it was now the Paris district, now the Anglo-Flemish, which predominated in pushing the sale of its raw materials to the factories of Picardy and Artois or which invaded the towns and villages of the plain in the form of colonies of artisans or individual workmen who brought with them the industries of their respective lands. The monotonous surface of the Somme country, interposing few obstacles to ready migration, favored economic penetration from both north and south; and the quiet struggles for economic and cultural control which inevitably resulted from the geographic situation were but preludes to those clashes of arms by which it was sought to crown the work of peaceful penetration by full political control.

The strategic position of the Somme battlefield becomes even more apparent when one considers the great lines of movement across the region, the principal roads, canals, and railways. On first thought one might suppose that the nearness of the sea on the west would result in a principal movement of people and products along east-west routes. The contrary is the fact. The great movement is north and south by roads and railways and by canals which occasionally must pierce the divides between different drainage basins by long tunnels. From the southwest, south, and southeast in France, from Great Britain on the northwest, Belgium and Holland on the north, and Germany on the northeast many of the more important transport lines converge to pass through the gateway between the barrier of Artois and the barrier of the Ardennes and hence to cross the historic Battlefield of the Somme.

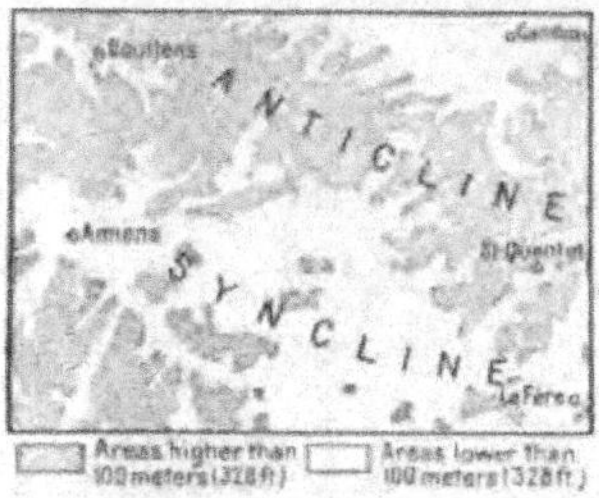

FIG. 28—Map showing the influence of the shallow northwest-southeast folds on the Somme battlefield. The slightly upfolded belts (anticlines) remain on the average a little higher than the down-folded belts (synclines).

When the war of movement in the summer of 1914 changed to the war of position in the autumn and the battle lines extended southward from the dunes near Nieuport to the vicinity of Noyon, there to make a right-angled bend toward the east, special elements of strategic and tactical importance characterized the region of the Somme. It was henceforth one of those

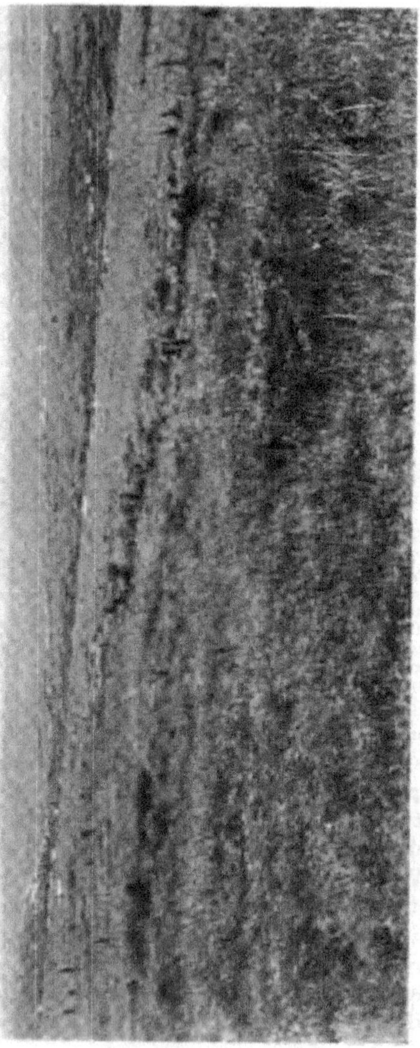

FIG. 29—Looking west across La Boisselle ravine, or Mash valley (just northeast of Albert), from a point near the village of La Boisselle, showing the gentle slopes of the late-mature plain (compare with Fig. 30). Here the loam rests directly on the chalk, and the latter is exposed on valley slopes and in the trenches.

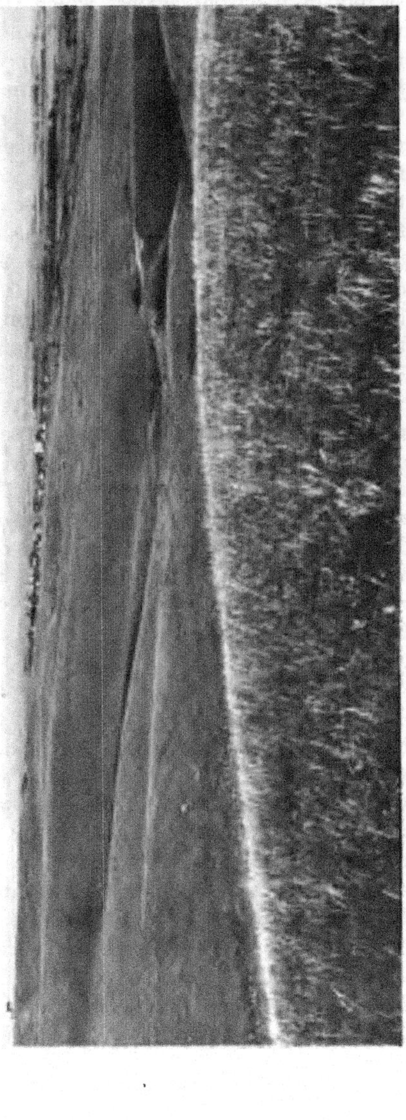

FIG. 30—Late-mature erosion surface of the Great Plains near Bismarck, North Dakota, showing gently rolling slopes similar to those of the Somme battlefield.

sectors of the front against which the Germans, profiting by the advantage of interior lines, could hurl the overwhelming forces they had concentrated at some central point within the bend before the Allies could determine whether those forces were to be moved west, southwest, or south. It was a sector which tempted such an attack, since it offered large rewards for success. An advance down the Somme would not only cut the lines of communication connecting the northern battle front with the rest of France, but if pushed beyond Amiens would drive a wedge between the British and French armies, thus destroying their unity of action and throwing the British forces back upon the coast, where naval communications would form their only effective connection with their allies. On the other hand, a successful attack by the Allies in the Somme region would have the double advantage of compelling the enemy to retreat from the dangerous Noyon salient and of closing the strategic gateway into Flanders, thereby forcing the Germans to depend upon the more limited lines of transport and communication passing through the narrow valleys of the Ardennes Mountains. Thus the precise form of this battle front gave to the historic field of the Somme a temporary and local significance additional to that which it already pcssessed for broader strategic reasons.

Surface Features of the Somme Battlefield

The most striking characteristic of the Somme battlefield is its monotonous succession of low, rolling plain. It is true that the faint northwest-southeast folds, which have produced in the region immediately to the west a marked parallelism of valley trenches and upland strips, are continued into the area now under discussion. A layered map of the battlefield (Fig. 28) shows a suggestion of the northwest-southeast alignment of topography in the form of alternate belts of slightly higher and lower land—the higher belts representing up-folds, or anticlines, and the lower belts down-folds, or synclines. In the 1916 Battle of the Somme the French and British were fighting to get from the lower land of the Somme syncline to the crest of the anticline next north and to force the Germans down into the next syncline beyond. But while slight variations of altitude are of critical military importance, these undulations are so faint and the differences in average elevation

FIG. 31—German camouflage of a road across the Somme battlefield. Roads on the white chalk of the barren plain are readily visible for great distances; hence the necessity for concealing them by strips of this type, which prevent effective observation from enemy captive balloons or other points near the horizon. (French official photograph.)

are so small that even the trained geographer would scarcely remark them. He would probably describe the Battlefield of the Somme as a plain of low relief, dissected by the branches of the Somme River and neighboring streams to a late-mature stage of erosion, in which valleys with gently sloping sides are separated by low, rounded hills or by remnants of the flat upland surface which descend gradually toward their margins to merge with the valley slopes (Fig. 29). The topography is not unlike that of the Great Plains in the vicinity of Bismarck, North Dakota (Fig. 30), which are there in a similar stage of erosion.

THE LOW HILLS

It is clear from what has just been said that there are two elements of prime importance in the topography of the Somme region: the low hills into which the plain has been dissected by stream erosion and the valleys of those streams. Let us first examine in some detail the form and character of the hills, to learn in what way they must have affected the military operations, after which we will turn our attention to the valleys.

Perhaps the most obvious result of the low elevations and gentle slopes of the so-called hills is the facility they afford to travel in every direction. No hills are so high and few slopes are so steep as to offer any real obstruction to road making. The main highways, like those from Amiens, Roye, and St. Quentin to Cambrai, and from Ham to Bapaume, could be built in perfectly straight lines for long distances with a minimum of cutting and filling, while less important roads avoided the necessity of any grading by very moderate adjustments of course to the gentle contours of the land (Fig. 31). Parts of the plain are so little dissected that the upland surface remains practically level, and here even the secondary roads have straight courses for remarkably long distances. In the very heart of the Somme battlefield is such an undissected remnant of the level plain, known as the Santerre, across which the road from Amiens to St. Quentin, although one of the east-west roads and hence not of the first rank of importance, runs for nearly forty miles in a straight line. It was along this road that the Australians battered in the German front in August, 1918.

In a region where road making is so easy the inhabitants find that it is cheaper to build new roads than it is to take roundabout courses over roads previously existing. As a consequence the plain is covered by an intricate network of national highways, departmental highways, good country roads, and passable lanes, such as is seldom found in other parts of France.

Railway construction is likewise comparatively easy and cheap, with the result that in addition to the main through-going railway lines there are numerous local railways which serve the multifarious needs of this rich agricultural region. Add to this the system of canals which follow the valley bottoms, where the slopes are uniform and the rivers easily controlled, and one can truthfully say that nowhere else is there a region more abundantly provided with cheap and easy means of rapid communication. The Battlefield of the Somme was ideally circumstanced for the transporting of those enormous quantities of men and materials which are essential to modern military operations of the first magnitude.

In other respects the form of the terrain in much of the Somme battlefield favored repetition there of military operations on an extensive scale. The general lack of marked topographic inequalities which might give to one side overwhelming defensive advantages, such as existed in the plateau scarps east of Verdun and Nancy, and the absence of such strong and rugged relief as discouraged major operations in the Vosges and the Trentino, tempted first one army and then the other to seek victory by a sudden massing of men, or by an unexpected concentration of fire, or by the use of novel methods of warfare, under conditions which might reasonably be considered as equal except for the advantage to be gained by the element of surprise. When the tank became an important weapon of offense, it was evident that, whereas it might not be utilized to the best advantage on the marshy soil of Flanders nor on the steep, forested slopes of the Vosges, the open, gently rolling plain of the Somme region lent itself admirably to the new method of attack. Hence it was here that tanks were first employed in battle (Battle of the Somme, 1916) and here that they won their most striking victory (Battle of Cambrai, 1917). The flat uplands and gentle valley slopes which invited armies to move at will in every direction also invited the French farmer to clear the forests of old and to extract from the soil abundant harvests; and armies, which fight on their stomachs, saw in the productive region of the Somme a battlefield where their stomachs could always be full. In a score of ways Nature had so fashioned the plain of the Somme as to make of it an ideal stage upon which the drama of war should be enacted.

There is, however, one element of a natural fortification very noticeable in many parts of the Somme battlefield. In defending important positions by artificial works the military engineer has often constructed low earth ridges or embankments called *rideau* (French for "screen" or "curtain"). Now, the hill and valley slopes of the chalk country are often interrupted by a series of terraces or ridges, sometimes faint and irregular, sometimes from 5 to 25 feet high (or even more) on their steep, downhill sides, and so closely spaced as to turn the slope into a giant stairway. These natural *rideaux*, as the French term the terraces, appear to be due in part to long-continued cultivation of the slopes, during which the soil has for centuries been worked down toward the lower part of each cultivated strip, and in part, especially in case of the higher ones, to fractures in the chalk.[2] To prevent the encroachment of higher on lower fields the crests of the *rideaux* were often required, by custom or by law, to be left uncultivated. Thus, with their steep slopes and crests covered with bushes, trees, and stones cleared from the fields, they form natural defensive positions of considerable value.

The northern termination of the Somme type of topography, where it gives place to the low, marshy plain of Flanders, is strikingly abrupt. Northwest of Arras, in particular, there is no transition zone from the rolling chalk upland to the flat clay marshland below. Instead, one of the northwest-southeast trending fault lines described on page 88 cuts across this part of the country, and the area to the northeast has been dropped downward so as to bring the sand and clay formations of Flanders opposite the chalk. As we should expect, erosion has washed away the sand and clay to produce a broad lowland,

[2] Albert Demangeon: La Picardie et les régions voisines—Artois, Cambrésis, Beauvaisis. Paris, 1905, p. 44.

while the more resistant chalk has merely been dissected into rolling hills and valleys (Fig. 32).

Vimy Ridge

The combined result of the faulting or breaking of the rocks, plus the later erosion, has been to produce a topographic feature of the highest military importance known as Vimy Ridge. A traveler advancing northeastward toward the crest of the ridge finds that the country rises very gradually as he proceeds.

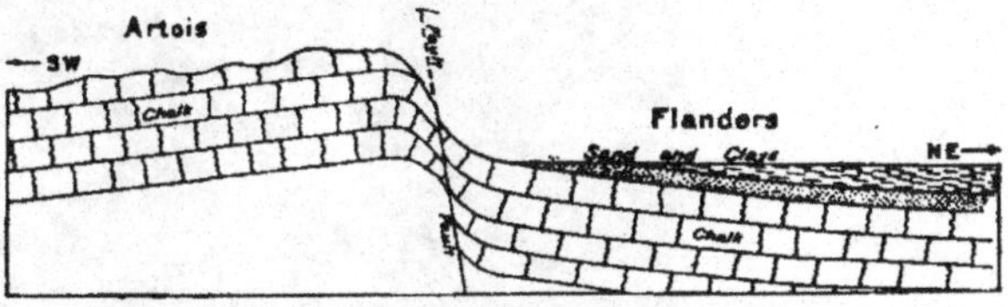

FIG. 32—Eroded fault cliff forming the steep northeast face of Vimy Ridge and separating the rolling chalk upland of Artois from the flat clay plain of Flanders.

Indeed, so gentle is the ascent that he might fail to give it particular notice, unless he turns and looks back. Then he cannot fail to realize that he is rising to an elevation which gives him a commanding view westward to the ruins of Mont St. Eloi and beyond (Fig. 33) and southward far past the towers of Arras. Continuing on his way, he reaches the crest. Instantly there bursts upon his vision a magnificent panorama of the Flanders plain (Fig. 35). The northeastern side of the ridge, the "fault scarp," as the geologist would call it, is steep, the ground dropping abruptly from the traveler's feet to the edge of the plain below. To the northwest, beyond the Souchez River, the ridge is somewhat higher, forming the heights of Notre Dame de Lorette; to the southeast it sinks lower and dies away in the plain of Arras. But toward the northeast nothing blocks the vision until the eye faintly glimpses on the horizon the dim outlines of Mont Kemmel far away in Belgium (Fig. 36).

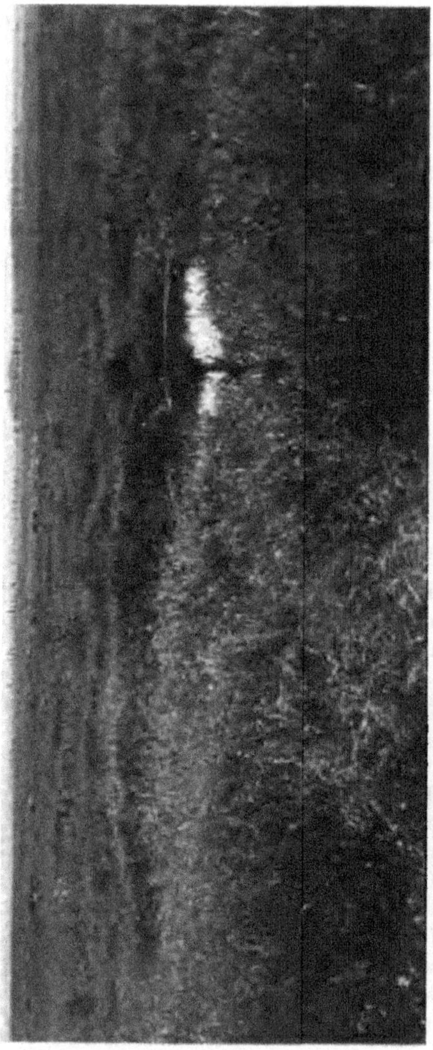

FIG. 33—View southward down the backslope of Vimy Ridge, showing the shell-torn surface, the line of poplars along the road in distance, and Mont St. Eloi with its ruined tower rising against the sky line. This is the view into the back areas behind the Allied front by which the Germans profited before they were driven from the ridge.

FIG. 34—Looking northeast up the backslope of Vimy Ridge, with the ruins of Souchez in the foreground. Note that direct observation of the German back areas was completely denied the Allies by the ridge barrier which blocked their view. Compare with Fig. 35. The white chalk is exposed where the cover of vegetation is lacking.

FIG. 35—View northeast over the Flanders plain from the crest of Vimy Ridge. This commanding observation of the German back areas was gained by the Allies through their advance from the position on the backslope shown in Fig. 34. Three great battles were fought to win this natural observatory.

It is not difficult to understand the great military value of Vimy Ridge. So long as the German trenches lay on the gentle southwestern slope of the ridge, German observation posts commanded an extended view behind the Allied lines (Fig. 33). They possessed the inestimable advantage of direct control of artillery fire, and a surprise attack by the Allies was difficult to prepare when roads and railways were under continuous observation and subject to instant and accurate bombardment. The Allies, on the other hand, from their trenches on the northeastern slope, had all direct observation cut off by the rising crest immediately in front (Fig. 34). The Germans could maneuver at will over a broad stretch of the Flanders plain beyond the ridge, providing the weather or their own airmen kept the aërial eyes of the Allied armies partly or wholly closed. Not until the Allies should drive the Germans beyond the crest would the situation be reversed and the Allies enjoy the topographic advantages, the Germans suffer the topographic disadvantages, inherent in the Vimy Ridge position. To make an advance of but a few hundred yards the Allied commanders in a series of terrific battles sacrificed their men literally by the tens of thousands. Let the reader compare Figures 34 and 35 and he will see that the sacrifice was justified; for the gain is to be measured not in linear yards of advance, but in the increased depth and breadth of observation behind the enemy's lines.

GEOLOGICAL STRUCTURE OF THE HILLS

In a war of movement it is the surface forms of the land alone which play a principal rôle in the military operations. But in a war of position, where the opposing armies "dig in" and for long periods remain rooted to a given piece of ground, the character and structure of the soil and rocks beneath the surface exercise a profound influence upon the condition of the armies and the nature of the fighting.

Throughout the entire area now under discussion there extends a layer of *chalk* (Fig. 37) practically horizontal and so thick that ordinarily its bottom is neither exposed in the deepest

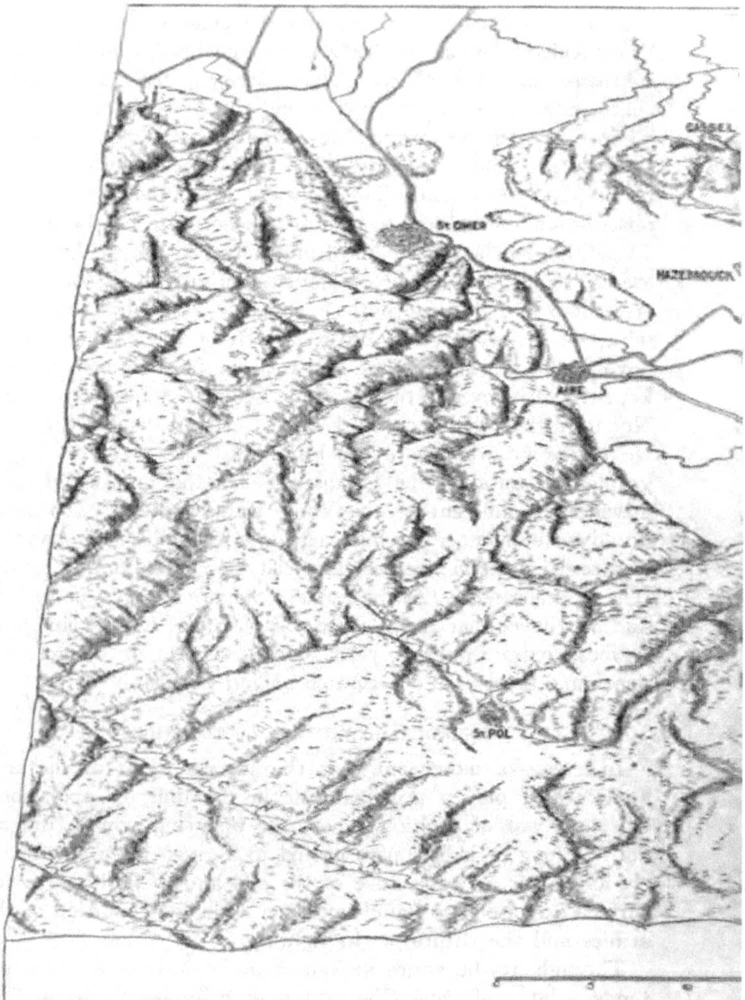

FIG. 36—Block diagram of the Mont Kemmel-Vimy R

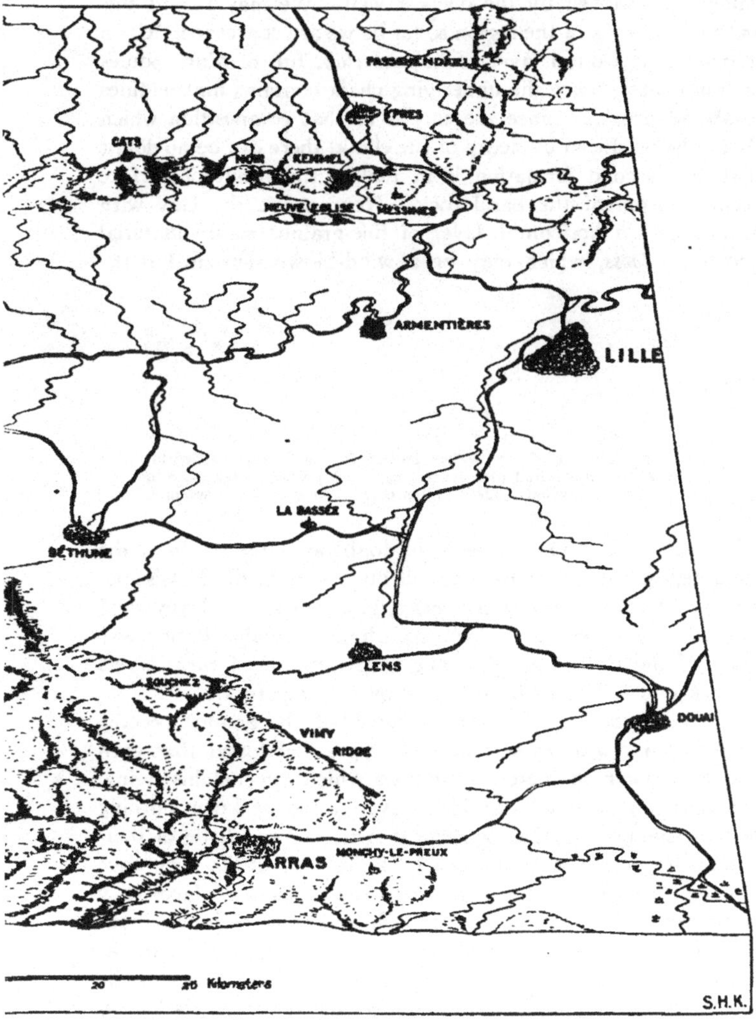

rion, showing the salient features of the terrain.

valley nor reached by the deepest well. We may regard this as the basal rock of the region so far as we are concerned. Upon it rests a horizontal layer of impure *clay*, full of flint nodules or fragments. Since the underlying chalk contains flint nodules in abundance, and since the clay is of that composition which should be produced by decay of the chalk, there can be no doubt that this second formation is in reality the result of surface decomposition of the basal rock. Next above the clay with flints comes a horizontal layer of fine-grained sandy material known as *loess*, which may be a wind-blown deposit, but the

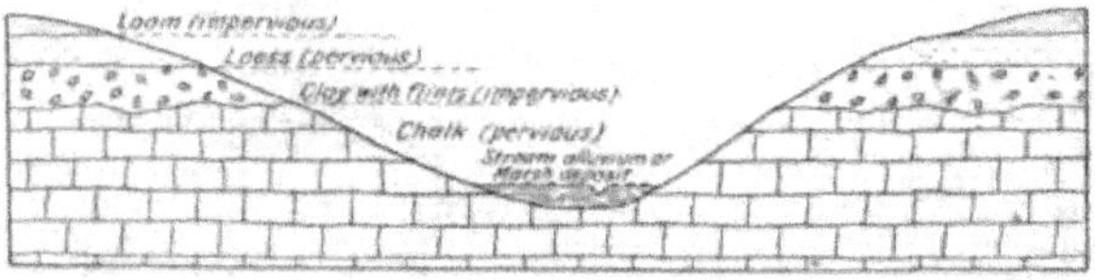

FIG. 37—Typical hill and valley section in the Somme battlefield, showing normal soil and rock succession, which gives an alternation of pervious and impervious beds of much consequence in trench warfare.

origin of which is just as much in doubt as is the origin of the loess deposits found along many of our rivers in the Mississippi valley. Finally, capping the series, is a deposit of clayey sand called *loam*, similar to the loam soils found abundantly in many parts of the United States. The usual relation of these layers of rock and soil to each other is shown in Figure 37.

The man who fought on the Battlefield of the Somme became very familiar with these four rock types and found that each had a peculiar character of its own which affected his living, his marching, and his fighting. Even when he knew nothing of their names and their geological structure and little of their areal distribution, he knew much of the kind of life they gave him in trenches and dugouts and of the kinds of surfaces they gave him to fight over. Had his officers known as much about the effect of these rocks and soils on military operations in the beginning of the war as some of them did at its close, his life in the trenches would have been easier, and fewer of his fellows would have died there. No commander in modern warfare can fully understand his battles unless he fully knows his battlefields below, as well as on, the surface.

In considering how the chalk, clay, loess, and loam left their indelible stamp upon the character of the Somme campaigns, we must first note how the different types are exposed at the

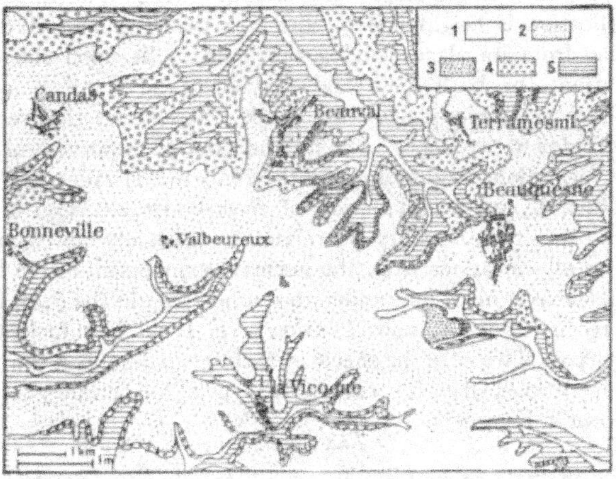

Fig. 38—Geological map of a portion of the Somme battlefield north of Amiens, showing the loam and loess on the hilltops and divides, the chalk in the valleys, and the clay with flints on intermediate slopes. Both the loess and the clay with flints are frequently missing from their normal positions. (Amiens sheet, geological map of France, 1:80,000.)

1, loam and other deposits washed into the ravines; 2, loam and loess covering the uplands; 3, loess or sandy beds; 4, clay with flints; 5, chalk.

surface (Fig. 37). As might be expected, the chalk, being the lowest of the four, is most frequently exposed in the valleys, where streams have cut down into it, or on the valley walls, where erosion has washed away the overlying beds. The loam, being the uppermost layer, naturally forms the hilltops or upland surface of the plain. The two intermediate beds, the loess and the clay with flints, must therefore outcrop on the gentle intermediate slopes or in the bottom of ravines which are not deep enough to cut into the chalk. A detailed geological map showing the actual surface distribution of each formation in a part of the region badly dissected by streams (Fig. 38) will therefore appear somewhat complicated despite the simple horizontal structure of the beds, for each valley and its branches and each hill and its spurs will show the several different formations exposed on its slopes at different levels. The apparent complication is heightened by the fact that the loess is often missing from its place, leaving the loam to lie upon the clay formation; or both loess and clay may be wanting, when the loam lies directly upon the chalk. Sometimes the loam is washed down the slopes to bury and conceal exposures of the lower formations; and frequently the chalk of the valley bottoms is hidden by an accumulation of river-borne alluvium or by marsh deposits. In reality the relations are usually quite simple, and, as all variations from the normal arrangement of the beds are easily recognized and understood when seen in the field, they need receive no further consideration here. It will be sufficient for our purpose if we note the effects of each formation upon the military operations, when it occurs in the typical hill or valley section as shown in Figure 37.

The Chalk

The chalk is so uniform in character wherever encountered that it gives to the whole plain of the Somme a unity of aspect that is very pronounced. Its presence is usually revealed by white patches exposed on hill or valley slopes, which give rise to the frequent repetition of the expression "white" in the place names of this part of France (Blanche Tache, Blancfosse, Blanc Nez, Blanc Mont). Where Nature left the chalk covered with but a thin coating of loam, loess, or clay, trenches and shell holes inevitably brought the fresh rock to the surface to form glaring white lines or patches in the landscape (Fig. 34). In the vicinity of Amiens the chalk over broad areas is only concealed by a very thin cover of soil and vegetation, and when in the spring of 1918 the British built a vast complex of trenches about the city to check a possible second German attempt at driving a wedge to the sea between the French and British armies, the magnitude of the defensive works was revealed to the German commanders by an imposing network of snowy ridges which nothing could camouflage.

It is because the chalk breaks and crumbles readily under the action of the weather and dissolves easily in rain water charged with carbonic acid that nowhere in the plain of the Somme do we find flat-topped plateaus or mesas sharply bounded by steep walls or cliffs such as characterize the landscape of the Verdun region (Fig. 88) but, instead, those rounded, gentle, monotonous slopes which we have already described. Only where recently and rapidly undercut, as by the lateral erosion of some meandering river, does the chalk stand up in really steep cliffs. We shall see later that the meanders of the Somme between Péronne and Amiens have produced some such cliffs, which acquired tactical significance in the Somme battles.

There are two features of the Somme battlefield which quickly impress themselves upon the observer and which materially affected the character of the fighting. One of these is the generally arid aspect of the landscape; the second is the marked concentration of the population in compact villages and towns, separated by broad stretches of open country in which isolated farmhouses or other structures are seldom visible. The transition from the deserted, arid plain to the crowded community is sharp and striking. Curiously enough, both of these characteristic features of the Somme region are due to the same geological peculiarity of the chalk. The rock is so extensively fissured that the rain which falls upon it, entering the vast network of crevices, quickly descends to depths so great that neither the roots of vegetation nor the shallower valleys can reach the level where the water comes to rest. Hence the arid appearance of the country, especially where the chalk forms the surface rock—an appearance which is accentuated by the removal of most of the forest cover, not merely from the chalk areas, but also from those belts where overlying deposits of loam or clay temporarily retain a portion of the rainfall near the surface. Hence, also, the abundance of dry valleys, which we shall later have occasion to consider more fully.

The concentration of the population in compact communities is equally explained by the fissured character of the chalk. The population must have water, and, since the fissures permit the water to descend to depths varying from 50 to 300 feet or more throughout much of the Somme region, deep wells must be sunk to obtain the desired supplies. But the cost of such a well is too great for the individual peasant. Only the rich can afford private wells. For the rest the wells must be the joint property of the community, many persons contributing to the cost and enjoying the benefits of each one. And, since the transport of water to any great distance from the communal well is a heavy burden, all the peasants seek to build their homes close about the particular well upon which they depend. Thus are explained the two strongly contrasted but closely associated types of culture (Fig. 12) which give rise to two equally contrasted types of defensive operations in the Somme campaigns: the open country of unobstructed vision and broad fields of fire, where lines of trenches alone could afford protection to the defense; and the dense clusters of houses, whose walls and cellars formed veritable forts which became the strong points of the whole defensive system.

Unlike limestone, the chalk does not contain any considerable number of natural subterranean caverns. But the rock is easily quarried, with the result that extensive artificial excavations are found under many of the towns and villages. These added greatly to the tactical strength of the battle front, providing secure refuge for large bodies of troops during heavy bombardments and hidden points from which machine-gun detachments could issue to sweep with murderous fire the lines of attacking forces. In preparation for the Battle of Arras in April, 1917, the great cellars and caverns under the town were connected by a series of tunnels driven for that purpose, and sufficient underground space thus provided for three infantry divisions with their headquarters, dressing stations, and other facilities. From these subterranean chambers there issued to the attack parts of two British army corps, which thereby escaped the slaughter awaiting them on their anticipated exit from the narrow streets of Arras, kept under accurate and deadly fire by the German artillery. One of the strongest points which long resisted British attacks in the 1916 Battle of the Somme was Beaumont-Hamel, 15 miles southwest of Arras, the secret of its resistance lying mainly in the extensive quarries and underground excavations for which it was noted.

The Clay-and-Flint Formation

The upper surface of the chalk is usually quite irregular, owing to the uneven progress of solution and decomposition, which proceeds from above downward. It is upon this irregular surface that the residual clay filled with flints, the decomposition product of the chalk, is found when the normal succession of beds is complete. As might be expected, the thickness of such a deposit is quite variable—from a few feet to as much as 20 or 30 feet, the thicker deposits commonly being restricted to the belts of higher country (Fig. 28). On the other hand, the deposit is often entirely missing over broad areas, especially in the shallow down-warp, or syncline, along which the Avre and lower Somme valleys take their northwest course to the sea.

In whatever phase the clay-and-flint formation manifests its presence, it gives to the landscape an appearance notably different from that developed on the chalk. Usually the clay is present in sufficient quantity to prevent the ready escape of water, thus giving a cold and wet soil which is not easily cultivated. Sugar beet culture, the capital industry of the Somme region, especially avoids this formation, not only because of the clay soil, but also because the flint nodules cause bifurcation of the beet roots. It follows that trees are more apt to be left growing on this unfriendly soil; and one not infrequently sees a hill whose treeless lower slopes on the arid chalk and treeless summit on the much cultivated loam and loess are separated by a girdle of trees growing on the more humid intermediate clay-and-flint formation. Villages on this formation are apt to boast numerous shade trees among the houses and larger groves in the outskirts, although an effort may have been made to clear and cultivate the rest of the country. In earlier days marshes were interspersed with the woodland, and even today unimproved roads through the remaining groves are apt to be muddy and at times almost impassable. Fortunately the clay-and-flint formation, which gives a landscape contrasting so strongly with that which characterizes either the chalk or the loess and loam, covers but relatively small portions of the Somme battlefield. Nevertheless it made itself felt as an unpleasant factor in the warfare of position in this region. East of Albert the Fricourt Wood, Mametz Wood, Railway Copse, the two Bazentin Woods, Bernafy Woods, Trônes Woods, and others whose names are forever linked with memories of some of the bloodiest struggles in the 1916 Battle of the Somme, are partially or wholly rooted in the unfriendly clay-and-flint soil. Had this formation not been present on the battlefield, fewer men would have had to be sacrificed in the hard task of clearing enemy troops from peculiarly difficult forest positions.

The Loess and Loam Formations

The two uppermost formations, the loess and the loam, are frequently classed as one under the French name *limon*, because they have certain characters in common and because sometimes, especially where neither is strongly developed, it may be quite impossible to separate them. Indeed the *limon* formation as a whole is quite complex, comprising seven different beds, of which two are gravel layers; but in the Somme region the loess and the loam are usually the only members of the series present over appreciable areas, and they alone need receive our attention. Together they constitute a fine-grained covering deposit on the hilltops and level uplands, or "plateaus," which reaches thicknesses of 25 to 30 feet or more in the east but thins out towards the west to 15 feet, 10 feet, and even less. It is the typical hilltop deposit, whether the hills be high or low;

but of course it is not wholly absent in the valleys, since rain and streams readily wash some of the material from the uplands down into the depressions, there to constitute much of the valley alluvium. On little dissected, broad, flat uplands, where washing has been at a minimum, the deposit is preserved in the greatest thickness. The broad and level plain of the Santerre southeast of Amiens, across which the British launched the great offensive of August 8, 1918, is so heavily coated with loess and loam that the underlying chalk and clay-flint formations are completely concealed over extensive areas.

Where typically developed the loess is light yellow or light gray in color, very fine-grained, and crumbles to a fine powder in the hand. It is sufficiently porous to permit the ready passage of water and thus ranks with the chalk as a pervious formation. The loam is more of a reddish brown and contains a considerable proportion of clayey matter. Like the loess it is fine-grained, easily cultivated, and comparatively fertile. Its fertility has been artificially enhanced by the addition, throughout long periods, of certain elements lacking in its natural state. Because the loam is the uppermost bed of the series it is exposed over a greater area than any of the other three formations. It and the chalk determine the character of the Somme region as a whole, the chalk forming most of the surface in the valleys, the loam most of the upland surface. The intermediate clay-flint formation and the loess show in limited areas on intermediate slopes.

The clay content of the loam, while not so great as in the clay-and-flint formation, is sufficient to render it comparatively impervious. As a result natural roads on the loam are so apt to be boggy in rainy seasons that the peasants in some sections have the proverb: "Good soil, bad roads."[3] If the loam is very thick, or if it rests directly on the clay-and-flint formation, the unimproved roads may become almost impassable, while the soil is then so cold and wet that its agricultural value is appreciably reduced. At the battle of Agincourt it was the loam

[3] Demangeon, La Picardie, p. 83.

surface of the plateau, soaked by the rain of the previous night and tramped into a bog by the feet of 60,000 men and their horses, into which the French, heavily weighted with their armor, sank so deeply that they could scarcely move, while the more lightly equipped English made good use of their advantage. "Artillery in such a sea of mud could not be brought into position on either side."[4] In the Franco-Prussian War the attack of the Germans on Villers-Bretonneux was materially crippled by the fact that their cannon, although drawn by six horses, could be moved only with the greatest difficulty in the loam of the plateau, then soaked by successive rains, and by the further fact that many of their shells failed to explode when striking in the mud. In the World War of 1914-1918 the unfavorable character of the loam continued to exercise its baleful influence, now to the disadvantage of one side, now to the disadvantage of the other.

Where the loam is thin and rests upon the pervious loess or chalk, under-drainage takes place with sufficient rapidity to eliminate the disadvantageous features mentioned above. Then one finds both good soil and good roads. Indeed, it is the extensive areas of good loam soil in the Somme region which from time immemorial have attracted to this plain one of the most numerous agricultural populations of France, led to the almost complete deforestation of the region, made necessary the construction of numerous roads and railways, induced the building of villages on the upland where water is difficult of access and costly to secure, raised the price of lands to a high figure, and yielded harvests so rich that poorer lands to the north and south have eagerly looked to it for a portion of their foodstuffs. It is the loam which is responsible for the fact that in the Somme region it is, except in wet weather, easy to move, and always easy to support, a great army.

In the more arid parts of the Somme upland continued traffic along the unmetaled roads pulverizes the chalk, loess, and to

[4] J. W. Fortescue: A History of the British Army, 7 vols. to date, London, 1899–1912; reference in Vol. 1, p. 60.

some extent the loam; and, when the dust is removed by winds or rain wash, or by that primitive manner of road mending which consists in scraping away the loose débris until firmer soil is reached, the surface of the roadway is progressively lowered until it forms the bottom of a narrow, steep-sided trench. These sunken roadways are sometimes 10 to 20 feet deep below the enclosing banks and constitute military obstacles of which both sides made excellent use. Some of the most formidable German strongholds southeast of Arras encountered by the British in their advance of April, 1917, derived their strength from a series of sunken roads in that region which had been abundantly fortified by numerous German machine-gun detachments.

Relation of Geological Structure to Trench and Dugout Locations

One effect of the increasing power of heavy artillery fire is to compel opposing armies to dig ever more deeply into the earth in search of protection. Shallow trenches are deepened, and dugouts near the surface give place to elaborate underground chambers reached by long descending flights of stairs. Extensive galleries and tunnels are run for the purpose of sheltering large bodies of troops assembling for an attack, or to place and explode mines under the enemy's trenches. It is evident that the geological formation of some battlefields, consisting of well-drained pervious rocks, may be highly favorable to such military engineering works; whereas in other battlefields, underlain by strata saturated with water which cannot drain off, the construction of tunnels and dugouts and even the digging of deep trenches may be practically impossible. The Somme battlefield falls neither in the one class nor the other, for we have already seen that it is composed of both pervious and impervious formations in alternating layers. Trenches, dugouts, and tunnels will be dry and habitable if excavated in the right formation; but if ignorance of the geological structure of the battlefield leads to excavations in the wrong formations, an army may find the waters beneath the earth more dangerous than the fire above. In a series of defensive works located on a hill slope in the Somme region (Fig. 39), a trench (No. 1) which is dug wholly in the relatively impervious loam will accumulate water, especially if the loam is of considerable thickness above the loess. But a trench (No. 2) located in such manner as to cut through into the underlying loess, or one (No. 4) located wholly in that pervious bed, will be dry, since water entering such trenches will drain off into the pervious formation. Trenches (Nos. 3 and 5) whose bottoms reach through the loess into

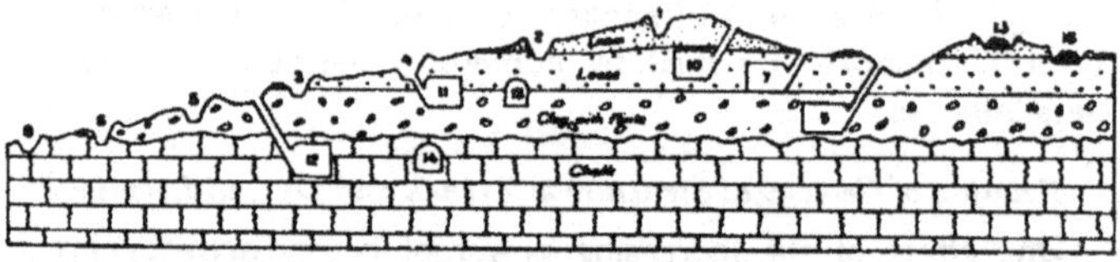

FIG. 39—Geological cross-section of a typical hill in the Somme battlefield, showing possible locations of trenches, dugouts, tunnels, and munition bunkers in the different formations. Odd numbers show poor locations, even numbers good locations as regards rock structure and underground drainage. (Tactical considerations are ignored in order to condense many features within convenient space.)

the impervious clay-and-flint layer, or which are cut wholly in that layer, will hold water so effectually as to make their occupation extremely trying to the men; but if a trench (No. 6) in the clay cuts through into the pervious chalk the water will quickly drain away. Trenches cut only in the chalk (No. 8) will likewise be dry.

A dugout (No. 7) excavated wholly in the loess but with its bottom close to the contact of the underlying clay will be very wet, because the flow of underground water is at a maximum through the pervious beds just over the surface of the impervious formations; while a dugout (No. 9) wholly in the clay will also be wet, since, although the water drains in more slowly through the dense clay, it has no chance to escape. Dugouts (Nos. 10 and 12) wholly in the loess and chalk, and not close to contacts with underlying impervious beds, will be comparatively warm and dry. Dugout No. 11 is in a particularly bad location, since waters flowing through the loess just over the clay contact will pour into the excavation from the upper walls and will be effectually impounded in the clay basin formed by the bottom of the excavation. Of the two tunnels No. 13 will give endless trouble for the same reason, whereas No. 14 will be dry. Even munitions bunkers may be located wrongly, as when they form basins in the impervious loam (No. 15) and hold the water from every rainfall; or correctly located (No. 16) so as to permit rainfall to drain off through the pervious loess.

The hazards of war do not always permit an army to select the terrain on which it will fight. Nevertheless it is evident that in a region like the Somme the terrain on which battle is delivered may be intelligently or unintelligently utilized. Unfortunately it happened during the great war that a higher officer, even a general of engineers, would sometimes remain blindly unaware of the vital relation of geological structure to the fighting power of his army, while his subordinates complained bitterly of the wasted effort and the unnecessary suffering which such ignorance entailed. Dugouts were excavated and then found useless because it was impossible to drain them. Tunnels and mines driven at great cost of time and labor were abandoned before attaining their object, because a water-bearing zone above or below was cut into and the workmen "drowned out." Men suffered misery in trenches deep with mud and water. And all this when changes of but a few feet in surface location or differences of a few feet in depth of excavation, perfectly feasible from the tactical point of view, would have avoided the damage incurred. Even with no change of location, drainage could often be effected by sinking in the bottom of the wet trench or dugout a narrow subsidiary trench reaching into an underlying pervious formation, or by driving pits or wells down into that formation so as to let accumulated waters escape below. With the lapse of time and at the cost of bitter experience these facts became more and more widely known. Geologists were attached to the different armies, and the results of their studies were utilized by the army engineer.

The Germans perfected an elaborate geological staff for service at the front, and documents captured by the Allies showed that this staff prepared and distributed instructions on the siting of trenches, dugouts, and other defense works which were abundantly illustrated by diagrams similar to Figure 39, showing correct and incorrect methods of excavating in the rock formations of the Somme region. Experience had demonstrated that in underground structure, as in surface topography, the terrain of a battlefield may offer advantages and disadvantages, which, when the first are properly capitalized and the second avoided, give to a resourceful commander that superiority over a less intelligent adversary which in an otherwise equal struggle might determine the issue of the conflict.[5]

Variable Effects of Artillery Fire

The four different formations of the Somme battlefield react differently to artillery fire. A shell exploding in the mud formed by the loam after a heavy rain, or in the water-soaked clay, does less damage than one bursting on contact with the hard and brittle chalk, as the Germans first found to their cost in the battle of Villers-Bretonneux in 1870. On the other hand, a wet battlefield on the loam or clay may be drained if the shell craters penetrate into the underlying pervious loess or chalk, thus establishing numberless points for the downward escape of accumulated waters. Or shell holes in the loess may be turned into permanent ponds of water if they penetrate to the underlying clay. So important were these effects of the heavy bombardment that the army geologists were required to prepare in advance of an attack maps showing the kind of terrain which would exist after a proposed bombardment and over which infantry, tanks, and other arms would have to advance to their objectives.

[5] For an excellent discussion of the value of geological studies to military operations, see A. H. Brooks: The Use of Geology on the Western Front, *U. S. Geol. Survey Professional Paper 128*, pp. 85–124, 1920. See also W. B. R. King: Geological Work on the Western Front, *Geogr. Journ.*, Vol. 54, 1919, pp. 201–221, and D. W. Johnson: The Rôle of the Earth Sciences in the War, from "The New World of Science," New York, 1920, pp. 177–217.

Tertiary Erosion Remnants

In addition to the ordinary rounded hills and flat uplands of the Somme region, composed of the normal series of four formations described on preceding pages, there are within the area a few scattered remnants of another geological formation (Tertiary sands and clays) which acquire military importance because they rise above the general level of the plain or because they are forested. It is evident that the Tertiary formation once covered the whole region but that erosion has removed all of it except isolated patches which now remain as mute witnesses to the former condition of things. Sometimes the erosion remnants rise above the plain as buttes or mesas which effectively dominate the surrounding country, as in the southeastern part of the Somme region, where groups of them represent outliers of the Marne (or Parisian) plateau from which they have been separated by erosion. One of the groups is the so-called Noyon-Lassigny massif, comprising several tablelands northeast and southwest of the town of Noyon. With their commanding view over surrounding plain and valleys, their precipitous rimming scarps and forest-covered uplands, these mesas constitute military obstacles of the first importance. Once solidly entrenched in the Noyon-Lassigny massif, the Germans were never expelled from it by direct frontal attacks.

In other instances the Tertiary deposit rises to only a moderate elevation above the general level and is mainly noticeable because it supports an island of forest in the midst of the plain, where the trees find abundant moisture held in the sands by layers of clays. This is the case with the famous Bourlon Wood west of Cambrai, which commands a long stretch of the marshy valley of the Scheldt, and the Holnon Wood west of St. Quentin. Between St. Quentin and Cambrai there are other wooded knolls of the same origin, many of which have determined the location of villages because the layers of clay keep abundant supplies of good water close to the surface, where it is easily accessible. Both knolls and villages have played a significant rôle in military operations in this theater of war from very early times, and in the present war were the occasion of some of the bloodiest conflicts occurring in the Somme area.

THE VALLEYS

While the rolling upland of the Somme battlefield is for the most part arid and treeless, the principal valleys are exactly the reverse. It would be difficult to imagine a more striking contrast than that which the valleys of the Somme and its

FIG. 40—The marshy valley of the Somme River near Amiens, one of the most important natural defensive barriers of France. From Roman times to the present it has again and again with its morasses blocked the passage of hostile armies.

chief tributaries present to the general aspect of the landscape. Instead of aridity the broad valley floors show a maze of marshes, ponds, and sluggish streams. Countless springs pour their waters into the valleys from either side and come bubbling to the surface in the stream channels themselves. Reeds and grasses wave over the waters, affording cover to wild fowl; while far-spreading peat bogs offer treacherous footing to man and beast. Groves of poplars and other trees add their verdure to the ribbons of green winding amid barren hills of yellow and brown (Fig. 40).

The secret of so striking a contrast is not difficult to discover. The uplands are arid because the rain water quickly descends through the porous loess and chalk to considerable depths. The main valleys are humid because the accumulated waters pour forth in springs from the fissured chalk upon their level floors. In other words the deeper valleys are the paths by which the accumulating waters escape to the sea. And just as the peculiar character of the chalk is primarily responsible for much of the aridity of the upland, so also is it responsible for the marsh-covered peat bogs which clog the valley bottoms.

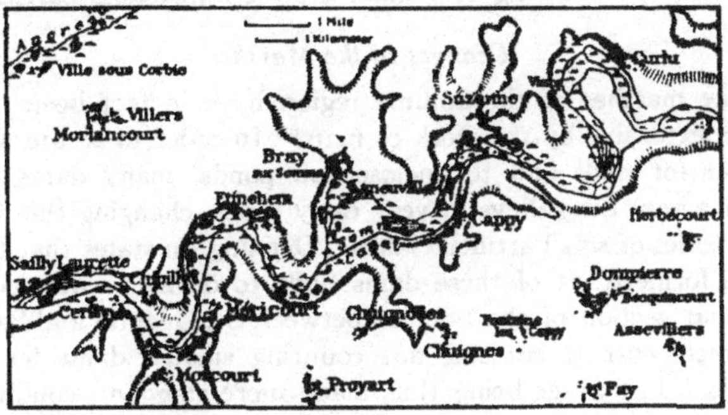

FIG. 41—Typical portion of the marshy valley of the Somme, showing interlacing stream channels, *clairs* of open water left by the extraction of peat, and the Somme Canal.

In regions where rainfall runs off over the surface, carrying much surface soil downhill into the streams, rivers have floods in rainy seasons and low-water stages at other times, and the waters are heavily charged with sediment. Both the great ranges in stream volume and the abundance of sediment are unfavorable to the formation of peat deposits. Flood plains of alluvium are more apt to result; and, while these may be marshy in their natural condition, such marshes are easily turned into meadows, like those along the Oise which make La Fère and Chauny famous for their hay. But in the chalk region the régime of the rivers is very different. The rain sinks into the ground, traverses the porous loess, and moves slowly along the fissures in the chalk. Surface run-off and surface soil wash are greatly reduced, and the rivers are supplied by a great underground reservoir which feeds perennial springs with crystal clear water. Hence the Somme and its branches, flowing on faint gradients with fairly regular volumes, unusually free from floods and sediment, offer conditions highly favorable to the growth of aquatic plants and to their preservation in the form of peat. As this type of vegetation makes poor forage for stock and the peat bogs are not easily reclaimed, the river bottoms are apt long to remain almost impassable barriers.

Changes in the Marshes

The marshes of the Somme region have in fact been made more extensive by the work of man.[6] In order to secure water power for mills and to enclose fish ponds, many dams were built across the valleys in very early times, changing the rivers into series of small artificial lakes. Demangeon states that there were formerly 31 of these dams built to develop water power on that section of the Somme between St. Quentin and Sailly-Laurette east of Amiens, not counting smaller dams for fish ponds. The water being thus made more stagnant, conditions were unusually favorable to marsh growth and peat formation. In the eighteenth century, however, man began in earnest the reclamation of the marshes he had helped to increase and which his desire for hunting and fishing grounds had helped to preserve. A bad reputation attached to the swampy lands because of the fevers suffered by those who came to work near them, and called for a reduction of the cause of the evil. The extraction of peat in great quantities for fuel, fertilizer, and other economic uses replaced in the aggregate a large total area with rectangular ponds of open water which the local population call *clairs* (Fig. 41). The construction of the Somme Canal and lateral ditches abolished the dams which interfered with the flow of the main river. In the vicinity of the larger towns, like Amiens and Péronne, peat bogs were transformed into market gardens, or *hortillonnages*, yielding rich harvests of vegetables

[6] Demangeon, La Picardie, p. 139.

which were carried to town in quaint boats adapted to the narrow ditches and interlacing channels of the aquatic habitat in which they were the only means of transport. But Amiens with its "water market" and its islands protected by branching channels of the river was still called the "Venice of the North," while even today the area of uncultivated marshland and treacherous peat bog is large.

The Marshes as Military Barriers

The depth of the peat varies from 10 or 15 feet or less in the upper valley of the Somme to more than 30 feet in parts of the lower valley. In the valley of the Ancre northeast of Amiens the peat is 20 to 25 feet thick and consists of many beds alternating with layers of calcareous sand washed in from the surrounding chalk. The Hallue, Avre, Noye, Celle, and other tributaries of the Somme are likewise floored with peat deposits of appreciable thickness and constitute in each case a military obstacle of no mean importance. The abundant springs serve to keep the water from freezing, even in very cold weather, so that winter does not afford any guarantee that the barrier can be crossed on foot. The Somme continues to flow when the Seine farther south is covered with ice; and in 1870, when the Germans reached the Hallue in the expectation of crossing it on ice to attack the French, they were surprised to find the stream still running despite an extremely low atmospheric temperature.

As we have seen on an earlier page, the Somme has throughout military history been the great marsh barrier of this part of France. Likewise the marshy valley of the Scheldt has repeatedly played an important military rôle. In 1793 the French under General Kilmaine entrenched themselves in the famous quadrilateral known as Caesar's Camp, having their front protected by the Scheldt from Cambrai to Bouchain, their left by the Sensée and its marshes, their right by the wooded heights of Bourlon, and their rear by the river Agache (Fig. 42). The passages over the Scheldt were closed and the valley floor was flooded. So formidable was the French position that the Allies under Coburg were compelled to force its evacuation by a turning movement to the south.[7] In 1870 the French Army of the North under General Faidherbe deployed along the line of the river Hallue, using the marshes and peat bogs of this valley as a protection against the Germans moving out eastward from Amiens. The Germans failed in their frontal attacks against the strong position and found it necessary to call up reinforcements and execute a turning movement before the inferior French forces could be dislodged. Similarly the Oise above and below La Fère and other marshy valleys of this part of France have served as defensive barriers at critical periods in French history; and in the great war so recently ended history in this respect repeated itself many times over. The "line of the Somme," the "line of the Oise," the "line of the Scheldt," and many like expressions appear again and again in the despatches of each campaign.

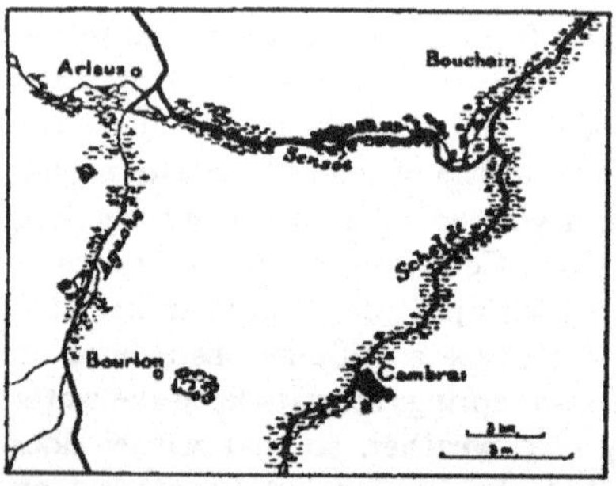

FIG. 42—The quadrilateral of Caesar's Camp, protected on three sides by marshy valleys and on the fourth by the wooded heights of Bourlon.

A marked peculiarity of many valleys in the chalk region is their asymmetrical cross-section. The eastern or northeastern walls are notably steeper than the opposing sides. Demangeon[8] explains this contrast on the theory that rain driven by southwesterly winds striking more directly against the eastern and northeastern valley walls, and the sun's rays falling more directly upon them, cause them to weather back with comparative rapidity into steep slopes; whereas the better protected, more humid, and less actively washed western or southwestern slopes

[7] Fortescue, Vol. 4, Part I, p. 119.
[8] Demangeon, p. 42.

are more slowly weathered and hence maintain a gentle inclination. Whatever the proper explanation, the fact may be observed along many of the valleys, as for example the Selle and portions of the Ancre. It is evident that such asymmetry of valley slope might have significant consequences in military operations. When in 1870 the Germans advanced from Amiens to attack the French forces standing behind the marshy barrier of the Hallue, as related above, they were compelled to assault the steeper eastern wall of the valley under French fire and were repeatedly driven back by charges of the French infantry, which swept down the steep slopes upon them. The asymmetry of the valley walls, by giving to the French a distinct advantage, accounted in part for the failure of the Germans to dislodge the defenders by frontal attacks and for the fact that the German losses far exceeded those of the French.

Dry Valleys

It is only the deeper valley bottoms which receive the outflow of countless springs and are covered with marsh and peat deposits. The shallower valleys and small ravines are as arid as the upland below which they are entrenched (Fig. 43). This is because the water in the chalk finds ample opportunity for escape without rising to the level of the shallower depressions. Following an exceptionally wet season the ground-water level may rise high enough to cause some dry valleys to carry temporary streams; and an exceptionally dry period may cause the upper parts of wet valleys to lose their springs and streams for a time. Many French authorities agree that in addition to these temporary fluctuations there has long been in progress a steady lowering of the average ground-water level, owing to progressive solution of the chalk, by which the underground channels are widened; to progressive deforestation, with consequent increase of surface run-off; and to continual increase in the number of wells, especially those pumping large quantities of water for sugar refineries and other industrial establishments. There is historical evidence of this progressive lowering of the groundwater level, for springs like the big Saint Firmin spring at Roye have failed, and valleys formerly wet, like the upper Omignon northwest of St. Quentin, the upper Tortille, and certain branches of the Ancre, are now dry or nearly so. This means that some of the smaller valleys and ravines which were military obstacles of some significance fifty, a hundred, or several hundred years ago, are less formidable in character today, and that the students of a later century must in turn remember that what were marshy barriers in the World War of 1914-1918 may appear to them as quite dry valleys.

It would be wrong to imagine that the dry valleys have no military significance. As places of concealment below the upland level, as channels directing the advance of troops and determining the location of lines of communication and supply, and in other respects, the dry valleys not only offer the same advantages as wet valleys of similar form but the further advantage that all of the valley floor is available for use. The German tactics of "filtration" profited greatly by these dry valleys, sometimes with serious consequences to the Allied line, as in the 1917 Battle of Cambrai, when the German penetration to the upper reaches of Ravine 22 south of the city caused the loss of much that had been gained by Byng's surprise attack with massed tanks. As military obstacles the dry valleys are of course far less formidable than those carrying permanent streams or containing marshes. But in wet weather some of the dry ravines of the Somme region, known as *riots*, temporarily become as terrible as mountain torrents. The *riot* of Herclain east of Cambrai, in which St. Vaast, Inchy, and Maurois are located, is noted for the devastating floods which have repeatedly swept through it, carrying destruction to everything within reach of the turbulent waters. Such floods are most apt to occur when the ground is frozen and a sudden thaw deluges the surface with water which cannot sink into the hardened soil. In such a case even the Somme itself, as well as its main tributaries, may suffer from floods. Local floods in the *riots* are also produced in warmer seasons when the ground is hardened by drying and a heavy rain causes much surface run-off. In order to prevent the damage incident to these flooded *riots*, ditches are dug to lead off the excess waters or successive dams are built across the valley to check the rush of the torrent and to retain its waters in a series of reservoirs or ponds. The sudden appearance of a flood in one of these dry valleys is capable of working havoc with military

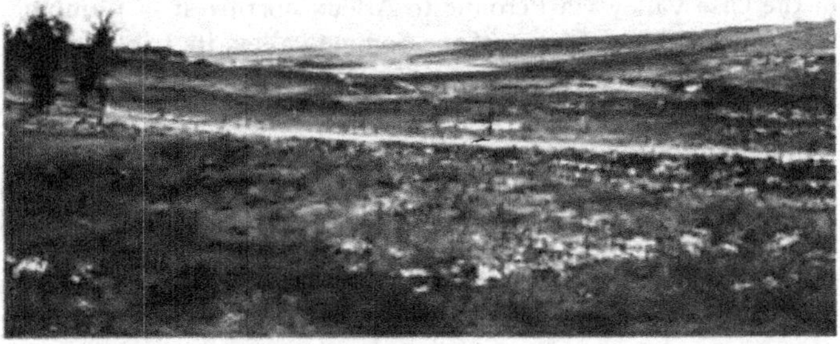

FIG. 43—Typical shallow, open ravine or dry valley in the chalk of the Somme battlefield just north of Bray-sur-Somme.

forces and their equipment which may have taken refuge there, while forces expecting to cross the valley would find their plans seriously deranged.

The Canals

The valleys, both wet and dry, have been utilized in constructing a series of canals across the Somme battlefield (Pl. II). One of these, the Somme Canal, follows the valley of the Somme River to connect St. Quentin with the sea. It was designed to be the great trade route of this region; but because it runs in a general east-west direction, whereas physical geography ordained that the Somme region was to be a gateway between the north and the south, all the great hopes entertained respecting its future were blasted. Today it is little more than a feeder for the north-south canals. Of these the Crozat-St. Quentin-Scheldt canal, connecting Chauny, St. Quentin, and Cambrai by utilizing parts of the Oise, Somme, and Scheldt valleys and certain of their tributaries, constitutes one of the most important trade routes in all France. The traffic over this water route, which not only brought the coal region of the north into touch with the industrial region of Paris, but also bore trade from as far south as Lyons to the ports of Dunkirk and Amsterdam, soon became so congested that another north-south canal was a pressing necessity. The Canal du Nord runs from Noyon in the Oise valley via Péronne to Arleux northwest of Cambrai, taking advantage of parts of the Somme valley, its tributary the Tortille, and a branch of the Sensée valley known as the Agache. As a second north-south waterway of capital importance it is destined to relieve the Crozat-St. Quentin-Scheldt system of some of its overburdening traffic. The Oise Canal follows the valley of the Oise River across the southeastern corner of the Somme battlefield, and the courses of the Scarpe and Sensée in the northeast have in part been canalized.

A canal forms an artificial moat in the bottom of the natural moat of the valley which contains it, and is a serious military obstacle (Fig. 44). The Somme Canal adds to the obstruction formed by the Somme River and thus increases the difficulties of armies moving north or south across this region. The lines of the Crozat-St. Quentin-Scheldt canal and of the Canal du Nord provide two north-south barriers of prime importance, especially where they traverse small valleys or dry ravines which otherwise might oppose little difficulty to the passage of troops. Only at one point, on the Somme-Scheldt divide between Bellicourt and Le Catelet, does the Crozat-St. Quentin canal pass underground for any considerable distance. There is a short tunnel at Le Tronquoy which pierces a minor divide within the Somme drainage basin; but at Bellicourt the canal disappears underground for a distance of more than three miles, and the surface of the broad divide is devoid of topographic obstacles. Here, then, is a strategic point where troops may pass the long north-south barrier without even seeing it. The Canal du Nord presents two such points: near Cachy east of Roye, where the Oise-Somme divide is pierced by a short tunnel; and near Bertincourt north of Péronne, where a longer tunnel

FIG. 44—The Oise Canal at Chauny, part of the Oise valley barrier. Note that the bridge has been destroyed and that material has been lost in attempting the passage. (French official photograph.)

carries the canal through the Somme-Scheldt divide. The great military importance of these two north-south canal barriers, often of quadruple strength (valley, marsh, river, and canal) will be apparent from a reading of any detailed account of the Somme campaign. Both the Allied and Teutonic armies used them as principal lines of resistance.

STRATEGIC POINTS ON THE SOMME BATTLEFIELD

It is interesting to note to what a marked degree the strategic aspect of the valleys of the Somme region has affected the location and character of towns in the past. From Roman times Amiens has guarded one of the principal passages of the Somme barrier, at a point where a narrowing of the valley and the presence of several islands facilitated a crossing. On the other hand, advantage was early taken of the divided channels of the river and of the marshes to make of Amiens a fortress which was so well protected by natural obstacles that an enemy always found it most difficult to reduce. Péronne was first located on the left bank of the river, but at the time of the Norman invasion the inhabitants moved out into the marshes and there built for themselves a fortified town which acquired special significance because it defended an important passage of the river at the point where the latter turns sharply from a northerly to a westerly course. The natural barrier of the marshes was so effective a protection that the fortress was long deemed impregnable. Even in later wars the town, dominated on the north by the commanding plateau spur called Mont St. Quentin, has been considered of special military value, and in 1536, 1870, and twice during the recent war German guns belched fire from this stronghold on the Somme.

Ham took advantage of a peninsula of the chalk upland which, projecting out into the Somme valley, so narrowed it as to make a place of easy crossing and which at the same time was so surrounded by marshes as to make a strong point for the defense of that passage. St. Quentin stands at the eastern end of the Somme barrier, where roads converge from all directions. It occupies a strong position on a peninsula of the chalk upland, protected on the east by the Somme marshes and on the south and west by a deep ravine. Roye guards the southeastern end of the rectilinear Somme-Avre depression, where passed a Roman road from Rheims to Amiens. Montdidier was first located in the valley of the Doms, but in the fifteenth century, when the wars of the Burgundians began, the inhabitants took refuge on a steep-sided spur of the chalk upland which was protected by the marshes of the Doms on the west and by deep ravines on the north and south.

La Fère has been, from earliest times, one of the chief northern defenses of the Paris region. Located on an island in the midst of marshes, with branches of the River Oise on all sides, it has well been called the type example of a fortress or natural citadel isolated in a marsh barrier. It guards one of the river gateways giving access through the Laon-Lassigny escarpment to the district of Paris. Cambrai, the center of many important converging roads, guards a passage over the marshy barrier of the Scheldt upon a spur of chalk upland protected by marshland on the west and north, and on the south by a ravine. In the marshes the river pursues an uncertain course, splitting into several branches which readily shift their positions or may easily be shifted by man. Doullens figured early in French history as a strong point guarding the eastern end of the natural trench of the Authie. Arras stands near the southeastern end of Vimy Ridge and was protected by the formidable marshes of the Scarpe. In early times its fortified walls formed a valuable connecting link between the Canche and Authie valley barriers on the west and the Scarpe and Sensée marsh barriers on the east.

Clearly the valleys and marshes of the Somme battlefield, through their effect on early military history, exerted a profound influence on the location of those cities and towns which were to figure most prominently in the Somme campaigns of 1914–1918.

CHAPTER IV

MILITARY OPERATIONS ON THE BATTLEFIELD OF THE SOMME

THE INVASION OF THE SOMME

Such, then, is the country across which the tide of battle flowed and ebbed in the tragic years 1914–1918. In the latter part of August, 1914, the German First Army under Von Kluck, after swinging far to the west through Belgium, turned southwest through the "gateway of Vermandois" and debouched upon the open plain of the Somme battlefield. German detachments swept rapidly across the gently undulating surface, routed small French forces between Arras and Bapaume after a stiff fight, dispersed them again at Albert, and drove them to seek protection behind the Somme barrier.

The French Sixth Army (sometimes called the Army of the Somme) had meanwhile formed behind the line of the Somme eastward and southeastward from Amiens, while its left was covered by additional troops holding the Somme barrier west from Picquigny, 8 miles northwest of Amiens, to the sea. "The invaders were arrested for two days on the Somme between Amiens and Péronne, the Allies holding a strong position behind the marshes through which the river here flows."[1] But by August 30 the Germans in great numbers had crossed the barrier east and west of Ham, and August 31 Amiens fell into their hands. The line of the Somme had been forced, and the French Sixth Army now fell back on Paris, soon to play a vital rôle in the critical struggle on another battlefield—that of the Marne.

In the meantime German hordes farther east had occupied Cambrai on August 26 without difficulty and violently attacked

[1] G. H. Perris: The Campaign of 1914 in France and Belgium. New York, 1915, p. 269.

the small British army in the vicinity of Le Cateau. The British forces, at that moment retreating from Mons, were in an extreme state of exhaustion; but, wrote their commander Sir John French, "I determined to make a great effort to continue the retreat until I could put some substantial obstacle, such as the Somme or the Oise, between my troops and the enemy."[2] After a further defeat at Le Cateau the British fell back to the line of the Oise between Noyon and La Fère, where they paused for a few hours of rest on August 28 and 29, while the French Fifth Army on their right took up a defensive position behind the Oise from La Fère northeastward to a point beyond Guise. Although the Oise marshes form less of an obstacle than the peat bogs of the Somme, and Turenne in the campaign of 1653 had rejected the proposal to stand behind that stream in opposing the advance of the Spaniards, on the ground that it was impossible long to defend the passages of such a river,[3] it was nevertheless a military barrier of real value. Under its protection the Fifth Army was now able to undertake the delicate maneuver of shifting the bulk of its forces from its right to its left wing, in preparation for a counterattack upon the Germans. This counterattack, known as the Battle of Guise-St. Quentin, was precipitated when the Fifth Army debouched from behind the line of the Oise between La Fère and Guise and attacked toward St. Quentin. About this time, however, the Germans attacked the weakened French right and succeeded in forcing the Oise barrier east of Guise. General Lanrezac had therefore to turn his attention to the task of restoring security to his right wing by driving the enemy back over the river, and the attack on St. Quentin hung fire. The British having meanwhile retreated from the Oise to a line farther south, and the Germans having pushed the French left wing back behind the barrier and captured La Fère, the whole Fifth Army abandoned the line of the Oise and fell back August 30 to conform with the British retreat.

[2] Sir John French: Despatches: Mons; The Marne; The Aisne; Flanders, London, 1914, p. 10.
[3] Napoleon Bonaparte: Memoirs of the History of France, Historical Miscellanies, London, 1823, Vol. 3, p. 65.

In less than a week the German wave had swept the Battlefield of the Somme clean of Allied troops. The onrush had been so overwhelming, and the necessity under which the Allies labored to deliver their main counterblow farther south was so imperative, that no natural defensive line in the Somme region could be held in sufficient strength to retard for long the German advance. Yet even under such conditions the physical features of this battlefield left their impress upon the brief campaign of a week. The swift enemy advance, the temporary check at the Somme, the assembling of the French Sixth Army behind that barrier, the retreat to the line of the Oise, the delicate maneuver of Lanrezac's Fifth Army, the preparation for and the vicissitudes of the Battle of Guise-St. Quentin, all are fully intelligible only when studied with a clear understanding of the stage upon which these events were enacted, and more particularly of the striking contrast between the rolling, open surface of the upland, so easy to traverse, and the broad, marshy floors of the Somme and Oise valleys, so difficult to cross.

It is probable that the Oise barrier would have played a still more important rôle in the last days of August, 1914, but for the failure of the British to support the operations on that line. Sir Douglas Haig had promised the aid of his artillery for the 29th and of his infantry for the evening of the 30th, subject to the approval of Field Marshal French, but was later compelled to withdraw this promise, saying that on account of instructions received he was unable, to his great regret, to participate in the operation. This lack of unity in the face of the enemy, which was emphasized by the British Minister of War, Lord Kitchener, in his instructions to Field Marshal French to the effect that his command was entirely independent and never, in any case or in any sense, would he be under the orders of an Allied general,[4] was to deprive the Allied armies of the tactical and strategic value of many a natural barrier before a long series of disasters would teach them the folly of divided counsels in

[4] Joseph Mangin: Comment finit la guerre, *Rev. des Deux Mondes*, Vol. 56, 1920, pp. 481–520, 721–762; Vol. 57, 1920, pp. 241–285, 481–537, 774–815; Vol. 58, 1920, pp. 74–101; reference in Vol. 56, pp. 502, 506.

the presence of a dangerous foe. The strength of Nature is no counterpoise to the weakness of man.

Nevertheless, it is along the marshy barrier of the Oise that one must seek the first influence of topography upon the maneuvers which led to the victory of the Marne. We have already seen that it was under the protection of this barrier that the Battle of Guise-St. Quentin was prepared. As the Oise flows from northeast to southwest in this part of its course, the effect of the French Fifth Army in debouching from the barrier was to force the army of Von Kluck farther toward the west, thus tending to separate it from the other German armies marching southward on his left. Furthermore, the excellent lines of road, railway, and canal following the natural trench of the Oise in some measure constrained commanding officers of the German First Army to take advantage of the line of least resistance toward the southwest in moving different units of their forces. Von Kluck in his "March on Paris, 1914" (p. 76) reports that as late as August 28 he was under orders to "march west of the Oise towards the lower Seine," where a concentration of fresh Allied forces along the Seine barrier was considered by the German high command to be a possibility; but there can be little doubt that the causes mentioned above operated to force his army farther west than was safe. The result was that Von Kluck arrived upon the Battlefield of the Marne with a gap of many miles intervening between his army and the German Second Army on his left—a perilous gap imperfectly filled by German cavalry. We shall later see that the creation of this gap resulted in the defeat of the isolated German First Army in the Battle of the Ourcq, a defeat which involved all the German forces eastward to Verdun and made of the Battle of the Marne a brilliant Allied victory.

When the German high command found its armies entangled in a dangerous struggle on the Marne battlefield, it faced the necessity of concentrating all its energy, first in an effort to meet the great Allied counterblow, and later to check the Allied pursuit at the Aisne barrier. It accordingly drew in the German right wing by abandoning places occupied on the Somme battlefield during the advance on Paris. As a result we find the Somme region for a few days practically devoid of troops of both armies, for during the bitter struggle along the Aisne neither contestant could turn his eyes to the vacant spaces of the rolling plain.

The Maneuver of the Somme

Nevertheless the plain of the Somme was a valuable prize, and there could be no doubt but that the first to recover breath and assume the initiative would reach forth to seize it. Its rich fields offered life to the armies which should possess it. If the Germans could occupy it they might also control the Channel ports, threaten England more closely, and endanger the short lines of communication crossing the Channel to the British armies in France. Furthermore, the prolongation of the German line on the Aisne westward along the Avre-Somme valley to the sea would not merely give them a natural defensive barrier in place of an arbitrary and artificial front for their extended right wing, but would in addition give them a line about 40 per cent shorter than the one they actually secured from Noyon to the sea and at the same time prevent the formation of such a dangerous strategic salient as that with its apex near Noyon which so long threatened the security of the German armies in France. For the Germans merely to gain and hold the strong, short line of the Somme would thus of itself constitute a great German victory.

On the other hand if the Allies could seize and hold the entire region of the Somme, retain control of the Channel ports, bend back the German right wing so as to deprive it of the protection of the Somme barrier, and impose upon the German front a dangerous salient, their victory would be incontestable. If in addition they could push the lengthening German right wing back to the line of the Oise the victory would be even more complete, for the Oise valley carried the chief line of supply of the German western armies, and to bring this line, including the vital railway junction at Tergnier, 3 miles west of La Fère, under Allied guns would spell disaster to the enemy. There could be no danger that the Allies would find their maneuver blocked by a firm German stand behind the Oise barrier, for that river makes so acute an angle with the Aisne that the resulting salient in the German line would be absolutely untenable.

About the middle of September, while the struggle on the Aisne was still in full swing, began the contest for possession of the unoccupied Somme region. In this struggle the Allies manifested the superior skill in maneuvering and, while failing to secure all they hoped for, won a victory of the first magnitude. Joffre announced a "new plan," which was to attack and envelop the German right flank, pressing it back toward the line of the Oise. The German right was firmly established on the wooded Tertiary mesas of the Noyon massif at the southern margin of the Somme battlefield (see p. 121). This group of imposing highlands constituted for the Allies a formidable military obstacle and for the Germans a solid *point d'appui* upon which their front was to hinge throughout a large part of the war. After vain attempts on the part of the French to dislodge the enemy from this stronghold, a new French army under De Castelnau moved around the western side of the obstacle, advanced northward into the Somme battlefield, and faced eastward along a north-south line from Péronne to Lassigny (Fig. 24), with the double object of attacking toward the line of the Oise and the rail junction at Tergnier, while at the same time holding as much as possible of the Somme region, covering Amiens, and making secure the rail routes across the Somme plain to the Channel ports. The Germans promptly sought to checkmate this maneuver by breaking through to the west and driving the French clear of the Somme area. A violent combat raged from September 21 to the end of the month, the net result of which was that the French held fairly close to their north-south line but failed to reach the Oise and seize Tergnier. The Germans had escaped disaster but had failed to break the wall of steel which was stretching steadily northward to confine them within the fast forming Noyon salient and to bar them from the prizes they should have moved to seize many days earlier.

The Germans were now fully alive to the peril of the Allied maneuver and sped troops northward and westward to turn back the growing left wing of the Allied armies. So effective was the German pressure that at one time the local command of one of the Allied flanking armies considered falling back on the line of the Somme.[5] But such an admission of defeat was refused, Foch was sent to the spot, and French and British troops were rushed northward to support and extend the Allied left wing and if possible to force back and envelop the German right. The great battle of maneuver thus shifted ever northward over the Somme plain, with clashes first in the Lassigny-Péronne sector; then in the region of Albert, where the Germans succeeded in establishing themselves in a strong position behind the marshy valley of the Ancre for a distance of five miles and along the crest of the arch (Fig. 45) east of that river; next about the formidable bastion of Arras, which remained to the French, while south of the town the contesting armies finally rested on their arms facing each other for a stretch of seven or eight miles along the small valley of the Crinchon, a stream less formidable than the Ancre but possessing a broad, flat valley floor bounded by walls in places fifty feet high; and still later, about October 1, in the district of Lens, where the Germans with a keen appreciation of the great advantages conferred on the holder of Vimy Ridge seized and held that important topographic feature at the northern border of the Somme battlefield. The feverish contest in which each army sought to outflank the other came to an end only in mid-October when the two fronts had been extended across the Flanders plain to the sea near Nieuport.

The historic struggle for position on the plain of the Somme during the latter half of September, 1914, might well be called the first great Battle of the Somme in the World War, had not usage decreed that the battle of 1916 should be known by that name. It ended in an Allied victory which in many respects was more far-reaching in its consequences than the first victory of the Marne. The Allies emerged from the contest with the

[5] Mangin, *Rev. des Deux Mondes*, Vol. 56, p. 514.

short sea route to the Continent saved, with the Channel ports and their rail connections secure, and with at least half of the fertile plain of the Somme in their possession. The enemy had been forced to accept a front unprotected by natural topographic barriers save for short stretches along the Ancre, the Crinchon, and at Vimy Ridge, a front so long that in 1917 he would be forced to shorten it in order to economize man power, and of such a form that, while it would give him the advantage of interior lines, it must at the same time constantly threaten the bulk of his forces with irreparable disaster. The bloody battles for the Channel ports were attempts by the Germans to gain what had been lost by their failure to checkmate the Allies' maneuver and would undoubtedly have been followed by further efforts to open yet wider the Noyon salient, to shorten the German front in France, and to secure a position behind an unbroken natural topographic barrier. As has been suggested by General Berthaut,[6] the battle cry "Nach Calais!" might more correctly be rendered "Nach der Somme!" and the price paid by the Germans in the unsuccessful bids for the Channel ports is a measure of the Allied victory in the Maneuver of the Somme.

It is not necessary to follow in detail the local fluctuations of the battle front newly established across the plain of the Somme. Albert, protecting the southern end of the Ancre front; Péronne, guarding the critical angle of the River Somme where it turns sharply from a northerly to a westerly course, and where the valleys of the south-flowing Tortille and the west-flowing Cologne join and so prolong the valley barriers of the two segments of the Somme; and other important points changed hands more than once before equilibrium was established. In its final form the line of battle stretched across the plain from the solid pillar of the Noyon-Lassigny massif in the south to the equally solid pillar of Arras-Vimy Ridge in the north, with no strong topographic features determining its position except along the Ancre and Crinchon valleys and the crest of the arch east of the Ancre. Possession of each of the two supporting

[6] Général Berthaut: De la Marne à la Mer du Nord, Paris, 1919, p. 87.

pillars was shared by the two opposing forces. The Germans held the northeastern part of the Noyon massif, the French the southwestern part. Vimy Ridge and a short portion of its continuation as the ridge of Notre Dame de Lorette had been seized by the Germans; but Arras, with its protecting marshy valleys and underground labyrinth of quarries and cellars excavated in the chalk, was held by the French. Thus for each contestant it was necessary to dislodge the enemy from one or both of the terminal buttresses of his Somme line if anything more than a local victory was to be secured.

We have already noted the failure of the French in frontal attacks against the German positions on the Noyon massif and their failure to flank the Germans out of those positions by the advance of De Castelnau's army. Early in October the Germans violently attacked the strong point of Arras, but, notwithstanding the fact that German troops were able to seize the heights of Vimy Ridge on the north and to reach the valley of the Crinchon on the south, Arras itself remained securely in French hands and projected like a bastion into the German lines. Later in October the French began the long struggle to wrest Vimy Ridge from the Germans, and bitter fighting continued at intervals until May, 1915, with varying fortunes but no great changes in the positions of the combatants. Meanwhile the Germans had dug themselves deeply into the gentle backslope of the ridge, adding to the natural topographic advantages of the position a formidable network of trenches, caverns, and tunnels excavated in the chalk. Every effort was made to hold what the German high command rightly regarded as one of the strongest and most vital points in their whole battle line. The French attached equal value to the peculiar topography of Vimy Ridge, and Foch was now carefully and methodically organizing a tremendous effort to capture it, rupture the German front, and drive the enemy northeastward over the Flanders plain. Could he succeed in this, the German line across the Somme plain, robbed of its northern support, must give way under Allied pressure.

The First Battle of Vimy Ridge

The first of the great battles for Vimy Ridge, often called the Battle of Artois, began on May 9, 1915. For a month and a half there raged one of the bloodiest combats of the war. Twenty days were required for the single operation of taking the network of trenches known as the "Labyrinth." In the defense of the ridge the Germans engaged sixteen divisions, and with the aid of the natural advantages of their position were able to stop the French advance before it had reached the crest. The battle ended with the Germans still in possession of the Vimy buttress, although their hold upon it had been measurably weakened.

The Second Battle of Vimy Ridge

The second battle of Vimy Ridge began on September 25 of the same year, in connection with the British offensive at Loos, northwest of Lens, in the Flanders plain, and continued into October. Souchez was taken by the French in the violent struggle, and the crest of the northern part of the ridge finally attained after the Prussian Guard had been dislodged from an almost impregnable position. But the second great battle ended with the major portion of the crest still in German hands.

During the months which followed the Germans repeatedly made determined efforts to regain a firmer grip upon the natural stronghold from which they had been partially dislodged. In April, 1916, the British took over the front in the Arras-Vimy sector, and on May 21 and June 2 the Germans recovered some terrain from their new antagonists in this region, during combats sufficiently important to rise above the general level of the endless struggle which ever kept the chalk slopes of Vimy Ridge stained with blood. British counterattacks regained part of the lost ground, but other parts of no great strategic or tactical importance were left in German hands in order that offensive preparations which were in progress farther south might not at this time be weakened.

The First Battle of the Somme

A new operation of impressive magnitude was about to be launched in the center of the Somme plain. Here the Germans had devoted all their skill to the task of rendering impregnable a front traversing a rolling plain which offered them outstanding topographic advantages at only a few points. But even where natural advantages were equal between the two antagonists, superior ability in utilizing them could give to one side a line of superior strength. It cannot be doubted that the Germans, particularly in the early years of the war, usually showed greater skill in adapting their defensive organization to the surface form and underground structure of the battle zones than did their enemies. On the Somme plain they not only turned the marshes of the Ancre and the valley of the Crinchon to good account but in addition organized the low rolling hills and flat uplands into a system of mutually supporting positions, across which trenches and wire entanglements were sited with rare ability and machine-gun positions so disposed as to sweep with a grazing fire every approach across the smooth, open slopes characteristic of the Somme topography. "The ground lent itself to good artillery observation on the enemy's part, and he had skillfully arranged for cross-fire by his guns."[7] Ravines and valleys of the plain furnished hidden gun positions from which flanking fire could be concentrated on attacking troops advancing across the open. The towns and villages, together with their underground quarries and cellars characteristic of a chalk country, were transformed into individual fortresses, and new chambers and galleries were cut in the easily excavated rock. These deep underground shelters were later to prove "fatal man-traps" which Hindenburg and Ludendorff would order destroyed;[8] but for the present they were properly regarded

[7] Except where otherwise stated, all quotations in this section are from Sir Douglas Haig's Despatches. These have recently (1919) been published in book form in London and Toronto.

[8] Erich von Ludendorff: Ludendorff's Own Story, August, 1914–November, 1918: The Great War from the Siege of Liège to the Signing of the Armistice As Viewed from the Grand Headquarters of the German Army, 2 vols., New York, 1919; reference in Vol. 1, p. 321.

as well-nigh unconquerable strongholds. Open pits and quarries provided place for machine guns, trench mortars, or heavier artillery. The scattered patches of woodland left growing on the inhospitable clay-and-flint formation and other infertile spots were transformed into strong points which were rendered almost impregnable by the combination of water-soaked soil, labyrinthine wire entanglements, trenches and redoubts hidden in a tangle of underbrush, and innumerable concealed machine-gun nests both on the ground and in the trees. Woodland fortresses and village-cavern fortresses were linked together by trenches to make as formidable a defensive system as had ever defied the power of a determined assailant. At Verdun the French enjoyed topographic advantages far superior to those which the Germans possessed in the Somme plain; but when compared with the stupendous military engineering works by which the Germans had fortified the Somme terrain, the "poor defenses of Verdun were literally mere child's play."[9]

Against this system the French and British armies launched on the first day of July, 1916, an offensive which was destined to be pushed with terrific energy for many months and to be known in history as the First Battle of the Somme. Whereas the Maneuver of the Somme had involved a flanking operation on a grand scale, the First Battle of the Somme was designed to pierce the German line by direct frontal attack. The limited objectives announced in the official despatches, after the larger objective had failed of realization, do not correctly represent the importance of the operation, which was in reality the first great attempt by the Allies to break through the German defensive system by sheer weight of prolonged, intense artillery fire. It is true that there were other objectives, chief among which was the relief of the dangerous German pressure on Verdun; and the attempt to discredit the objects avowed by Sir Douglas Haig, particularly the argument that there was no crisis at Verdun after early February,[10] is sufficiently answered

[9] Victor Giraud: Histoire de la Grande Guerre, Paris, 1919, p. 408.
[10] W. L. McPherson: The Strategy of the Great War, New York, 1919, p. 264.

by the revelations of General Mangin, which show that late in June the situation at Verdun was so grave that General Pétain was for the third time urging the necessity of beginning a retreat to the left bank of the Meuse,[11] an obstacle which might cause disaster to the defenders isolated on the right bank if the enemy broke the last dike in front of Verdun, against which he was then ready to strike. The French Government was urging the hastening of the Somme operations, "in order to relieve Verdun as quickly as possible." To these appeals Joffre replied on June 26 that the preparation for the Somme offensive had already begun (the bombardment started June 24) and that the defenders of Verdun should hold firm on the right bank. On the other hand, General Mangin leaves little doubt that the breaking of the German line was Joffre's chief objective in the Somme operations.

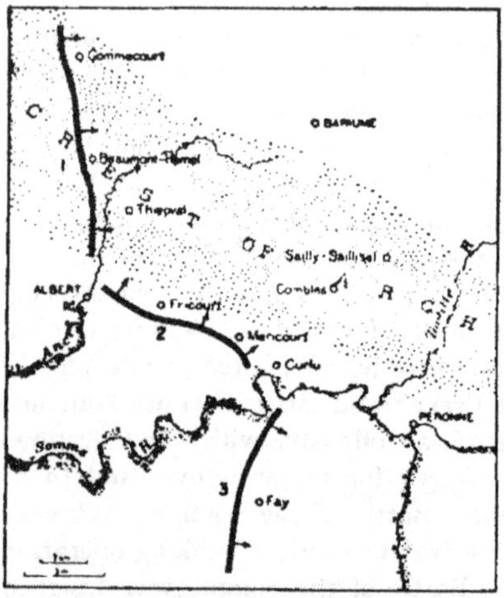

FIG. 45—Three sectors of operations in the 1916 Battle of the Somme. 1, subsidiary attack eastward along the crest of the arch; 2, main attack northeastward to gain the crest; 3, subsidiary attack eastward to drive the Germans back on the Somme barrier.

Surprise could not be relied upon to bring success as in certain later offensives, for the enemy's excellent observation of the Somme plain from the slightly higher land of the anticlinal arch (Fig. 45) gave him ample warning of the extensive preliminary preparation for the attack, while the moment of assault was foretold by the long bombardment. The line was to be broken by unprecedented artillery fire, despite the massing of

[11] Mangin, *Rev. des Deux Mondes*, Vol. 56, p. 737.

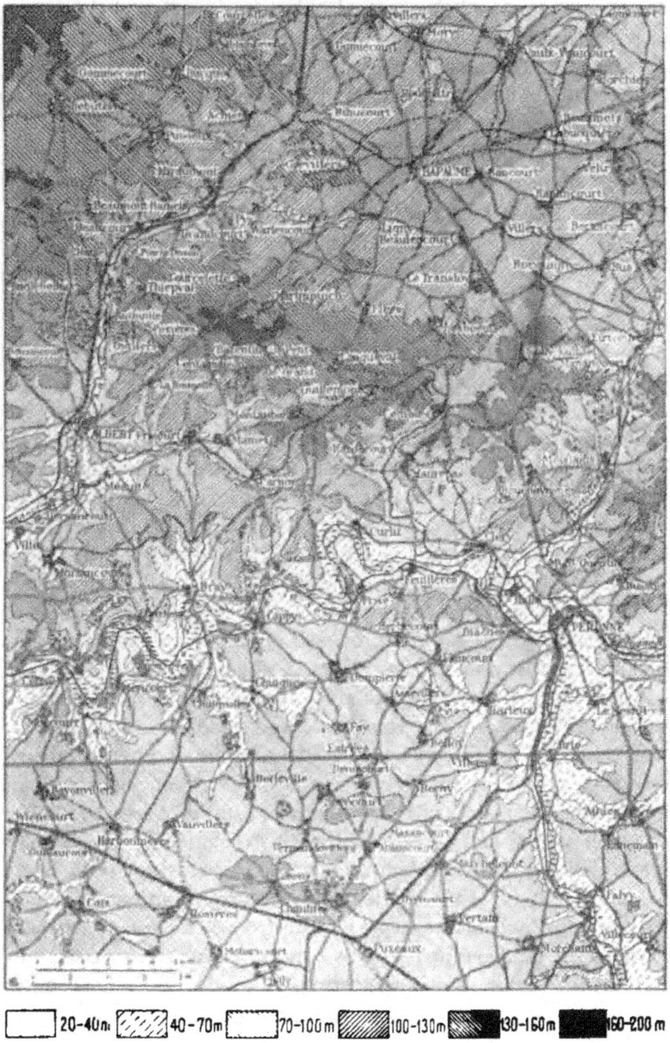

FIG. 46—Field of operations of the First Battle of the Somme. (From the Lens and Amiens sheets of the contour map of northern France and Belgium, 1:100,000, by the Geographical Section of the General Staff, London, 1916.)

enemy reserves, and the operation has been well styled "*l'offensive de rupture.*"[12] Could the rupture be accomplished, the Arras-Vimy bastion would be so threatened from the rear as to become untenable, and the whole line across the Somme battlefield would crumble.

The topography of the Somme plain (Fig. 46) was in many respects especially favorable to such an enterprise. The enemy's front was protected by no continuous natural barrier of formidable proportions. The Ancre made such an acute angle with the Somme that it was impossible for the Germans between the two rivers to utilize both streams to their junction without creating a narrow and vulnerable salient in their line. Their front left the eastern slopes of the Ancre valley near Albert and ran along the southern face of the low anticlinal arch (Fig. 45) to reach the Somme half a dozen miles west of Péronne. As the course of this river is in general east-west, while the front imposed on the Germans trended north-south, it was not practicable for the latter to take advantage of the marshy valley for any great distance, although the marshy meander at Curlu and the abrupt cliffs of the meander scarp known as the "Gendarme's Hat" were skillfully utilized.

To feed the Allied guns with the enormous quantities of munitions required for the proposed offensive and to move and supply rapidly the masses of men engaged necessitated the construction of a network of new standard-gauge and narrow-gauge railways, trench tramways, and roads; and this labor was greatly simplified in a region of low relief and gentle slopes. For the purpose of supplying water to the advancing troops in ample quantity and with the necessary promptness 120 miles of water mains had to be laid, and here again the topography made the task relatively simple as compared with the same operation on certain other battlefields. "Many additional dugouts had to be provided as shelter for the troops, for use as dressing stations for the wounded, and as magazines for storing ammunition, food, water, and engineering material. Scores of miles

[12] Les étapes d'une victoire, *Rev. des Deux Mondes*, Vol. 47, 1918, pp. 419–456.

of deep communication trenches had to be dug, as well as trenches for telephone wires, assembly and assault trenches, and numerous gun emplacements and observation posts. Important mining operations were undertaken, and charges were laid at various points beneath the enemy's lines." Some of these operations would have been practically impossible, others extremely difficult in the water-soaked clay of Flanders or in the hard crystalline rocks underlying parts of the Italian front; but the dry chalk of the Somme plain with its covering of loam and loess was well adapted to such engineering works.

On the other hand, the marshy valleys of the Somme, Ancre, and smaller streams were serious obstacles across which to supply an army on the offensive, and many miles of causeways had to be constructed and maintained under heavy enemy fire. Furthermore, the fissured character of the chalk, by permitting the ground water to sink to low levels, made it extremely difficult to secure adequate water supplies over the uplands between the principal stream valleys. "To meet this difficulty many wells and borings were sunk, and over one hundred pumping plants were installed." It should be noted, also, that while the excellent subdrainage of the chalk insured a dry upland surface during fair weather, except where the clay-flint formation and very thick loam retained moisture unduly long, in rainy weather the extensive covering of relatively impervious loam must transform unmetaled roads and shell-torn battlefields into quagmires. As the Germans held the higher land of the anticlinal arch east of the Ancre, the Allies suffered a disadvantage as to observation which no preparation could overcome and which superior work in the air could only partially alleviate.

It was planned that the Allied offensive should pivot on the Arras-Vimy bastion at the north. The principal immediate objective was the crest of higher land formed by the anticlinal arch mentioned above. This crest rose from 325 to 350 feet above the valleys of the Somme and the Ancre, and its dissection by ravines tributary to those rivers gave stronger relief and somewhat more pronounced slopes than are found in other parts of the plain. In the sector east of the Ancre the German front line trenches were located near the base of the southern slope of the arch. One and a half to two miles farther back, along the critically important crest, ran the German second line of defense. If this crest were reached, nothing but lower land would lie beyond, clear into the plain of Belgium. The prime advantages of effective observation and direct artillery control would shift to the side of the Allies; Bapaume, well down the northern slope of the broad arch, would be closely threatened; and a large section of the German line must give way.

The low arch (Fig. 45) which was to play so important a rôle in the First Battle of the Somme is cut across from northeast to southwest by the Ancre River north of Albert and by the Tortille north of Péronne. At one place between these two valleys the crest is rather deeply trenched by a dry ravine in which the town of Combles is located. Other smaller ravines, usually dry, head against the crest and extend southwest, parallel to the larger valleys mentioned. As the German front ran southward from the Arras-Vimy bastion to the crest of the arch near Beaumont-Hamel and continued along the east side of the Ancre valley across the arch to its southern base, then turned eastward along the base to the Somme west of Péronne, and then once more ran nearly due south some distance west of the upper Somme, it is evident that the character of the terrain naturally divided the operations for the capture of the crest into three distinct undertakings. In the center British and French forces would make a direct frontal attack northeastward up the ravines to clear the main portion of the crest from the Ancre to the Tortille; this principal operation would be aided on the left by British forces attacking eastward, roughly parallel with the axis of the arch, on a front of some ten miles from near Albert northward to Gommecourt; while on the right the French were to strike eastward south of the Somme on a front of equal length and drive the Germans back on the line of the upper Somme.

The attack was launched on July 1, after a week-long hurricane of artillery fire had plowed the battlefield into a shapeless mass of earth in which German trench systems could scarcely be distinguished, and mines exploded under the enemy's lines had thrown the defenders into confusion. On the northern of the three sectors above described the attack was nevertheless a failure. The German position behind the Ancre at Thiepval was exceptionally strong. In front of and below the village ran the marshy valley, through which the sluggish river took its course, canal-like, between two dikes. Behind the village the land rose as a smooth glacis, effectively swept by enemy fire, to the Thiepval "plateau," a typical portion of the crest of the broad arch from which the Germans had direct observation of British movements in the valley, on the farther slopes, and for a long distance to the north and south. German artillery fire, controlled by direct observation, cut the dikes of the Ancre and flooded the valley floor, harassed troops assembling for attack, destroyed assembly trenches, and swept all lines of communication with deadly accuracy. Along the crest lay the Schwaben, Stuff, and Zollern redoubts, deep trench strongholds skillfully sited so as to increase the natural difficulties of the terrain. All combined to give a system of defenses "which may fairly be described as being as nearly impregnable as nature, art, and the unstinted labor of nearly two years could make them."

Beaumont-Hamel lay in a steep-sided, flat-bottomed ravine which continued the natural defensive position of the Ancre valley toward the northwest, just at the point where the main valley bent eastward behind the German lines. "The position was immensely strong, and its holders—not without reason—believed it to be impregnable. All the slopes were tunneled deep with old catacombs—many of them made originally as hiding places in the French Wars of Religion—and these had been linked up by passages to constitute a subterranean city, where whole battalions could be assembled."[13] In the outskirts of Beaumont-Hamel were a number of open quarries, and just south of the village was a ravine (Y Ravine) nearly a mile long and having vertical walls in places 30 feet high, into which had

[13] John Buchan: The Battle of the Somme, New York, 1917, p. 153.

been excavated German dugouts connecting by tunnels with the catacombs mentioned above. "The four successive German lines were so skillfully linked up subterraneously that they formed virtually a single line, no part of which could be considered to be captured till the whole was taken." Against these strong positions the British Fourth Army under General Rawlinson hurled itself with determined fury. Farther north on the upland some progress was made in the initial assault; but the valley positions held firm despite the explosion of a great mine under Beaumont-Hamel, and, in the words of Field Marshal Sir Douglas Haig, "The enemy's continued resistance at Thiepval and Beaumont-Hamel made it impossible to forward reinforcements and ammunition, and in spite of their gallant efforts our troops were forced to withdraw during the night to their own lines." More than four months of sanguinary conflict was to ensue before these strong positions would be wrested from the hands of the invaders. Ludendorff had reason to believe the places impregnable and to be much surprised by their ultimate capture.[14]

In the center, from the Ancre eastward to the Tortille, success was achieved, but at a terrible price. The bloody struggle continued, with only imperfect respite in December and January, from July 1, 1916, until early in March, 1917. The enormous strength of the village-cavern fortresses and of the woodland strongholds led to numberless special actions which in former wars must have ranked as important battles. To preserve the great advantage which they enjoyed in direct observation from the plateau tops, especially from the crest of the main ridge from Thiepval eastward, the Germans fought with a ferocity which was only equaled by the grim determination of their assailants. When the British finally conquered the Thiepval plateau on the ridge summit, the German commanders sacrificed thousands of their men in one furious counterattack after another in the vain effort to cling to the vitally important upland. To regain a single point on the upland the Germans launched eleven big assaults in twenty days. A German order,

[14] Ludendorff, Vol. I, p. 343.

issued prior to one of these counterattacks, read: "Men are to be informed by their immediate superiors that this attack is not merely a matter of retaking a trench because it was formerly in German possession, but that the recapture of an extremely important point is involved. If the enemy remains on the ridge he can blow our artillery in the Ancre valley to pieces, and the protection of the infantry will then be destroyed."[15] At the eastern end of the ridge and overlooking the Tortille valley, the counterpart of the Thiepval plateau was found in the Sailly-Saillisel plateau (Fig. 46). In speaking of the necessity of capturing the village fortress of Sailly-Saillisel, the Commander-in-Chief of the British forces wrote to his Government: "Possession of the high ground at this latter village would at once give a far better command over the ground to the north and northwest, secure the flank of our operations towards Le Transloy, and deprive the enemy of observation over the Allied communications in the Combles Valley."

To lessen the consequences of assaulting under enemy observation, the British Commander-in-Chief sometimes resorted to the perilous expedient of beginning major attacks with large forces under the cover of night, although this involved the movement of incompletely trained men across the open and their assembly in the darkness on a specified line close to the enemy's positions. To secure as much cover as possible from observation and fire, the dry ravines were much utilized for the advances of attacking columns; and as a result inconspicuous and formerly unnamed depressions became known the world over by the names given them by the soldiers. Prominent among these were the "Mash Valley" (Fig. 29), along which was made the heroic advance past La Boisselle* and Ovillers on up to Pozières and the famous windmill on the plateau crest; the smaller ravine farther east known to the British as "Sausage Valley," from which was launched one of the main attacks on Contalmaison; "Caterpillar Valley" north of Montauban, in which troops assembled

[15] Buchan, The Battle of the Somme, p. 142.
* For the villages mentioned in this section, see Fig. 46.

for attack on the two Bazentins and Longueval; while the main valley of the Ancre where it bent eastward into the German line was utilized in a most important penetrating movement by which British forces advanced from Hamel to Beaucourt under protection of the high valley walls, later continuing the penetration as far as Grandcourt. Combles ravine could not be used as a line of advance, but on the contrary constituted a serious obstacle to the attackers, because in the ravine itself the village of Combles with exceptionally extensive catacombs was a strong fortress from which a deadly fire could be poured up and down valley, while other village fortresses on the uplands swept the valley floor with flanking fire. The valley of the Ancre was sometimes a dangerous obstacle for the Germans as well as a line of penetration for the British, as in the operations about the little village of St. Pierre Divion north of Thiepval on November 13, when so many Germans were hemmed in between the attacking force and the marshy barrier that the number of prisoners taken exceeded the number of those attacking. In general, however, one might say that the dry ravines, and occasionally the larger valleys, were the means by which the Allies moved forward for the conquest of the upland.

The third sector of the attacking front, from the large meander of the Somme at Curlu southward, was held by the French. In the main attack on the crest of the arch French forces also co-operated, holding the territory from Combles ravine eastward to the valley of the Tortille and conquering the plateau of Sailly-Saillisel. Their subsidiary operation on the right wing of the attack gained much ground. North of the Somme it was necessary to move eastward so as to clear the enemy from strong positions on the scarps cut by the river in its meanders, positions which were sometimes very difficult to attack because protected on one side by the valley marshes at the base of the cliffs. A stiff engagement took place at the scarp known as the Gendarme's Hat, near Curlu, on the opening day of the offensive. The French advance was only stopped, however, when the attackers faced the formidable position of Mont St.

Quentin, which is the key to the defense of Péronne. Between the Tortille and the Cologne valleys the southern slope of the anticlinal arch forms a southwestward projecting spur, which ends in a bluff overlooking Péronne on the north. It is this spur end which is dignified by the name Mont St. Quentin, and, since it is bordered on the north by the marshes of the Tortille, on the west by the marshes of the Tortille and Somme, on the south by the marshes of the Somme and Cologne, and affords commanding views northward up the Tortille valley, westward down the Somme valley, and southward up the valley of the upper Somme, it is clear that it is a geographic feature of considerable military importance. Possession of it must give possession of Péronne in the marshes close below, standing at the intersection of the important north-south valley depression (Tortille-Somme) with the equally important east-west depression (Cologne-Somme)—a point of such value that in 1870 the Germans ordered its capture at all costs. The French now were unable, however, to advance beyond the Tortille marshes, which they reached west of the Mont, and an expectant world waited day after day for the communiqué which never came, telling of the fall of this topographic stronghold. South of the river the French made notable advances across the little dissected surface of the remarkably flat plain known as the Santerre, where there are few topographic obstacles until the valley of the upper Somme is reached. Among the few, one of the most important is a low mound formed by one of the Tertiary erosion remnants (see p. 121) rising slightly above the general level; and Chaulnes, situated on the mound and near an important railway junction, was the key to the German position in this sector. Repeated efforts of the French to take the position were unsuccessful, but farther north the Germans were forced eastward, until in the vicinity of Péronne they were so hemmed in against the marshy obstacle at their backs that their only course was to retreat to the eastern bank to avoid a catastrophe.

Throughout the whole of this great battle, and on all three sectors of its front, the troops had to deal for the most part with two of the four formations typical of the region—the basal chalk and the surface loam. The degree to which the chalk influenced the character of the fighting is obvious from the frequent references to the chalk pits, quarries, cellars, catacombs, subterranean galleries, tunnels, and similar excavations which fill the pages of every published account of the battle. Even more striking, if possible, was the influence of the loam upon the military operations and their consequences. In every detailed description of the Somme campaign the author will be found to have paid his respects many times over, often without realizing it, to the uppermost geological formation of the battlefield. In the pages of a single popular history of the battle [16] we may read that "the soil of the place was the best conceivable for digging, for it cut like cheese, and hardened like brick in dry weather;" while the pits and brickyards in more clayey parts of the loam (*terre à briques* of the French) were organized into strong points of resistance. But when "there was a deluge of rain . . . the sodden ground and flooded trenches crippled the movement of our men," "the whole land became a morass, . . . every road became a watercourse, and in the hollows the mud was as deep as a man's thighs." "Off the roads the ground was a squelching bog, dugouts crumbled in, and communication trenches ceased to be." "Trenches . . . were often three feet deep in water," and "it was no light job to get out over the slimy parapets." "The roads were past praying for," and "the bringing up of supplies and the evacuation of the wounded placed a terrible burden on our strength. Under conditions of such grievous discomfort an attack on a comprehensive scale was out of the question, the more when we remember the condition of the area behind our lines." Even "the dusty hollows became quagmires," because they were floored with loam washed down from the uplands.

If it be feared that our observer has exaggerated the military consequences of the character of the loam in rainy seasons, turn to the sober despatches of the British Commander-in-Chief.

[16] Buchan, The Battle of the Somme.

"Unfortunately, at this juncture, very unfavorable weather set in and continued with scarcely a break during the remainder of October and the early part of November. . . Constant rain turned the mass of hastily dug trenches for which we were fighting into channels of deep mud. The country roads, broken by countless shell craters, that cross the deep stretch of ground we had lately won, rapidly became almost impassable, making the supply of food, stores, and ammunition a serious problem. These conditions multiplied the difficulties of attack." Throughout the despatches we read repeatedly of preparations for further advance "hindered by bad weather" and of "awaiting better weather for further operations." But on November 9 a dry spell began which made possible the launching of delayed attacks along the Ancre, although the scope of the operations had to be restricted because "the ground was still very bad in places." Even then, "opposite Serre the ground was so heavy that it became necessary to abandon the attack at an early stage."

In the end the water-soaked loam proved the best ally of the German. The British and French had won a great victory over their enemy by driving him from the important crest north of the Somme and by backing him into a perilous position south of the river. With his morale seriously impaired, with the Allies threatening him from a crest which gave them enormous advantages in observation and artillery control, and with the danger of a forced retreat across a marshy barrier adding to his embarrassments, he was forced to the construction of the Hindenburg Line (Siegfried Stellung) far in his rear, and to fall back upon it at the earliest opportunity. But he had escaped the greater peril. The terrible condition of roads and trenches in the loam, and the morass formed by the shell-torn zone, slowed down the action of the Allied battering ram. It stuck in the mud at the critical moment, just when the last of the original German strong lines was in the grip of the attackers and when a break-through might have been hoped for. The pressure on Verdun had been relieved and other important objectives gained; but the piercing of the German front, although attempted with all possible power of guns and men and aided by the first use of tanks in the history of war, had failed of accomplishment. Napoleon's "fifth element"—mud—had, in the sober words of the British Commander-in-Chief, "prevented full advantage being taken of the favorable situation created by our advance, at a time when we had good grounds for hoping to achieve yet more important successes." Joffre had urged that the British, in company with the French, should continue the pressure in spite of all difficulties; and General Mangin is of the opinion that had unity of command existed this would have been done, the German line would have broken, and final victory would have been achieved in 1917. As it was, the German line was merely bent, not broken, and the Arras-Vimy bastion stood firm.

The loam which had mired the French at Agincourt, and worried the Prussians at Villers-Bretonneux in 1870, had saved the Kaiser's armies from complete disaster. At the same time it had, in combination with the subterranean fortresses of the chalk, rendered the Allied advance so difficult and costly and the apparent results attained so limited, as to contribute greatly to the "semi-disgrace" into which certain commanders on this front now fell, and to bring about the replacement of Joffre by General Nivelle.

The Hindenburg Retreat

Further results of the Somme battle were soon to make themselves evident. Possession of the crest of the low arch between the Ancre and the Tortille gave the Allies such command of the enemy terrain to the north that broad areas became untenable, and during the month of February the Germans evacuated one position after another.

Late in February and early in March these local withdrawals became merged in the great Hindenburg Retreat, the main movement of which began March 16, and which was the sequel to the two great struggles on the Somme plain. The Maneuver of the Somme had left the Germans holding a front which was longer than they could effectively man, insufficiently protected by natural obstacles, and forming a salient which might some day prove a trap. The First Battle of the Somme had lengthened the front while at the same time diminishing German man power, deprived the Teutons of the strongest of their natural defenses, and demonstrated that the salient could in time be pushed in. Furthermore, the German high command had now definitely decided upon a purely defensive policy in the west, while awaiting the effects of the submarine campaign and exploiting their gains in the east.[17] Such a policy would be safer if a great Allied offensive on the Somme front were rendered for a time impossible by a change of position from the old line in front of which had been painfully elaborated all the machinery for launching such an offensive—roads, railways, tramways, shelters, assembly trenches, subways, dressing stations, hospitals, wells, water mains, munition depots, etc.—to a position in front of which none of these things should be left in existence. All these considerations demanded withdrawal to a shorter and better defended line across the Battlefield of the Somme. It was to such a line that the enemy was now retreating, pivoting on the impregnable Vimy buttress in the north.

Unfortunately the Allies were unable to press the enemy and so disorganize his retreat. Sir Douglas Haig reported to his Government that "when the thaw commenced in the third week of February the roads, disintegrated by the frost, broke up, the sides of the trenches fell in, and the area across which our troops had fought their way forward returned to a condition of slough and quagmire even worse than that of the previous autumn. On the other hand, the condition of the roads and the surface of the ground behind the enemy steadily improved the farther he withdrew from the scene of the fighting. . . Over such ground and under such conditions rapid pursuit was impossible." Again the loam was proving to be the best ally of the Germans.

The only serious difficulty encountered by the Allies in the early stages of their pursuit was at the marshy valley of the Tortille-upper Somme, which is likewise the line of the Canal

[17] Ludendorff, Vol. 2, p. 2 *et seq.*

du Nord (p. 130) and is backed by the commanding position of Mont St. Quentin. After overcoming the resistance of German rearguards posted along this natural trench to delay the pursuers, British troops, who had taken over a further portion of the front formerly held by the French, occupied Mont St. Quentin and Péronne and succeeded on March 28 in getting across the Somme farther south at Brie. The British Commander-in-Chief found that "the River Somme, the bridges over which had been destroyed by the retreating enemy, presented a formidable obstacle," and in his despatch covering the operations pays a special tribute to the Royal Engineers for the able manner in which they bridged the barrier, here of triple strength—marsh, river, and canal. Bridges had to be constructed at six points before the pursuers crossing to the east side would have their communications established behind them. Foot-bridges for the infantry to pass in single file were completed by the night of March 18, medium-type bridges for horse transport and cavalry by the morning of March 20, the heavy bridges for all forms of traffic not until the afternoon of March 28, or ten days after the arrival of the pursuers at the obstacle. Such are the embarrassments which a river barrier imposes upon an army, even when not defended by an enemy in force.

East of the Somme-Tortille barrier there is no serious north-south obstacle until one reaches the Crozat-St. Quentin-Scheldt canal, which follows in part the marshy valleys of the Somme headwaters and upper Scheldt. Between Cambrai and St. Quentin the new "Hindenburg Line," really a zone of defensive works several miles in width, was based on this obstacle; but south of St. Quentin it swung southeastward to follow behind the Oise valley and canal to a point south of La Fère. Here, therefore, the depression followed by the Crozat Canal, consisting of the headwater portion of the Somme from St. Quentin to St. Simon, and of two minor valleys from St. Simon to the Oise at Tergnier, constituted a single topographic barrier capable of serving as "a very solid position of German defense,"[18] in

[18] Berthaut, De la Marne à la Mer du Nord, p. 144.

advance of the main Hindenburg Line. The French pressed the enemy back from the advanced position and continued to the main barrier at the Oise, where the Germans had flooded the broad marshes to make their new front the more impregnable. It had been part of the German plan to deliver a serious counterattack upon the pursuers when they should have the Crozat Canal obstacle at their backs;[19] but the attack was not pushed with sufficient vigor, and by the end of the first week in April the Allies were close to the main Hindenburg Line from the Scarpe to the Aisne, having in some places captured the outer defenses of that position.

Like the former line around the Noyon salient, the new German front was based on two solid natural buttresses. In the north the Vimy buttress still held firm. At the southern end of the line the Noyon massif had been abandoned in favor of the St. Gobain massif just south of La Fère in the southeastern corner of the Somme battlefield (Fig. 24). This outlier of the Tertiary plateau is the largest area of high land bordering the plain of the Somme. Its flat upland and steep sides are in good part covered with forest, and on three sides it is surrounded by a natural moat formed of the marshy valleys of the Serre, Oise, and Ailette Rivers. Only the northeastern border lacked a valley protection, and this was secure because it lay well within the German lines. The St. Gobain buttress was even more solid than the one which had been abandoned near Noyon.

The new front between the terminal buttresses was not only much shorter, but was also far better protected than the old line farther west. It was not merely that it was based on the Oise valley, on the Crozat-St. Quentin-Scheldt canal system for considerable distances, on the Canal du Nord for a few miles southwest of Cambrai, and was supported by such formidable intermediate buttresses as the old marsh-girded fortress of La Fère and the city fortresses of St. Quentin and Cambrai, whose suburbs and subterranean caverns were organized into impregnable positions; but with a skill which can only command

[19] Ludendorff, Vol. 2, p. 7.

the highest admiration the Germans had sited their lines of trenches and barbed wire across the rolling hills in such manner that every line of approach was across open surfaces swept by machine guns and by fire from artillery concealed in ravines behind the German front, while as few concealed artillery positions as possible were left to the Allies. So adroitly was the front adjusted to the topography that the advantage of observation from higher parts of the plain, especially from such dominating points as Bourlon hill west of Cambrai, lay with the Germans. No one who studied the line on the ground could fail to be impressed with the ability of the Germans to turn every element of surface form to their advantage. Sir Douglas Haig testified that "the line had been . . . sited with great care and skill to deny all advantages of position to any force attempting to attack it."

The Third Battle of Vimy Ridge

Allied strategy now had as its objective the breaking of the German front by prying it loose from the terminal buttresses of Vimy Ridge and the St. Gobain massif. Both of these points could be attacked from the old Allied positions, without the long delay incident to preparing an offensive in the devastated areas abandoned during the Hindenburg Retreat. Two principal operations were accordingly planned to deal with these two strong points. Vimy Ridge had already defied two great Allied offensives and a continuous series of minor struggles. In order that the new attack on the buttress should not be defeated by the natural and artificial obstacles of this highly important position, the same extensive preparations which had preceded the Battle of the Somme in 1916 were undertaken, including the building of reservoirs, the installation of numerous pumping plants, and the laying of many miles of pipe line, to insure abundant water supplies to every part of the arid chalk hills, and, in addition, the construction of plank roadways which would make traffic possible when rains turned the loam into a sea of mud. As the region east of Arras was to be included in the field of operations, the large system of underground quarries in the town and its suburbs was linked together by tunnels, fitted up for occupation by a great number of troops, and connected with the trench system to the east by long subways. The attacking forces were assembled in these subterranean chambers, safe from enemy fire, and moved out to the assembly trenches by the underground routes, thus defeating the German intention to smother any attack by concentrated artillery fire upon the accurately located surface exits from the town. It was part of the Allied plan to attract to the Arras-Vimy stronghold as many as possible of the German reserves before the attack on the extremely difficult terrain defending the St. Gobain buttress should be launched; and the great advantages enjoyed by the holder of the heights of Vimy made it reasonably safe for the Allies to assume that their enemy would put forth every effort to hold that part of his front, rather than seek to dislocate Allied strategy by another withdrawal like that from the old Somme front farther south. "No such withdrawal from his important Vimy Ridge positions was likely. He would be almost certain to fight for this ridge."[20]

"The great strength of these defenses demanded very thorough artillery preparation," which extended over three weeks, culminating during the last few days in a terrific bombardment. The third great Battle of Vimy Ridge, often called the Battle of Arras, was fully engaged when the infantry attacks, led by tanks, were launched on April 9, 1917. The struggle was fierce and bloody, and the defenders launched many counterattacks from tunnels and dugouts. But "the whole might of Canada" was assembled for the assault, and in two days the Germans were driven from the entire ridge and down into the plains to the east. In the expressive football vernacular of the British soldier, the Germans had been "kicked into the hole," a saying which evidences full appreciation of the value of topographic position in modern warfare. The excavations with which the defenders had honeycombed the chalk ridge proved to be man-

[20] Sir Douglas Haig's Despatches, London, 1919, p. 82.

traps when the attacking flood swept past, and such subterranean passages as the Völker and Prinz Arnault Tunnels were found packed with Germans.

Before the attack the Allies' observation had been blocked by the ridge crest close in front (Fig. 34). Now, after an advance of but a few hundred yards, they commanded a magnificent panorama of the Flanders plain (Fig. 35) from a crest rising almost 400 feet above it. The solid Vimy buttress which had supported the new German front on the north was wholly in British hands. The way was paved for a future Allied advance across the plain. From the crest of the ridge Allied control of artillery fire over a considerable zone immediately to the east was so effective that the Germans on April 13 began the evacuation of that part of the plain nearest the ridge. In the words of Ludendorff, "the situation was extremely critical," "a day like April 9 threw all calculations to the winds" and "was a bad beginning for the decisive struggle of this year."[21] The loss of a supposedly impregnable and critically important topographic position was sufficient ground to make Ludendorff admit that April 9 was one of his "black days."

The Allies were correspondingly elated. The British Commander-in-Chief rejoiced that "the capture of the Vimy Ridge had removed a constant menace to the security of our line." His troops would now enjoy a higher and drier foothold, with all which that means to the health and morale of an army; and in addition to excellent observation over German movements on the muddy plain of Flanders he would have his own back areas concealed from direct view.

It is interesting at this point to pause a moment and note that the bloody battles of Flanders were fought to gain possession of the Messines-Passchendaele ridge and its continuation westward in the Mont Kemmel ridge, the most important high ground in the plain of Flanders; the first great Battle of the Somme to gain the crest of the anticlinal arch forming the most important high land in the plain of the Somme; and the terrific

[21] Ludendorff, Vol. 2, pp. 22–23.

struggles for Vimy Ridge to gain the most commanding position on the borderland between the two plains. "Commanding positions" and "dominating heights" have not yet lost their military significance.

East of Arras the attack was also successful. Here the southeastern terminus of Vimy Ridge is dissected into several low hills, and for one of these, Monchy-le-Preux, particularly hard fighting took place. The hill is protected on the north by the valley of the Scarpe and on the south by that of the Cojeul. Its crest dominates a wide stretch of country to the east and west, as well as the valleys north and south. In 1654 the French army under Turenne, in order to threaten the Spaniards besieging Arras, had taken up a strong position on this same dominating high land, with its right resting on the marshes of the Scarpe and its left on the Cojeul River, "its flanks being thus perfectly supported by these two natural obstacles."[22] The Germans now occupied this naturally strong position, and their "commanding positions on Monchy-le-Preux Hill blocked the way of advance along the Scarpe." Large numbers of machine-gun detachments concealed in a series of sunken roads converging on the village of Monchy-le-Preux obstinately defended every approach to the summit. After heavy fighting the position was carried, and from the dominating crest the conquerors turned a murderous fire upon the Germans retreating eastward over the open plain.

Farther north the Arras-Lens railroad, because of its succession of cuts and fills necessitated by the rolling character of the chalk plain, was a serious obstacle to overcome. But here, as in the great Battle of the Somme farther south, the loam, soaked by melting snow and rain, was a chief embarrassment. It gave a special character to the fighting and impressed all who recorded the events of the battle. In Conan Doyle's history of "The British Campaign in France and Flanders" one reads of men plodding ankle deep in the mire, of repeated attacks impeded by the deep mud, of men nearly buried alive

[22] Napoleon Bonaparte, Memoirs of the History of France, Vol. 3, p. 74.

in the dreadful morass; while the despatches of the Commander-in-Chief speak of the difficult going over wet and sticky ground, of troops suffering great hardship and movements seriously hampered, of delays in bringing up guns and operations postponed till the ground should be drier, and of the incalculably greater results which the offensive might have achieved if unfavorable weather had not made the ground so bad as to cause various delays by which the enemy profited to bring up needed reserves and strengthen his positions. The loam was the persistent enemy of complete Allied success, even when it could not save the Germans from what was in some respects the most serious military reverse they had yet suffered in the war.

The operations continued slowly throughout April and well into May. Their net result was to place the entire Arras-Vimy buttress wholly in the grip of the British, where it was to form an impregnable bastion about which the waves of new German offensives would lash in vain, even when the lines to the north and south of it were overwhelmed. It would be difficult to overestimate the gain to the Allied cause represented by the victory achieved in the third and successful battle for Vimy Ridge.

The St. Gobain buttress was so formidable a natural position that it was decided to pry the German line loose from it by an attack delivered farther east, along the heights north of the Aisne River. The terrain of this offensive and its effects upon the operation are considered in another chapter, for it falls within the limits of the Battlefield of the Marne. Suffice it to say here that the operation was a failure, and that the St. Gobain buttress stood firm until the autumn of 1918.

The Battle of Cambrai

During the summer of 1917 the battle line across the Somme plain remained essentially stationary. The Germans had retreated to the Hindenburg Line for the express purpose of standing on the defensive there, while the British were forced to consume the time in re-establishing their trench systems, shelters, roads, railways, and other equipment in the new areas which the Germans had abandoned to them. Only at Cambrai was the usual monotony of dogged positional warfare broken by an important operation, designed to pierce the enemy's line by a surprise attack. Sir Douglas Haig had decided to try the effect of massed tanks as a means of destroying the elaborate wire entanglements of the Hindenburg Line (Fig. 47), thus

FIG. 47—The main wire defenses of the Hindenburg Line on the chalk plain of the Somme battlefield, southeast of Arras. It was the skillful siting of the defenses on the topography of the plain, more than the formidable entanglements themselves, which gave the famous line its great strength.

eliminating the long artillery bombardment which always gave the enemy ample warning of an impending assault. Just as the gently undulating Somme plain had witnessed on November 15, 1916, during the First Battle of the Somme, the first use of the tank in warfare, so now, a year later, it was to witness its first use as a major offensive weapon. The Cambrai sector was selected as the point of attack because that part of the plain was specially favorable for the employment of tanks, and preparations for the attack could be well concealed there under the cover of Havrincourt Wood and other natural protection. The dominating height of Bourlon, which commanded not only the surrounding plain and long stretches of the Agache valley carrying the Canal du Nord on the west, but also the Scheldt River and canal on the east, was the main objective of the attack, rather than the city of Cambrai.

On November 20 at dawn, screened by artificial smoke clouds but without artillery preparation, a veritable army of the iron monsters crawled across the rolling plain, smashing their way into the German lines. It was a clean break-through, and infantry and cavalry pushed on for a number of miles until the obstacles formed by the Scheldt valley with its river and canal, the rolling hills of the open plain, and the wooded Tertiary erosion remnant of Bourlon hill began to break the force of the blow. At Masnières a tank trying to cross a partially destroyed bridge over the Scheldt River and Canal broke through, completing the destruction of the bridge so that cavalry could not cross the barrier in sufficient strength to overcome the enemy's resistance at that point. After some delay a squadron succeeded in crossing on a hastily constructed temporary bridge and after heavy losses took up a position in a sunken road where it defended itself until nightfall. No tanks could cross on the temporary structure, and heavy traffic was restricted to the one bridge seized intact at Marcoing. As a result this passage of river and canal was so slow that the Germans had time to reorganize resistance on the low hills immediately to the east. Even where the canal was successfully crossed by infantry the river sometimes, as near Crèvecoeur, effectually checked further progress. The Scheldt barrier had proved an insuperable obstacle to any effective advance to the eastward.

Elsewhere on the battlefield difficulties were encountered. The open plain rendered the slow-moving tanks excellent targets. One German artillery officer, serving a field gun single-handed until killed, put 16 of the monsters out of action. Sunken roads delayed the bringing up of guns, without which some of the German positions could not be cleared. The northern slope of a small valley south of Flesquières which was effectively swept by German machine-gun fire, together with the village on the upland crest, proved a difficult barrier to pass. When it was successfully negotiated, the British positions north of it "were completely commanded by the Bourlon Ridge, and unless this ridge were gained it would be impossible to hold them." On the other hand, "possession of Bourlon Ridge would enable our troops to obtain observation over the ground to the north, which sloped gently down to the Sensée River. The enemy's defensive lines south of the Scarpe and Sensée Rivers would thereby be turned, his communications exposed to the observed fire of our artillery, and his positions in this sector jeopardised. In short, so great was the importance of the ridge to the enemy that its loss would probably cause the abandonment by the Germans of their carefully prepared defense systems for a considerable distance to the north of it."[23] The southern side of Caesar's Camp (p. 125), which had so often opposed its natural strength to the genius of man, was again to write an important page of history.

The fight for the Bourlon height was of a most desperate character, attack and counterattack following each other in quick succession for five days, at the end of which time neither side controlled the whole ridge. Two days were then spent in preparation for a renewal of the struggle. Meanwhile the Germans were gathering, for an assault on a wide front, forces "whose secret assembly was assisted by the many deep folds and hollows typical of a chalk formation." The German counter-offensive was launched on November 30, and south of Cambrai made such rapid progress up the "natural runway" formed by a deep gulley known as Ravine 22 that the British positions on this part of the front were turned. Farther north, where the plain was more open and surprise therefore less possible, the attack was stopped short, although in supporting their line the British were much hampered by the Canal du Nord. This was a deep trench 80 feet wide with steep, brick-lined walls, without water at this point, across which the soldiers had to

[23] Haig's Despatches, p. 159.

scramble as best they could, aided by ropes, since all bridges were swept by German gunfire. Meanwhile the British troops on the eastern side of the Scheldt barrier were in a perilous situation, hemmed in between the enemy in front and the river behind them, while their hold on Bourlon ridge was very precarious. Both positions were therefore abandoned, the British falling back to the western side of the canal and to the Flesquières ridge during the first week of December.

The "surprise battle" of Cambrai had failed because the valley of the Scheldt, with its river and canal, had blocked the eastward advance of the right wing of the attacking force; and Bourlon hill, the isolated remnant of Tertiary sands and clays rising above the rest of the plain, had as effectually checked the northward advance of the left wing.

The Second Battle of the Somme

Throughout the winter of 1917–1918 the opposing forces on the Somme battlefield faced each other along the strong Hindenburg Line. Then, on March 21, 1918, with dramatic suddenness the German flood burst from its confines on a front 45 miles wide and swept westward over the plain. The Hindenburg defensive had given place to the Ludendorff offensive. The Somme front had been selected for the first of the great German drives designed to end the war before American aid could become effective, in part because the gently rolling surface of the plain was favorable to military operations on a grand scale.[24]

Except along the Oise valley the British front was unprotected by strong natural barriers, for the Germans maintained outlying defenses west of the Scheldt and Somme valleys from Cambrai to St. Quentin. Indeed, for many miles between these two cities the Germans had left the British on the western crest of a divide from which the land slopes down southwestward to the Somme. The topography certainly tempted the Germans to "kick the British into the hole," and here if anywhere the Allies should have organized strong defensive positions along natural

[24] Ludendorff, Vol. 2, p. 220.

obstacles in their rear, in anticipation of such an event. True, such obstacles were only too few in number; for the valley routes for the most part trended toward the west, pointing the way to Amiens and the sea. Thus again the topography invited the Germans to a great offensive, the ultimate objective of which should be to drive a wedge between the French and British armies and force the British back upon the coast.

The second great Battle of the Somme opened with an attack in which Ludendorff sought to secure the advantages of surprise not so much by the use of tanks, the "Cambrai method," as by the employment of overwhelming quantities of toxic gases. Preliminary preparations for the attack had been carried out largely by night, and by day the accumulating masses of men and materials lay concealed in the villages, in the subterranean caverns so abundant in the chalk, and in the scattered patches of woodland. A thick white fog covered the whole battlefield during the entire morning of the attack. Blinded by fog, stifled by gas, and with their communications cut by the heavy bombardment, the British were unable effectively to defend their line, even where it lay behind a strong natural barrier. The German wave overwhelmed every obstacle, and the British retreat westward across the Somme plain was quickly under way.

Between La Fère and Moy the British Fifth Army front was protected by the broad, marshy valley of the Oise with its river and canal. Although in some measure forewarned of and prepared for the attack on other parts of the line, the British command apparently thought this sector so strong naturally that the enemy would make no serious effort against it. Trusting too much to the protection of the terrain, it reduced the defending forces to a thin line of troops, wholly inadequate to meet a strong attack. As this part of the front had been taken over from the French only a few weeks before, its organization was not as perfect as it might have been. The British commander had laid out but not completed a three-zone defensive system. Evidently the French, and possibly also the British, had not adequately entrenched themselves. The greatest military genius in history gave expression to the truth that "the natural positions commonly met with are insufficient to shelter an army from another that is more numerous, without the aid of art."[26] It has been widely stated that one cause of the successive Allied disasters was the reluctance of both French and British troops to dig into the earth sufficiently. In his account of "The British Campaign in France and Flanders" Conan Doyle vigorously defends the British army against this criticism, but his opinion is not supported by some of those most responsible for the British defensive systems. One highly placed general of engineers admitted the truth of the charge, and illustrated his difficulties by saying that on making a tour of inspection to an army front where he had prescribed a certain system of urgently required trenches, he found only two hundred men digging, but two thousand watching a football match.

Be that as it may, the Germans did attack the Oise barrier with very heavy forces. An exceptionally dry spring had rendered the marshes passable for infantry at many more places than usual, thus detracting from the effectiveness of the obstacle. The attackers picked their way across the valley, concealed by the dense fog. When the British outposts, who could not see 50 yards in any direction, discovered the assaulting columns, it was too late to defend the crossings. Even after noon the fog was still dense, and one officer is reported by Conan Doyle to have sent back this message at 12:30 p. m.: "Boche all round within 50 yards, except rear. Can only see 40 yards, so it is difficult to kill the blighters." Telephone and telegraph wires were generally cut, and, as the S. O. S. signals of the outposts could not be seen by the artillery and machine gunners, it can hardly be said that the initial passage of the barrier was really opposed. The small advanced groups fought to the last man, with that grim determination which is the glory of the British soldier on whatever field he fights; but they were no match for the numbers which overwhelmed them, often before they could pass back word of the

[26] Napoleon Bonaparte, Memoirs of the History of France, Vol. 3. p. 72.

attack. The Oise barrier fell, not because it was not strong, but because it was not strongly defended.

Farther north the British struggled valiantly to preserve their front west of St. Quentin, where it based on the vantage point of Holnon Wood, located, like the Bourlon Wood west of Cambrai, on one of the isolated Tertiary erosion remnants rising slightly above the general level of the plain. This position was held all day and was only abandoned when outflanked by enemy forces advancing over less difficult parts of the terrain.

The first strong natural position behind the British front was the line of valley depressions followed by the Crozat Canal from Tergnier to St. Simon, the marshy Somme River and the Somme Canal from St. Simon to Péronne, and the Tortille River and Canal du Nord from Péronne northward. "The Fifth Army Commander decided on the evening of the 21st March, after consultation with the G. O. C. III Corps, to withdraw the divisions of that Corps behind the Crozat Canal. The movement involved the withdrawal of the 36th Division . . . to the line of the Somme Canal." These withdrawals took place that same night, and the next day found the southern end of the British line protected by the Crozat-Somme barrier. Unfortunately the full value of the obstacle was not to be realized, because the charges placed under the numerous bridges often failed to explode, or only partially wrecked the structures.

On March 22 there was a fierce battle along the Crozat Canal. Aided by partially destroyed bridges and by rafts, German forces succeeded in crossing the canal at four points between Tergnier and St. Simon but were thrown back at two. That evening the troops of the XVIIIth Corps, farther north, opposite St. Quentin, where the main force of the German blow had fallen, were ordered to fall back behind the Somme barrier, in touch with the IIIrd Corps on their right, while a large bridgehead east of Péronne was to be held by the XIXth and VIIth Corps next in line to the north. The danger of attempting to defend this bridgehead in the face of increasing German pressure and without the aid of any natural obstacle resulted in orders being issued the next morning for the XIXth Corps to retire behind the Somme marshes, while the VIIth Corps later in the day withdrew behind the Tortille valley-Canal du Nord line. Thus the entire Crozat-Somme-Tortille barrier, from the Oise valley to the source of the Tortille, was utilized as the first strong position behind which the defeated British armies sought protection.

Despite the fact that unusually dry weather made the Somme marshes less of an obstacle than usual, permitting the crossing

Fig. 48—The small erosion butte of Mont Renaud commanding the Oise valley route to Paris just south of Noyon. The natural strength of the position has been increased by artificial fortifications. (French official photograph.)

of infantry between the recognized points of passage, the situation was for the moment reassuring on that part of the front. Most of the bridges had been destroyed, and attempts of the enemy to cross were repulsed with heavy losses. Throughout the night of the 23rd and all day of the 24th the north-south sector of the valley south of Péronne was held by the defenders, except at Pargny, half way between Péronne and Ham, where the enemy succeeded in establishing a bridgehead on the western bank. At Ham, however, the bridges were not completely demolished, and the enemy crossed in force. The two enemy crossings of the Crozat Canal had been maintained and others effected, so that this least formidable portion of the long line of valley and canal had to be abandoned. Its defenders, covered on their right by the Oise from La Fère to Noyon, fell back westward toward the Canal du Nord (Libermont section), which from the Oise at Noyon follows up one small marshy valley and down another to join the Somme west of Ham.

FIG. 49—The fortified summit of Mont Renaud, showing at the right how effectively it commands the valley of the Oise below. (French official photograph.)

North of that portion of the Oise valley between La Fère and Noyon, and parallel to it, the long outlying strip of the Parisian plateau which we have called the Noyon massif interposed a wedge of wooded, difficult country between the enemy forces advancing westward down the valley toward Noyon and those operating north of it, in the country south of Ham. It now became apparent that these northern forces, instead of pushing westward, were facing south with the intention of attacking the Lassigny-Noyon massif, constituting the first line of the northern defense of Paris, and, by breaking through the Noyon gateway, gaining that part of the Oise valley which from here turns more nearly southward and forms a natural pathway to the capital. The forces advancing down-valley along the southern side of the massif would ensure the fall of that part of the obstacle northeast of Noyon by outflanking it from the south. The French had now taken over this part of the front and were vigorously defending the route to Paris. But neither north of the Noyon gateway through the massif, nor in the Oise valley south of it, could they stop the German drive. The massif was outflanked, the town of Noyon fell, and the French withdrew southward behind the Oise barrier, pivoting on the St. Gobain buttress, of which they held the southwestern portion.

Two miles southwest of Noyon the new French line crossed the river to reach Mont Renaud (Figs. 48 and 49), a small erosion remnant left standing in the valley and commanding a good view of the gateway through which the German hordes were pushing southward. Thence the line turned northwest along a northern outlier of the Lassigny massif called Porquéricourt Mountain, which dominates the southern end of the Canal du Nord. The Oise valley-Mont Renaud-Lassigny massif was a formidable obstacle, and against it the Teuton armies hurled themselves in vain. For several days they delivered one attack after another against the Lassigny massif. An outlying mesa at Lagny was captured after a bloody struggle. Porquéricourt Mountain was at last overrun, and the town of Lassigny taken; but five violent attacks on Mont Renaud during the single day of March 27 were hurled back with heavy losses to the enemy; the valley of the Divette, a natural trench in front of the largest of the plateau remnants to the west, could not be crossed; the northwestern flank of the plateau, thrice assaulted in one day, withstood all attacks. Nearly a week of terrific onslaughts netted the Teutons a bare foothold on the northern ramparts. The formidable bulk of the main massif was unshaken. Attempts to pass the Oise at Chauny and other points between the St. Gobain and Lassigny pillars were likewise failures. What some authorities believe constituted from the first a vital part of the German plan of campaign, an advance toward Paris by the Oise valley route, was wrecked against the river-plateau barrier.[28]

After the loss of the Crozat Canal the defensive position of the Allied armies had consisted of the marshes, river, and canal in the valley of the Oise from south of La Fère westward to Mont Renaud; thence the tributary valleys of the Oise and Somme which carry the Canal du Nord from Noyon to the Somme, the southern end of this line being supported by the Lassigny massif; thence the main valley of the Somme with marshes, river, and canal to Péronne; thence the Tortille and Canal du Nord on northward. Even so early as March 22, when it was seen that heavy French reinforcements must be thrown into the battle to check the German onrush, General Fayolle received orders "to re-establish the situation, south of the parallel of Péronne, upon the line Péronne-Noyon."

This line was, however, already seriously breached. The enemy bridgehead on the west side of the barrier near Pargny had been extended and Nesle captured. South of Nesle the Canal du Nord sector had been flanked, and the enemy was beating against the Lassigny massif west of it. At the northern end of the line the enemy had discovered a gap between the VIIth and Vth Corps, poured through it, and compelled the VIIth Corps to abandon the line of the Tortille. The situation on this sector of the front was extremely serious, and the Third Army, now comprising the troops north of the Somme, was ordered to retire its center to the next natural barrier west of the Tortille, the River Ancre. "There seemed every hope that the line of the Ancre would be secured and the enemy stopped north of the Somme." This hope was justified, and from Albert to Beaumont-Hamel the center of the Third Army about March 26 took up a strong position behind the marshy trench from which it was never dislodged. Although the British in the confusion of the retreat first reached the line of the river without wire, bombs, and other necessary equipment, they were able with their rifles to hold much larger German forces at bay along

[28] La bataille de France de 1918, *Rev. des Deux Mondes*, Vol. 46, 1918, pp. 241–302.

the valley until their position could be strengthened. The Germans got across the river at Albert but could not get up the western valley wall.

The westward advance of the Germans north of the Somme seriously threatened the flank and rear of the forces holding the north-south sector of the river from Péronne southward. At the same time the enemy bridgehead on the west bank of the Somme near Pargny had been deepened and broadened by vigorous attacks until a dangerous gap existed between the British XVIIIth and XIXth Corps. Outflanked on both sides, the troops holding the river from north of Pargny to Péronne were ordered to fall back over the level surface of the Santerre, their left flank and rear protected by the east-west portion of the Somme valley, which was defended by small, hastily improvised forces facing north along the barrier. The British next endeavored to hold a north-south line across the Santerre just west of Chaulnes. But this line was still several miles east of the British front north of the Somme, and therefore exposed to a flank attack by the Germans north of the river. The 350 men hurriedly detailed to hold this part of the Somme barrier were utterly inadequate to the task, and before they could be properly reinforced the Germans pushed over to the south bank, compelling a further retirement of the main British front. During this latter operation the line was so bent that for several miles it followed from east to west along the little valley of the Luce. The successful defense of this marshy stream, and of the important bridge over it at Caix, by French and British forces, checked the enemy advance and secured the safety of several British divisions. The entire north-south belt of marsh, valley, and canal, extending in unbroken line across the Somme plain from the River Oise at Noyon to the River Scarpe northwest of Cambrai, had now passed into the hands of the Germans.

West of the Lassigny massif there is a stretch of rolling plain trenched only by small ravines until the head of the Doms valley is reached, near Montdidier, some 12 miles away. Northwestward from Montdidier the marshy valleys of the Doms and Avre, with their partially wooded slopes, constitute a topographic obstacle of much importance—the first west of the Canal du Nord (Libermont section) to trend even approximately north-south. The strength of this barrier is enormously enhanced by the fact that it is bordered on its southwestern side by the rising slopes of one of the anticlinal arches previously described (p. 96). Once established on these slopes the French would enjoy not only the protection of the river, but in addition some of the advantages of commanding position possessed by the Germans on the slopes of the other arch during the 1916 Battle of the Somme. North of Moreuil the Avre valley turns farther toward the west and becomes more of a pathway to Amiens than a barrier protecting it. But the Montdidier-Moreuil sector is rightly to be considered an obstacle, which gains in significance because it is the strong western pillar of a strategic gateway open toward Paris, the eastern pillar of which is the Lassigny massif.

As we have already seen, the Montdidier-Lassigny gateway is barred by no serious natural obstacle. One partially wooded Tertiary erosion remnant, the hill of Boulogne-la-Grasse, stands as a sentinel in the middle of the gap but does not effectively block it. Beyond, the way to Paris is dangerously inviting. It would, however, be a hazardous foe who should dare to pass through while the two strong lateral pillars were in enemy hands. Hence the strong efforts of the Germans to capture the Lassigny massif already mentioned, and the even more determined attempts to secure the Montdidier-Moreuil obstacle which must now be noted.

The main valley of the Avre, above its junction with the Doms at Pierrepont, trends almost exactly east-west. It is marshy and wooded, but ordinarily one would hardly think of it as a barrier against troops moving westward, or even southwestward, more or less nearly parallel with its course. The line of the Doms-lower Avre, described above, was alone capable of effective defense against such an advance. Yet it was the valley of the Avre from its source near Roye westward to Pierrepont, thence northwest to Moreuil, behind which the French were ordered to stand in an effort to halt the Teuton drive. While the enemy was still well to the eastward, General Debeney arrived on the ground with the advance guard of his army, hastily called from the Lorraine front, and issued orders to organize, without delaying an hour, the south bank of the Avre between Roye and Pierrepont, and the west bank from Pierrepont to Moreuil. "Between Moreuil and Roye one must, at any price, attempt to defend the Avre." Evidently the French, who expected a drive toward Paris and who had just seen the westward-moving Germans turn south to fling themselves against the Noyon-Lassigny massif, feared a similar movement southward toward the Montdidier-Lassigny gateway. But conditions were here very different, for the Germans were already southeast of the Roye end of the barrier in great force. The error of trying to hold the east-west section of the Avre was quickly demonstrated. On March 27 the enemy advanced westward both north and south of the river, outflanked the whole line, and threw the defenders into confusion. At the cost of considerable losses a hasty retirement behind the Doms was effected, and Debeney issued a moving appeal to his men to hold that line *à tout prix*. Other troops were hurriedly despatched to close the Montdidier-Lassigny gateway, taking advantage of the strong point furnished by the isolated hill of Boulogne-la-Grasse.

The storm now broke against the Doms-Avre barrier. In addition to the fourteen divisions originally launched on this narrow front, seven others were hurriedly transferred south from parts of the line as far north as Albert and Bapaume. In all 240,000 men delivered a terrific blow in the hope that the shock would fairly shatter the obstacle.[27] The heroic army of General Debeney was weak in numbers, not all of its forces having yet arrived in line. Their defenses were but partly organized. Nevertheless the natural strength of the terrain combined with the supernatural heroism of the devoted Frenchmen to render the position impregnable, even when attacked by greatly superior forces. The river barrier broke the shock of the initial onslaught.

[27] *Rev. des Deux Mondes*, Vol. 46, 1918, pp. 290–291.

Montdidier and Moreuil fell, and the west side of its marshy valley was reached at a few points. But as many as seven furious attacks in succession were beaten off at other places. For three days the battle raged at white heat. Then followed several days of local struggles and preparations for a new grand

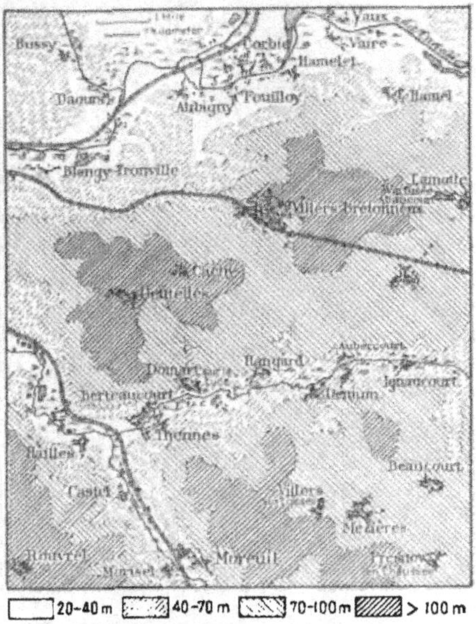

FIG. 50—Battlefield of Villers-Bretonneux, showing plateaus forming eastern defenses of Amiens. (From British Geogr. Sect., Gen. Staff, 1:100,000 map.)

attack, which was delivered on April 4. On this day "ten successive assaults in compact masses were to follow one another, formidable battering-ram blows by an enemy whom each check made furious." Two days longer the struggle raged. But the blow which the German high command had believed irresistible, had merely bent the line and forced it up the slopes beyond the river. The double strength of the barrier could not be over-

Fig. 51.—The open valley of the Somme River northeast of Villers-Bretonneux. In the distance is seen the gently undulating surface of the Battlefield of the Somme, north of the river.

come, and the Montdidier obstacle, like the Lassigny massif, remained unshaken.

East of Amiens the British had checked the German advance across the Santerre on a north-south line just east of Villers-Bretonneux, on the battlefield where the Germans in 1870 had fought for the control of Amiens. The village of Villers-Bretonneux lies at the western extremity of a low, broad mound of the loam which has great tactical importance (Fig. 50). Anyone who walks out north of the village for a few hundred yards and faces westward will have before him such a panorama as few places on the Somme plain can offer. The towers of Amiens ten miles away down the valley are in full view, and all that length of the marshy Somme barrier that lies between. Portions of the Hallue, Ancre, and Luce valleys, to say nothing of parts of the Somme to the north and east (Fig. 51), are likewise dominated by this "plateau," as the British called it in recognition of its commanding importance. Southwest of it lies the "Gentelles plateau," 40 feet higher and nearer Amiens, and rimmed by scattered patches of woodland which appear to owe their existence, in part at least, to the presence of the cold, wet soil of the clay-flint formation which lies just below the loam here and outcrops on the higher slopes of the upland.

Across the Luce valley, to the south, is the similar "Moreuil plateau," already held by the Germans and from which they directed a deadly fire upon the Gentelles upland. If the British could be driven from the Villers-Bretonneux plateau, Gentelles plateau, attacked from the south, east, and north, could be rendered untenable, and the Germans would have "kicked the British into the hole" formed by the valley junctions below. Amiens would be under artillery fire controlled by direct observation, would certainly be destroyed, and could probably be taken. The three plateaus were in fact highly important natural defenses of the city. Moreover, they would give the Germans an excellent view of all movements in the British back areas to the west and north and would conceal the German back areas from British observation.

As soon as the Germans recovered breath from their long pursuit of the British, they launched, on April 24, a violent attack on Villers-Bretonneux with four divisions supported by fifteen tanks. Practically all of the plateau and part of the Bois l'Abbé on the northern side of the Gentelles upland were captured. For a few hours the fate of Amiens hung in the balance. But this was the high-water mark of the German advance over the Somme plain. That night the invincible Australians, responding magnificently to the critical situation, hurled the Germans back to the east. The important western half of Villers-Bretonneux plateau was recaptured, Gentelles plateau was disengaged, effective observation and artillery control were denied to the Germans, and Amiens, "the key to the communications system linking the French front with the British front," was saved. Later the whole of the Villers-Bretonneux plateau was cleared of the enemy.

It is significant that the gently rolling plain of this region, which in the 1916 Battle of the Somme had witnessed the first use of tanks in warfare and in the 1917 Battle of Cambrai the first employment of tanks as a major offensive weapon, should in this 1918 Battle of Villers-Bretonneux witness the first occasion on which the Germans brought tanks into the battle line and the first use of the light, swift British whippet tanks.

In its final form the great salient produced by the German offensive had its southern side along the Oise valley for some 15 miles and along the Lassigny massif for nearly 10 miles; its apex along the Doms-Avre barrier for 15 miles; and its western side along the Ancre valley for 7 or 8 miles. At the southeast it pivoted on the St. Gobain buttress, which was shared by the opposing armies; but in the north it had no solid support on the German side, all of the old Vimy buttress remaining securely in British hands. To remedy this situation the Germans were already attacking the Arras sector with tremendous force.

We have seen that the valleys of the Somme region, with their rivers, marshes, and canals, played a very significant rôle in the third great struggle of the war on this historic battlefield. But it may be asked why these barriers, and particularly the one stretching from Noyon northward to the region of Cambrai with the Somme above Péronne as its central bulwark, did not serve to check the German drive. The answer is not in doubt. They were not properly organized as defensive lines. Napoleon's dictum that "the natural positions commonly met with are insufficient to shelter an army from another that is more numerous, without the aid of art," cannot be too often repeated. Full advantage of the opportunities for defense offered by the valley trenches could only be secured by clearing trees and other obstacles which might shelter an approaching enemy from as much as possible of the field of fire on the eastern banks, by making adequate preparations for the complete destruction of all bridges, by combining with the natural moat a proper system of artificial trenches, and by all the other measures which a wise prevision must dictate. And the lines thus organized must be adequately manned; for no natural position will hold itself.

In the present instance none of these necessary measures were properly executed. Although the line of the Somme had been in Allied hands for a year, the fringe of trees and underbrush bordering the stream had not been removed from the eastern bank, and during the defense of the valley it "afforded good cover to the enemy, and limited the field of fire of the defenders."[28] The destruction of the bridges was very imperfect, and at many points the Germans crossed river or canal on structures which should have afforded no footing. Conan Doyle[29] attributes the incomplete demolition in part to deterioration of the charges of explosives after the bridges had been mined and prepared for destruction by the French, and in part to a transfer of the work of destroying railway bridges from the control of the army, with the result that the work was poorly done and these bridges remained to weaken the defensive value of rivers and canals. Sir Douglas Haig emphasizes the effect of hostile artil-

[28] Haig's Despatches, p. 201.

[29] A. Conan Doyle: The British Campaign in France and Flanders: 1918, Vol. 5, London, 1919.

lery fire in blowing up some of the charges stored near the bridges ready for use at the last moment and in cutting the leads of others.

As for defensive works to strengthen the natural barriers behind the British front, little seems to have been done. Perhaps it was assumed that the Germans would not again attempt to secure a region which they had voluntarily abandoned, although such a belief would seem well-nigh impossible in view of the long-continued preparations for an offensive which had been observed behind this part of the German front. Nevertheless it has been claimed that many of the British commanders considered the long-heralded German offensive as mere "bluff." The excuse that too much of the distasteful work of trench digging injures the fighting power of an army is not wholly convincing when the safety of the army and the victory of a just cause, to say nothing of the lives of the men themselves, may depend more on "digging in" than on using firearms. Napoleon's castigation of the disinclination to entrench is worth repeating: "Both officers and privates dislike handling the pick-axe and spade; they therefore strive to outdo one another in echoing and repeating such notions as these: 'Field fortifications are more injurious than useful; there ought to be none constructed; victory belongs to him who marches, advances, and maneuvers; the soldier ought not to work; is not war attended with fatigue enough?'—Flattering, yet despicable arguments!"

When threatened with overwhelming disaster the British, "with their backs to the wall," did dig in magnificently. Within a few months of the time they were pushed back to the gates of Amiens, they had dug five thousand miles of new trenches; and a high officer of engineers gave it as his opinion that the mere sight of so formidable a system of defenses in the chalk country discouraged the Germans from ever again attempting a serious blow at Amiens. Certain it is that whereas this officer admitted the British were too weak in numbers properly to man so elaborate a system of trenches as that prepared in the Amiens sector, where the next German blow was expected—a numerical weakness which is emphasized by the British Commander-in-

Chief in his despatch of December 21, 1918—the German high command got the idea that this front was held in great strength; and Ludendorff gives this as one reason for abandoning the plan of an advance in the Somme area. Although at the time of Ludendorff's decision the defensive network was necessarily far from completed, it is probable that the elaborate system of partially excavated trenches would, by turning out on the ground the white chalk which lies close to the surface about Amiens, give an appearance of strength which actually did not exist. However that may be, had even a part of the labor expended in organizing the Amiens front been utilized in strengthening the north-south line of the Somme during the year it was in British hands, it is permissible to suppose that the big German push of March 21, 1918, might have been halted along that barrier. Direct comparison between the two cases is of course impossible, for the Amiens defenses constituted the forward zone of their period, whereas work on a secondary line like that of the Somme was necessarily subordinated to the needs of the forward zone of that time.

Whether or not the failure adequately to fortify the natural defensive positions of the Somme plain was justified by conditions known to the British Commander-in-Chief, we are here chiefly concerned with the fact that they were not so fortified; and hence that no criticism of the value of topographic barriers can be based on failure to check the Germans along those lines. We do learn of a general of engineers being sent back hurriedly to organize as well as possible the line of the Somme below Péronne, after the crash had come. But the time to undertake that task was in the preceding twelve months, not during a forced retreat. Sir Douglas Haig admits that "practically no work had been carried out with the object of securing the line of the River Somme" and gives as a reason the limited amount of labor available, all labor units that could be used in rear of the forward defensive zones having been allotted to the construction of the Péronne bridgehead defenses.[30] As we have seen, these defenses had to be abandoned without a fight, and

[30] Haig's Despatches, pp. 195, 217.

the attempt to check the enemy was actually made on the unorganized line of the river.

Finally, the number of men detailed for the defense of the Somme barriers was inadequate. The British Commander-in-Chief explains in some detail the circumstances which caused him, when advised of the impending German offensive, to concentrate his forces on other parts of his front and deliberately to assume the risk of having to yield ground in the Somme plain. And, while the advisability of making a determined stand on the line of the Somme River was debated during the first days of the retreat, it was decided that the reserves of fresh troops necessary to hold that line could not be brought up to it in time to check the enemy there. Ludendorff states that the left wing of his Second Army was "more delayed by having to cross the Somme than by the enemy."[31] A barrier insufficiently manned could at best only hold the enemy in check long enough to secure the orderly retreat of the defenders to another position farther back.

The Second Battle of the Somme gained for the Germans a vast area of the Somme plain, nearly 40 miles broad at the point of deepest penetration; brought prestige to German arms and hope to the German people; and profoundly discouraged the peoples of the Entente. But it failed to separate the French and British armies; it greatly lengthened the German line and reformed it into a dangerous salient; and, worst of all for the German cause, it forced the Allies to use common sense. Foch was appointed generalissimo five days after the Germans had smashed through the British front.

The Fourth Battle of Vimy Ridge

The German offensive in the west, on which the hope of victory was based, was soon being cramped and hindered by the Arras-Vimy bastion. Not until that formidable obstacle had been conquered would the armies immediately to the south have elbow room to develop their advance. Nor was it possible

[31] Ludendorff, Vol. 2, p. 231.

to press on to Amiens and beyond so long as the rear of the forces advancing down the Somme was threatened by an Allied attack launched from the Arras region. Accordingly, on March 28, the German Seventeenth Army was hurled against the buttress which blocked the way to victory. Ludendorff states that the principal objective was "to capture the decisive heights [Vimy Ridge] east and north of Arras; the next day the Sixth Army was to prolong the attack from about Lens and carry the high ground in that area [the Notre Dame de Lorette heights, forming the northwestern continuation of Vimy Ridge]. I attached the greatest importance to both these attacks."[82] Neither in the plain of the Somme nor in the plain of Flanders could the Germans push very far west until the Arras-Vimy bastion had been captured.

Nature had endowed the Arras region with a topography which well fitted it to play the rôle of an impregnable bulwark in the Allied line. We have already seen that north of the town Vimy Ridge trends from northwest to southeast. South of the town the streams flowing down the northeast slope of the anticlinal arch—the Sensée, Cojeul, Crinchon, Gy, and their branches—give a succession of ridges and valleys trending from southwest to northeast. Arras lies just within the angle formed by the intersection of these two topographic systems. With its labyrinth of subterranean caverns the city strengthened the apex of the eastward-pointing bastion. The British had so shaped their defensive works as to take advantage of the strong natural features constituting its sides. Hence their trench system, guided by the elements of the terrain, also formed an angle the apex of which was directed eastward (Fig. 52). This was the strongest point in the whole British line.

The Germans made a tremendous effort to storm the bastion. German infantry advanced almost shoulder to shoulder in six successive waves and were supported by a bombardment of the utmost violence. But the excellent observation from the higher ground, for which the Allies had paid so dearly, now

[82] Ludendorff, Vol. 2, p. 237.

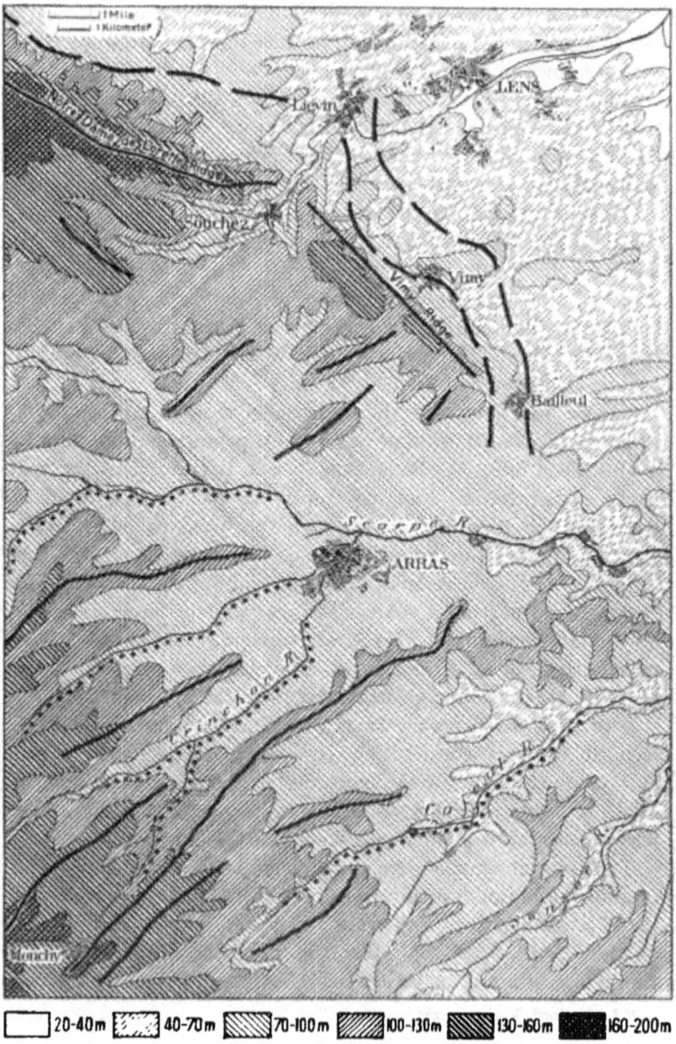

FIG. 52—For explanation see bottom of opposite page.

demonstrated its value. British artillery, controlled by direct observation, decimated the enemy forces as they concentrated for the attack and during their advance, while machine guns in favorable positions and field guns brought forward close to the British front line completed the slaughter. "In spite of employing extraordinary masses of artillery and ammunition, the attack was a failure," admits Ludendorff. And failure here was nothing short of disaster. "A German success in this sector," writes Sir Douglas Haig, "might well have had far-reaching effects. There is little doubt that the enemy hoped to achieve great results by this new stroke and that its failure was a serious setback to his plans." The German line bulged westward on the south and was soon to bulge westward on the north. But these bulges were necessarily limited so long as the bastion between held firm. Against its impregnable point the German armies suffered a rude and decisive check. They tried in vain to crush in its strong sides. Failing this, they gave up the struggle on the whole front north of the Aisne, and turned despairingly to another sector in the vain search for a decisive success.

On April 6, in the southeastern corner of the Somme battlefield, the Germans attacked the French on the line of the Oise between La Fère and Chauny and pushed south over the western edge of the St. Gobain massif, where they had lost a little ground after the retreat to the Hindenburg Line, to the valley of the Ailette carrying the Oise-Aisne Canal. Here their advance was halted. The operation put the entire St. Gobain massif and its surrounding moat again in German hands and thus

FIG. 52—Some of the major topographic elements utilized in constructing the defenses of the Arras bastion, showing the northwest-southeast trend of the Notre Dame de Lorette-Vimy Ridge crest and the southwest-northeast trend of ridges and valleys farther south. Note that the two systems meet at an angle near Arras. (From the Lens sheet of the contour map of northern France and Belgium, 1:100,000, by the Geographical Section of the General Staff, London, 1916.)

Solid lines show important ridge crests; broken lines show ridge slopes or outlying foothills important for observation and defense along lines parallel to the main crest, although erosion has made cross ridges and ravines more apparent; dotted lines show some principal drainage lines. The defensive system was controlled by these features of the terrain, even where trenches did not follow ridges or valleys continuously.

gave more solid support to the long southern side of the salient resulting from the push towards Amiens. Three days later Ludendorff started his westward push in the plain of Flanders (p. 77) which was to fail because the Arras-Vimy bastion on one side and the Mont Kemmel bastion on the other held firm. In June a short-lived offensive would give him practically full possession of the Lassigny massif but lead to no important results. His last great efforts to achieve victory were to be made on an entirely different terrain, in part because secret preparation and a surprise attack alone now promised any chance of success; and, as Ludendorff himself tells us, the open plain of the Somme "afforded too little cover for the necessary preliminary work."

The Third Battle of the Somme

The months which followed the German push across the Somme plain to the eastern defenses of Amiens witnessed not only the enemy offensive in the plain of Flanders referred to above, but his successful break-through at the Chemin des Dames, the collapse of his final offensive on the Marne, and the launching of the Allied counterblow which was to achieve complete victory under the supreme command of Marshal Foch. By August the time was ripe for Haig's main army to participate actively in the Allied offensive. The Germans saw the new storm coming and in the first days of the month paid a tribute to the defensive value of natural obstacles by withdrawing from the region they occupied southwest of the Avre valley and from their very limited holdings west of the Ancre, and by taking up positions behind those two barriers of river and marsh.

On the 8th of the month the British, to use their own expressive phrase, "kicked off" from the eastern end of the Villers-Bretonneux plateau. Ludendorff's expression had a different ring: "August 8 was the black day of the German Army in the history of this war."[33] The vital railway junction at Amiens was under long-range artillery fire, and one effect of an advance at this point would be to disengage it completely and to free

[33] Ludendorff, Vol. 2, p. 326.

the main Paris-Amiens railway from enemy interference. The advancing forces would be protected on their left flank by the marshes of the middle Somme (Fig. 41), while in front of them there was only the smooth surface of the Santerre, almost devoid of serious natural obstacles until the marshy valley of the upper Somme, above Péronne, should be encountered. The attack, launched in a fog without warning but accompanied by intense artillery fire and by the use of great numbers of tanks, for which the ground was most favorable, was immediately successful; and before nightfall the indomitable Canadians and Australians had advanced six or seven miles over the smooth terrain.

One hour after the British had launched their attack, the French First Army assaulted the Doms-Avre barrier immediately to the south. The obstacle was passed in a few places under conditions not dissimilar to those which aided the Germans in their passage of the Oise barrier on the 21st of March preceding, a dense fog preventing adequate defense of the crossings. Of still greater advantage was the fact that on the north the British advance aided the outflanking of the Avre line, and French troops filtering eastward along the *couloir* of the Luce could compel the evacuation of Moreuil; while at the south a flanking operation could be carried around the head of the Doms valley, thus making Montdidier untenable. Both operations were successful, and the line of the Doms-Avre, threatened at the rear from both ends, was precipitately abandoned by the Germans. Farther east the French Third Army next tried to outflank the Lassigny massif (which had been conquered by the Germans in their June 9 offensive) by passing around its western side and capturing the town of Lassigny. This operation, supported by a painful and costly advance over the massif itself and across the borders of the Parisian plateau east of it, was eventually to deprive the Germans of their hold on the entire Lassigny-Noyon massif and throw them back on the St. Gobain buttress, but only after the attacking forces had been aided by the Allied advance north of the massif and had themselves made the heaviest sacrifices.

By the evening of the 12th the Germans, pivoting on the Noyon-Lassigny buttress, had fallen back on their old north-south line of defense, the key to which, as we have seen, was Chaulnes, situated on one of the low isolated Tertiary mounds (p. 121) and guarding an important railway junction. Here the Germans put up such a stiff resistance that the British Commander-in-Chief decided to break off the battle and shift his attack to a sector north of the Somme. Chaulnes did not fall into the hands of the Allies until two weeks later.

Foch looked forward to the probability of encountering even stiffer German resistance along the natural barrier of the upper Somme, and directed an attack in the direction of Bapaume-Péronne which "would have for consequence the outflanking of the defense which the enemy would offer along the Somme and force him to a more or less general retreat."[34] Sir Douglas Haig chose the terrain between Albert and Arras for launching the attack, and states among the reasons for this selection the facts that "the ground . . . was suitable for the use of tanks," and "this attack, moreover, would be rendered easier by the fact that we now held the commanding plateau south of Arras about Bucquoy and Ablainzevelle," a spur of the anticlinal arch which played so significant a rôle in the 1916 Battle of the Somme. As the attack developed the forces operating at the north of the line would use "the River Sensée to cover their left, in the same way as the River Somme had been used to cover the left of the Fourth Army in the Battle of Amiens."[35]

The attack was a success, but in places encountered great difficulties. The marshy valley of the Ancre, backed by the dominating heights of the Thiepval plateau, opposed a serious obstacle to the troops advancing on the right. "The Fifth Corps, under General Shute, followed the curve of the River Ancre on a front of 9,000 yards. . . . It had no tanks, since the marshy valley and sluggish stream lay before it, . . . this

[34] Louis Madelin: La bataille de France, 21 mars–11 novembre, 1918, *Rev. des Deux Mondes*, Vol. 52, 1919, pp. 798–853; Vol. 53, 1919, pp. 59–99, 270–310, 533–569, 785–828; Vol. 54, 1919, pp. 64–108, 314–363; reference in Vol. 53, p. 550.

[35] Haig's Despatches, p. 264.

evil watercourse which had been flooded by the Germans and was 300 yards wide at one part. All bridges were gone, and the banks were low and boggy. The main stream was over 6 feet deep, and its channel could not be distinguished from the general flood. The whole morass was covered by a tangle of fallen trees, reeds, and artificial obstruction. . . The west bank was so overlooked that no one could move unscathed."[36]

Undismayed by the apparently impregnable barrier the dauntless British soldiers waded breast-deep through the waters under heavy fire and assaulted the heights to the east. Their task, which might otherwise have been impossible, was rendered easier by a concentric attack on that part of the arch east of the Ancre, delivered from the northwest and southwest by troops which had outflanked the main part of the barrier opposite the Thiepval plateau. Miraumont, in the upper Ancre valley, resisted for three days, but was finally captured. Difficult as was the taking of the Ancre position, it would certainly have proved even more costly had not a Prussian division (known to be of poor quality and hence given part of the river front to hold because of the protective value of marsh and stream) refused to fight, thereby throwing the whole line into confusion.[37] The crest of the anticlinal arch, which was taken by frontal attacks in the First Battle of the Somme, was much more easily secured by an advance parallel with its axis in this Third Battle of the Somme.

On the 26th of August the Germans south of the middle Somme began falling back to the north-south line of the Canal du Nord-upper Somme between Noyon and Péronne, the first strong natural barrier behind their former front, and which Ludendorff calls the "Kanal-Stellung." In electing to stand on this line Ludendorff gave much weight to the protection it would afford against tank attacks. Marsh-girt Péronne with its guardian Mont St. Quentin (p. 155) was the buttress upon which the northern end of this sector of the enemy's new front was

[36] Conan Doyle, The British Campaign: 1918, Vol. 6, p. 82.
[37] Ludendorff, Vol. 2, p. 342.

FIG. 53—The Somme valley at the Brie crossing, above Péronne. The importance of the marshy valley as a barrier is well shown by the presence of British tanks abandoned when it proved impossible to get them over the obstacle.

based. Its capture was essential, if the line was to be turned from the north. An attempt to force the river in front of Mont St. Quentin was a failure. The flooded valley, defended by heavy machine-gun fire, could not be negotiated. As the Tortille was a smaller obstacle, the Australian infantry selected for the attack were taken back to a point 4 miles west of Péronne and moved over the River Somme where its east-to-west course was under Allied control. Advancing again north of the river, the Australians forced the Tortille and stormed the mountain from the northwest, while the defenders were engaged with a portion of the forces directly in their front. The struggle was extremely violent, for the Germans fought tenaciously and counterattacked repeatedly, in order to hold this "most important tactical feature commanding Péronne and the crossings of the Somme at that town." In the opinion of the British Commander-in-Chief, "the fighting was exceptionally severe, and the taking of the position ranks as a most gallant achievement." The capture of Péronne quickly followed. Despite the loss of this critical point, the Germans continued to maintain their position behind the Somme-Canal du Nord obstacle between Péronne and Noyon for some days; and in the meantime their forces farther north took their stand behind the Tortille-Canal du Nord line from Péronne on northward, to Ytres at the northern end of the Tortille valley. Thus for a second time this striking north-south barrier (Fig. 53) determined the battle front of the contesting armies.

A new attack, designed to dislodge the enemy from this position, was already under way. Striking eastward from the Arras bastion, the British forces assaulted the German position at Monchy-le-Preux. As we have already seen (p. 165), this position "was one of great natural strength, well organized for defense, and commanded observation of much importance." Once before the British had overcome its resistance at great cost, only to lose it during the later German advance. Now they had to pay the price a second time. Fortunately the Germans did not fight with the vigor of former days. The position was stormed, and, pushing on across the parallel ridges and ravines descending the northeast slope of the anticlinal arch, which were skillfully utilized as defensive lines to retard the British advance, the attacking forces smashed the main German Drocourt-Quéant line of resistance facing the Arras bastion. This line was one of the strong positions on the western front and, because of the skill with which its defenses were adapted to the topography, entitled to rank with the Hindenburg Line, of which it was a northern continuation. When it collapsed, the German commanders ordered their troops in this section to fall back eastward on the Canal du Nord-Agache River line, and northward on the line of the Sensée River marshes, where they raised the water level by opening the sluices and by damming a tributary brook, thus converting the valley into a lake. By September 9 the British faced a continuous river, lake, canal, and marsh barrier 20 miles long, from Havrincourt to near Lécluse, behind which the Germans had sought protection. This was a part of the famous Siegfried Line, the position of which had in this sector been determined by the natural obstacles mentioned. It will thus be seen that the Agache River forming the western side of Caesar's Camp (Fig. 42) was again in use.

In the Bertincourt-Havrincourt sector the Canal du Nord offsets sharply to the east (Pl. II), taking advantage of the east-west valley of the small Exuette River. As a result, the advance of British forces to the northern section of the canal line would outflank German troops holding the Tortille valley section farther to the south and west, as well as those holding the continuation of that line along the Somme valley above Péronne. The defense of the Somme had already been compromised by the poor behavior of several German divisions, and in the face of a new threat the defenders, between the 4th and 8th of September, abandoned the whole of the Tortille-Somme-Canal du Nord line from the head of the Tortille southward and fell back toward the next valley-and-canal barrier to the east.

The high defensive value which the Germans attached to lines of rivers, marshes, and canals, was never more clearly evidenced than during this retreat. The intervention of the tank as a major offensive weapon had given a new and increased importance to these topographic features. Massive barbed wire defenses (Fig. 47), to which the German infantryman had once trusted for protection, were easily smashed through by the ponderous machines, and the whole German army was now suffering acutely from "tank fright." Ludendorff avows that mass attacks by tanks under cover of artificial fog were now his most dangerous enemy,[38] while Sir Douglas Haig reports that "so great has been the effect produced upon the German infantry by the appearance of British tanks that in more than one instance, when for various reasons real tanks were not available in sufficient numbers, valuable results have been obtained by the use of dummy tanks painted on frames of wood and canvas."[39] The German commanders reasoned that since wire might stop men but could not stop tanks, whereas water would stop tanks but might not stop men, the only safe line of defense consisted of a strongly wired position covered in front by river, canal, or marsh (Fig. 53). Hence arose the saying, so frequently heard during the last months of the war: "The boches are trying to get behind water."

At the south the French, both by direct assault and by a flanking movement, gradually pressed the enemy back over the Lassigny-Noyon massif to and beyond the Crozat Canal, capturing the outlying fort forming part of the western defenses of La Fère. Ludendorff placed great confidence in the strength of the canal line with its associated marshes and valleys. "It would no doubt have been possible to keep the left wing of the Eighteenth Army . . . on the Crozat Canal, but that required more men than the defense of the St. Quentin-La Fère line, with the broad valley of the Oise in front of it."[40] So the German front in the south was withdrawn behind the marshes of the Oise. At the north the enemy still stood behind the Sensée marshes and the Agache River-Canal du Nord line

[38] Ludendorff, Vol. 2, p. 340.
[39] Haig's Despatches, p. 302.
[40] Ludendorff, Vol. 2, p. 346.

as far as Havrincourt, a peculiarly difficult position to take, especially as the smooth slopes of the valley leading down to the canal were open and effectively swept by German machine-gun fire. From Havrincourt on south to the Oise the enemy had retreated close to his main line of resistance, the Hindenburg Line, which crossed the hills from Havrincourt to the Scheldt valley 8 miles south of Cambrai, then followed the great Scheldt River-Scheldt Canal-Somme River-St. Quentin Canal barrier to St. Quentin.

The strength of this line fully justified the repute in which it was held. The entanglements were of the most formidable character, broad belts of rigid iron posts closely set and intricately woven with extremely heavy barbed wire (Fig. 47), while the open space in front was swept by machine-gun fire directed from heavily armored concrete "pill boxes" skillfully concealed in the trench system (Fig. 54). Instead of being a single line of entanglements defending the usual trenches, the so-called "Line" was a zone 4 to 6 miles wide composed of a network of the heavily wired trenches linking up the town and village fortresses and their subterranean caverns in the chalk, and supplemented by numerous concrete fortifications. But the peculiar strength of the barrier, as already noted, lay in the remarkable skill with which the defenses were sited so as to take advantage of the natural features of the terrain. Not only did they command every approach by grazing fire across smooth, cleared ground, but they were in addition so disposed across the plain as to prevent the Allied artillery from obtaining any positions from which to bring an effective fire to bear upon them. The upper valleys of the Somme and Scheldt, which are followed by the canal system and which determined the general position of the Hindenburg Line in this region, were utilized as a natural fosse, to the west of which were the advanced defenses, including one or two of the continuous lines of trenches and heavy wire, while the double line of wired trenches, sometimes called "the Hindenburg Line proper" lay behind the canal to the east.

On the Somme-Scheldt drainage divide there is a break in the surface barrier, but the tunnel in the chalk which carries the canal from one valley to the other was turned to good account, serving as shelter for large bodies of troops who could reach their trenches above by shafts driven for that purpose. The open canal added materially to the shelter from enemy

FIG. 54—Partially destroyed concrete machine-gun shelter in a trench of the Hindenburg Line southeast of Arras. From this shelter a deadly grazing fire swept the exposed surface of the barren Somme plain.

fire naturally afforded by the valley depressions, and in its steep walls, as well as in the valley sides, were constructed numberless dugouts and concrete shelters. Machine guns in armored emplacements at the tops of the walls could sweep the approaches on the western side and the great trench itself in case it were entered by the enemy. Subways connected the valley-canal trench with the wired defenses east and west of it. For convenience one usually speaks of "the line of the canal;" but it should not be forgotten that, as Sir Douglas Haig has so well pointed out, it was not so much the artificial canal itself as the skillful use of the topographic features associated with it which gave to the Hindenburg Line its great strength.

Such was the well-nigh impregnable belt of intricately combined natural obstacles, artificial waterway, and newly executed military works behind which the badly beaten and much depressed German armies had sought protection and along which they held at bay for the moment their victorious pursuers. But the Hindenburg barrier had a greater significance than serving as a line of defense for the Kaiser's field armies. Back of it lay the railway cordon skirting the southern and western base of the Ardennes Mountains and passing through Maubeuge, Hirson, Mézières, and Sedan to connect the northern group of German armies on the western front with the southern group. From this central artery there branched out to the south and west the lines feeding the German front in Champagne, on the Marne plateau, and in the Somme plain. The Allied plan of campaign involved concerted blows at this vital railway artery, the British striking for the Maubeuge sector, the French and Americans at Mézières and Sedan. If the railway were cut the two groups of German armies would be separated, and one group forced back along the northern side of the Ardennes, the other along the southern side, with no effective lateral communication between them across the intervening mountain wedge. The Hindenburg Line was the most formidable barrier protecting the railway artery from the British blow. If that barrier were breached the whole German front would be endangered, while the political effect produced among the peoples of the Central Powers by the collapse of their strongest defensive position would be most far-reaching.

Local engagements having cleared the enemy from important outpost positions such as the Holnon Wood elevation west of St. Quentin, the way was cleared for a decisive attack on the main barrier of the Sensée River-Agache River-Canal du Nord-Hindenburg Line (Scheldt River and Canal-Somme River and Canal-Oise River and Canal). The Agache River-Canal du Nord sector west of Cambrai was attacked first on

September 27. Near the head of the Agache valley in the vicinity of Moeuvres the canal is the only serious obstacle to be encountered until Bourlon Wood some two miles farther east is reached. But downstream to the north the valley carrying the canal deepens and broadens and becomes marshy. The British Commander-in-Chief decided that "the northern portion of the canal was too formidable an obstacle to be crossed in the face of the enemy. It was therefore necessary for the attacking divisions to force a passage on a comparatively narrow front about Moeuvres, and thereafter turn the line of the canal farther north by a divergent attack developed fan-wise from the point of crossing." This was a very difficult maneuver, especially as the wooded height of Bourlon lay on the central axis of the advance and commanded ground to the north and south across which the diverging forces must pass. It was aided by the fact that the canal in this sector was dry as a result of the blowing up of the gates.

The maneuver was successful, but only after a bitter struggle where the crossings of the canal were forced. In the dim light of early dawn the attacking troops rushed down one wall and clambered up the other to capture the machine guns and forward field guns sweeping the depression. The engineers worked under heavy fire to bridge the obstacle, and to their gallantry and skill the favorable results achieved are in part attributed. When the fan-wise movement was threatening the defenders of the Agache line farther north, a single division succeeded in forcing a passage there; while the capture of Bourlon hill by Canadian troops, and of Flesquières ridge next south of it, assured the advantage of commanding observation to the British, and drove the Germans back behind the marshes of the Scheldt valley south of Cambrai. An attempt to cross the Sensée marshes and capture Arleux, guarding a historic defile through the morass where Marlborough had passed before, failed. The Germans had turned much of the valley into a lake by damming back the waters of the Sensée and one of its branches called the Trinquis. While this protected the Germans from an attack toward the north, it was an act which fitted exactly with the plan of operations of the British; since it was their purpose to use the Sensée barrier as protection for their left flank while advancing eastward across the Agache and Canal du Nord.

The front of attack was now greatly broadened, and on September 28 and 29 the whole line of the Scheldt valley and Scheldt Canal-Somme valley and St. Quentin Canal, from Cambrai to St. Quentin, was violently assaulted. A heavy bombardment, continued without intermission for two days, drove the defenders into their subterranean cellars, dugouts, and tunnels, and cut off their supplies of food and ammunition. Tanks crushed down the wire entanglements west of the canal and aided the advance of the infantry to the main barrier. Then the struggle for the canal and the Hindenburg defenses east of it began.

North of St. Quentin the canal crosses from the main Somme valley into the upper valley of the Omignon, a tributary to the Somme, through the Tronquoy tunnel; follows west down one branch of that tributary to the village of Bellenglise, then north up another branch to Bellicourt, where it enters the long tunnel under the main Somme-Scheldt divide already referred to. Bellenglise thus occupied the apex of a valley salient followed by the canal. "Equipped with life belts and carrying mats and rafts, the 46th Division stormed the western arm of the canal at Bellenglise and to the north of it, some crossing the canal on footbridges which the enemy was given no time to destroy, others dropping down the sheer sides of the canal wall and, having swum or waded to the far side, climbing up the farther wall to the German trench lines on the eastern bank." Tanks could not, of course, be used. The gallantry and skill with which this part of the difficult barrier was overcome is the subject of special praise from the British Commander-in-Chief in the despatch from which the foregoing quotation is taken.[41] An effective word picture of the heroic feat is painted by Conan Doyle, who describes the smooth sur-

[41] Haig's Despatches, pp. 282, 283.

face of the canal for a mile or more dotted with the heads of the English infantry, their officers trailing ropes behind them as they swam, while cables, broken bridges, extempore rafts, and leaking boats were pressed into service.[a] On the right of the 46th an American division gained that part of the valley and canal near Bellicourt and carried the higher land to the east. Thus a serious breach was opened in the Hindenburg Line just north of St. Quentin.

South of Cambrai a second breach was effected after a costly struggle. Here the valley of the Scheldt, with its river and the canal, proved an almost insuperable obstacle, just as it had done in the 1917 Battle of Cambrai. When the Germans were driven from their advanced defenses west of the valley, they fell back to the eastern side of the barrier. One division after another of the attacking troops flung itself against the position only to find crossing impossible, or else to discover that a precarious hold on the eastern side could not be enlarged because adequate support was blocked by river and canal. As along the canal south of Bellicourt, so here also the task was made still more arduous by the fact that tanks could not be used to advantage. Conan Doyle's account of the campaign gives one a good idea of the difficulties of the crossing. The Vth Corps of Byng's army reached the barrier September 30 without great opposition, as the Germans were purposely withdrawing behind its protection; but all bridges on the corps front were destroyed, "and for three days General Shute prepared for the difficult task of forcing this broad waterway." Not until October 5 did the corps get across and then only when the outflanking movement from the south compelled the Germans to retire. The IVth Corps was held up at all points except one, where New Zealanders reached Crèvecoeur on the eastern side but were driven out. They partially saved the bridge here, however, and maintained a foothold on the farther bank. On October 1 the New Zealanders again got into Crèvecoeur and began pushing troops over the incompletely wrecked bridge. "But it was

[a] Conan Doyle, The British Campaign: 1918, Vol. 6, p. 159.

desperate work, and the attacking brigade suffered heavy losses." The German retreat, forced from the south, soon lightened the task. The VIth Corps found several bridges intact, and part of its forces attempted to cross but found "nothing could be done without deliberate preparation." Small forces from one division "got across the canal during the night but were unable to establish any permanent bridgehead. In the morning of September 29, however, one brigade of this division made a lodgment upon the farther side and remained there, though with wire and machine guns before them. Pontoons were brought up during the day and many bridges thrown across." After very heavy fighting the heights at Rumilly, where the Germans stopped the attempt to debouch from the east side of the Scheldt in the 1917 Battle of Cambrai, were captured. But after a week of desperate struggle the corps was still in line along the western side of the barrier, although it had important bridgeheads established on the eastern bank. On September 28 the XVIIth Corps reached the river. "The orders were to push on and cross the canal. . . It was soon clear, however, that the line of the canal and river could not be easily rushed, for all the possible crossings were swept by a deadly fire." Under cover of Folie Wood two battalions endeavored to force a passage but failed. Later a crossing was effected on a broken bridge. Other units followed, and a bridgehead was opened out to the east. It had required a bloody struggle lasting a week, even with the aid of the outflanking movement from the south, to make an effective breach in the northern sector of the river-marsh-canal barrier.

Between Le Catelet and Bellicourt, on the divide between the Somme and Scheldt drainage basins, was the only portion of the line from Cambrai to the northern defenses of St. Quentin not defended by a valley-canal barrier. Here the canal passed under the divide in a remarkable tunnel over three miles long in the chalk, excavated in Napoleon's time. The unbroken surface of the upland afforded "the only place in the whole line where tanks could be used to advantage." It was decided to launch part of the main attack against this sector, and the 27th and 30th American Divisions, supported by Australians, were selected for the task. The fighting was of the most violent character. Realizing the special danger from tanks on this part of their front, the Germans prepared for their destruction by mines and traps, as well as by gunfire, as an offset to the absence of natural barriers. One brigade which had been well equipped with tanks is said to have lost all its heavy ones and most of its whippets in fifteen minutes. Americans smashed through the defenses and advanced well east of the tunnel; but groups of the enemy kept emerging from the ground behind them and renewing the battle in their rear. On barges in the tunnel there was room for two German divisions, and from numerous shafts seemingly endless reinforcements could be poured out on the surface. Advanced elements of the 27th Division were for some time cut off from support by enemy troops debouching from the subterranean stronghold in their rear. Enough of the divide north of Bellicourt was held by the 30th Division to pass tanks around over the unbroken surface north of the tunnel entrance and so down the eastern side of the barrier to help the troops engaged in extending the ground already gained there. Eventually the widening of this southern breach outflanked the defenses farther north, thus aiding the advance in the tunnel sector and compelling the retirement of the German forces still holding the eastern side of the Scheldt from Le Catelet northward.

Meanwhile the French First Army operating on the right of the British had encircled St. Quentin, by passing over the Tronquoy tunnel divide north of the city and by breaking the defense lines to the south where they crossed the level upland between the Somme and Oise valleys and were therefore unprotected by any natural obstacle. St. Quentin fell, and the victors pressed on to the line of the Oise, where they were effectively brought to bay. The entire Hindenburg Line, from St. Quentin to Cambrai, was now in Allied hands, after one of the greatest battles of the war. A strong natural position, to strengthen which every device of art had been most skillfully employed, had not

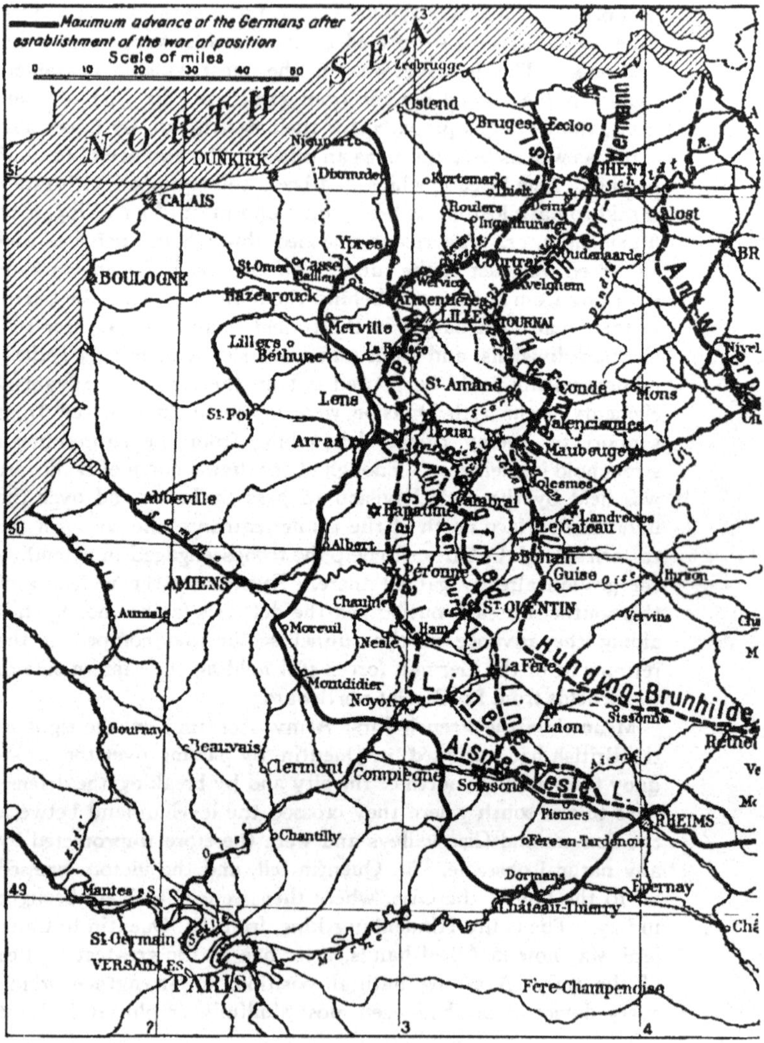

FIG. 55—Map of northern France with the main German defensive positions (dash lines).

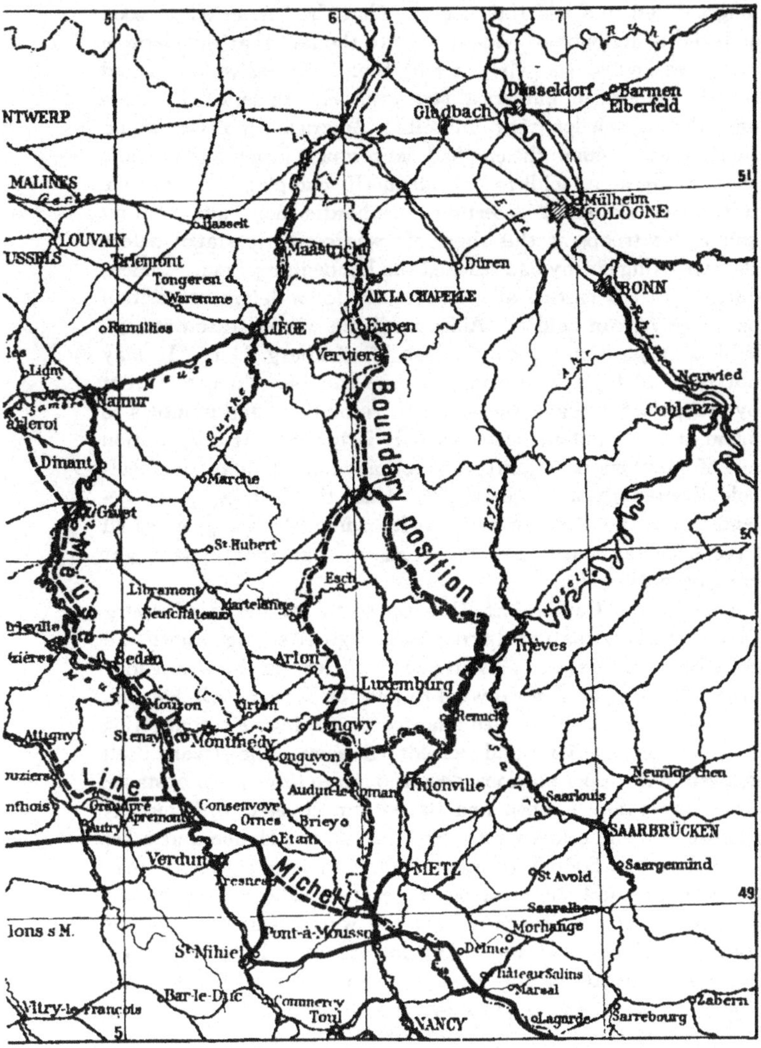

showing the use made by the German military leaders of the natural topographic barriers.

long sufficed to shield the enemy. That the barrier fell as soon as it did is undoubtedly due in part to the fact that the German army was no longer what it had been. Excessive losses had greatly reduced its numerical strength. A long series of defeats and the knowledge that unlimited American reserves would surely crush them in the end had dangerously lowered the morale of the German soldiers. Ludendorff complains of passive resistance, skulking, desertion, insubordination, and mutiny among his troops at this time. It was correctly stated before the retreating enemy had reached the Hindenburg Line: "Ludendorff . . . . is casting about anxiously for a defensive position on which he can hold the Allies until the winter season sets in. Will that position be the famous Hindenburg Line? It may not be, and for this reason. No line can be held now except by troops which have the stamina, resolution, and numbers to mend it by counterattacks when it is broken. It is a question not of numbers or natural obstructions, but of infantry of sufficient strength and quality."[43] Had the barrier been defended by the German army in its prime, the struggle would have been far longer and more costly, and the result might have been different.

East of the Cambrai-St. Quentin sector of the Hindenburg Line there is no natural barrier of any great strength until the Selle River is reached. This stream rises close to the Oise valley near Guise and flows northward to join the Scheldt near Denain. Its position is therefore well adapted to serving as a connecting link between the Oise and Scheldt barriers. The stream itself is, even in its upper portions, deep enough to form a real obstacle, while the valley is open and its bottom covered with marshy meadows. The eastern wall, like that of the Hallue and other streams in the Somme plain (p. 126), is in general steeper than the western, and the crest on the eastern side is often higher, dominating the approaches from the west. Dams across the river turned parts of the valley into lakes. Truly "it was a position of great natural strength." As soon as the security

[43] McPherson, The Strategy of the Great War, p. 404.

of the Hindenburg Line was imperiled, German General Headquarters ordered the preparation of a new defensive position along the Selle. This formed a part of the Lys-Hermann Line, extending from the Dutch frontier along the Eecloo canal, then the valleys of the Lys, the Scheldt, and the Selle, to reach the Oise near Guise. As the lower Selle makes an awkward angle with the Scheldt, the Hermann Line was here carried along the Écaillon River 2 or 3 miles to the east (Fig. 55).

On October 9, after the collapse of the Hindenburg Line, the German armies retreated across the plain to the Hermann Line, taking up their new position behind the Selle. In the Flanders plain to the north a withdrawal behind the Scheldt was ordered, while to the south the front crossed from the upper Selle to the Oise valley below Guise and continued on southward along the eastern bank of that river. The Germans thus held only a narrow strip of the Somme battlefield (Fig. 24) at the extreme east. The Allies promptly launched an attack to drive them from behind their new river-marsh barrier, force them back on the difficult country of the Ardennes, and capture Maubeuge. Advancing with their left flank screened by the Scheldt valley, they encountered determined resistance on approaching the line of the Selle. The Battle of the Selle began October 10 with attempts to force the passages with cavalry. These failed, and the task was given over to the infantry. Different army corps on reaching the western bank found it necessary to halt and organize for a general attack. One division attempting to establish a bridgehead north of Le Cateau gained the high ground east of the barrier, only to be thrown back to the western bank. Another reached the crest but was pushed back within two hundred yards of the river behind them. A third gained a hill on the east, but was likewise forced back. "Thus ended this weary day which . . . had exposed both the southern Corps of the Third Army to heavy losses with barren results. A week now elapsed, which was marked by very heavy artillery work on both sides."[44]

[44] Conan Doyle, The British Campaign: 1918, Vol. 6, p. 232.

Although the British occupied the west bank of the Selle throughout most of its length within a few days and by October 14 had established a number of more or less satisfactory bridge-heads on the farther side, it was not until the 17th that communications could be established and sufficient strength developed to attempt a grand assault on the strong position. The operations opened with a violent attack by French, British, and American troops against the short gap in the natural barrier existing at the divide between the source of the Selle River and the valley of the Oise. Three days of very heavy fighting were required to dislodge the enemy from the wooded country of the divide and drive him behind the Sambre-Oise barrier a few miles to the east. The southern part of the Selle defenses were now outflanked and the way opened for an assault on the main line of the river. This was launched in the darkness of night in order to reduce the advantages enjoyed by the enemy from his commanding positions. A heavy rain had soaked the loam on the upland and in the valley bottom, as well as the clay-flint formation outcropping along the higher slopes, changing them to a deep mire. The steep valley walls were slippery with mud. Everywhere the men suffered greatly from the nature of the ground, while tanks which had successfully been passed over the river could not operate to advantage. Men swam or waded the stream where bridges were lacking, those crossing the Écaillon finding the water up to their armpits. One field company of engineers lost fifty per cent of its effectives before it could complete its bridge over the Selle. Altogether the enemy resistance could not be overcome and his armies driven back to the eastward until three days more of heavy fighting had taken place.

Despite the fact that the line of the Selle had been hastily occupied and its organization was but little advanced; and the further fact that "the German infantry and machine gunners were no longer reliable, and cases were being reported of their retiring without fighting in front of our artillery barrage,"[45]

[45] Haig's Despatches, p. 293.

the barrier had held up the Allied pursuit for nearly two weeks.

Coincident with the operations against the Selle line, the French farther south launched an attack to drive the Germans from their strong position behind the Oise barrier between Guise and La Fère. After a very bitter contest footholds on the eastern wall of the valley were gained at several points in the vicinity of Ribemont, and the process of prying the enemy out of his strong and long-held lines was begun. Under steady pressure applied by the French north of the Aisne as well as along the Oise, the Germans began their retreat to the strong "Hunding-Brunhilde Stellung" (Fig. 55), withdrawing from the Oise valley eastward to the meridian of Guise in order to bring their north-south front into proper articulation with the east-west Hunding Line. The southward flowing portion of the Oise, below Guise, was now clear of the enemy.

Back of the Selle lay a number of shorter parallel rivers, among them the Rhonelle, Aunelle, and Honnelle, along which the Germans now fought delaying actions; while southward, beyond the heads of these streams, the enemy front was established behind the Sambre valley, part of a practically continuous northeast-southwest trench along which the Sambre and Oise flow in opposite directions from a point not far from Le Cateau. This marshy valley not only carries the Sambre-et-Oise Canal and the main Paris-Berlin railway, but also forms a military obstacle of some strength to troops advancing eastward. On November 4 the German front lay behind this obstacle from the vicinity of Landrecies on the Sambre to Guise on the Oise. Immediately to the north the Germans were on the western outskirts of the great Mormal Forest, ready to fall back through the woods to the line of the Sambre if necessary.

The main Battle of the Sambre opened November 4 on a front of nearly 40 miles, of which about two-thirds lay along the line of the Sambre or the Mormal Forest. The French on the right and the British in the center stormed the strong position, while the British left advanced across the Aunelle River farther north. The French reached the east bank at a number of points and dislodged the enemy from all his positions between Guise and Oisy. Guise itself was captured; and thus the valley of the upper Oise was opened to an Allied advance, and the Germans were compelled to evacuate the Hunding Line. A difficult crossing was effected by the British at Catillon and at the lock two miles south. "The 32nd Division . . . met strong resistance all along the river line. By hard fighting they forced a crossing at Ors." Other troops got across near Landrecies by means of rafts. On the left the Aunelle valley and the high ground east of it proved serious obstacles; and, when they were negotiated, the Honnelle River held the attacking forces in check for a time.

But the resistance of the demoralized German army was broken. The Teutonic hordes were now sullenly retiring eastward, cowed and whipped, to the Antwerp-Meuse line (Fig. 55), far in their rear, meanwhile begging an armistice from their conquerors. To such an army topographic barriers, be they never so strong, avail but little. The victorious British swept on to Maubeuge and cut the vital railway line which was the objective of Allied strategy, thus crowning with final disaster the whole campaign of the German northern armies. Before the armistice was signed the last of the invaders had been driven far from the Battlefield of the Somme.

CHAPTER V

THE BATTLEFIELD OF THE MARNE: THE PLATEAU-AND-LOWLAND BATTLEFIELD

In our study of the Battlefield of the Somme we found that in its southeastern part there appeared a number of high mesas or plateau remnants towering several hundred feet above the general level of the chalk plain. These were the Lassigny, Noyon, and St. Gobain "massifs," as the French call them, composed of more recent geological formations which once covered the older chalk but which have been removed from most of the Somme region by long-continued erosion. They were, moreover, mere outlying fragments of a continuous plateau occupying a vast area to the southeast.

If a traveler were to start eastward and follow along the base of the plateau scarp, keeping on the chalk himself but with the escarpment close at his right, he would make an interesting discovery. About Lassigny and Noyon he would see the chalk of the lowland disappearing southeastward under the plateau. Farther east, at Laon, the chalk would be found dipping southward beneath the escarpment (Fig. 57). The course of the plateau margin, swinging gradually southeastward, then southward, and finally southwestward, would lead him past Rheims, Épernay, and Sézanne. Near these points he would find the chalk always inclining inward toward the center of the circle described by his journey, and always plunging gently under the plateau formations (Fig. 58). Our traveler would discover, in short, that the chalk dipped in from all sides to form a shallow basin and that in the center of the basin rose a rudely circular plateau whose ragged margins he had been following.

NOTE.—For Chapters V and VI the reader should constantly consult the detailed map of the battlefield in the pocket (Pl. III) and the block diagrams (Figs. 62 and 64).

If the discovery prompted some curiosity as to why the central plateau should remain standing while the marginal areas had been worn low enough to expose the underlying chalk, an examination of the rock layers might suggest an answer. Toward the top of the escarpment would be found very resistant beds of limestone and other calcareous beds, more or less cavernous owing to solu-

FIG. 56—Generalized sketch map of the Battlefield of the Marne. Ruled areas represent uplands, white areas lowlands. For the topographic details and place names referred to in the text, see the detailed map of the battlefield in the pocket (Pl. III) and the block diagrams (Figs. 62 and 64).

tion of the rock along fractures and crevices in its mass. Lower down thick beds of soft, easily eroded sands would appear, alternating with fine-grained impervious clay or marl formations. At the base is the chalk. It is easy to see that when such a series of rocks was warped by movements of the earth's crust into a shallow basin and erosion attacked their upturned edges, a circular central plateau on the limestone, bordered by a rimming lowland on the chalk, must result. For the loose sand would quickly wash away, especially since rain falling upon it would find it already saturated by water sinking down through the cavernous limestones and held in the sand by the impervious clays beneath. The more resistant limestone would not erode so rapidly, but its margin would constantly be undermined by the washing away of the underlying sand, thereby developing a

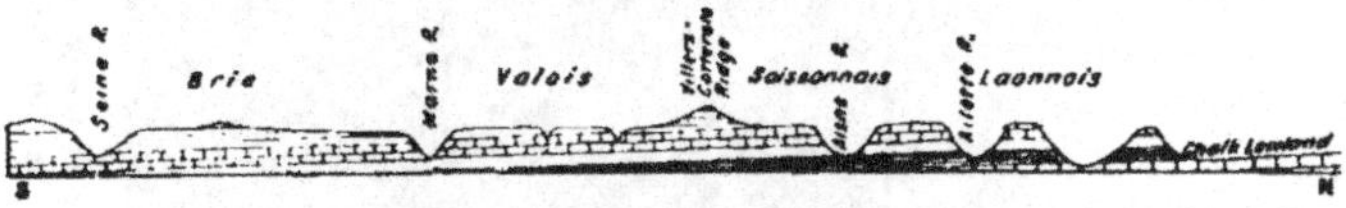

FIG. 57—Ideal north-south cross-section of the Marne plateau, showing beveling of different rock layers, with consequent development of successive east-west trending topographic belts (Brie, Valois, Soissonnais, and Laonnois).

steep scarp. As this process was continued on all sides, the limestone would form a more or less circular plateau, constantly shrinking in size, with steep marginal scarps; while the removal of the sand and the washing away of the unconsolidated clay would expose the chalk as a low plain belt, everywhere dipping in

FIG. 58—Ideal east-west cross-section of the Marne plateau, showing how different layers of the cap rock come to the surface in this direction, thereby varying the terrain of the battlefield. The key shows the relation of this cross-section to the cross-sections of the Battlefields of Verdun (Fig. 78) and Lorraine (Fig. 99) and forms a continuous section from the Marne plateau to the Black Forest.

toward the plateau (Figs. 57 and 58). The basin structure is known to geographers and geologists as the "Paris Basin," because Paris lies near its center; the central plateau is sometimes called the "Parisian Plateau," sometimes the "Marne Plateau" because its most typical portion is drained by the Marne River and its tributaries; while the rimming chalk lowland is called "la Champagne" (the open country) because of its vast expanse of barren plain.

FIG. 59—View northeast across the Champagne, showing the vineyard-covered lower slopes of the plateau escarpment in the foreground, the chalk plain to the right, the Mountain of Rheims bastion projecting far eastward in the left distance, with the Épernay curtain in the left middleground.

FIG. 60—Vineyard-covered slope near the base of the Marne plateau escarpment in the foreground; infertile areas of chalk in the distance, covered with thin grass or plantations of pines. (West of Châlons.)

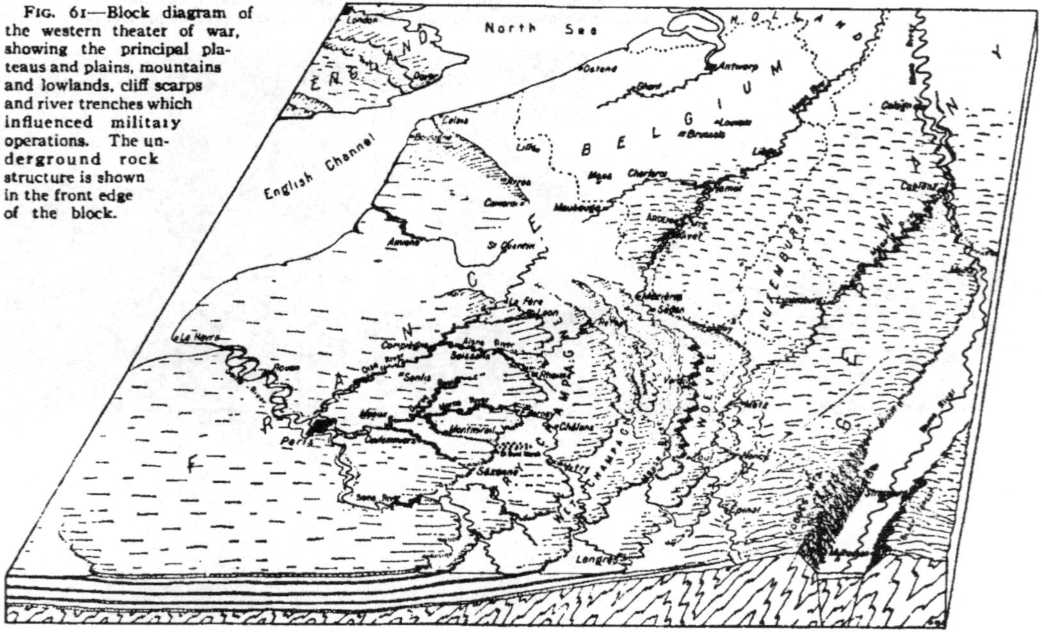

FIG. 61—Block diagram of the western theater of war, showing the principal plateaus and plains, mountains and lowlands, cliff scarps and river trenches which influenced military operations. The underground rock structure is shown in the front edge of the block.

Let our traveler climb to the crest of the escarpment at some convenient point and he can seize in an instant the remarkable contrast between the two regions. Looking westward over the plateau, the eye sweeps an expanse of flat upland where fields of grain are frequently interspersed with areas of woodland. Here and there the woods merge into a vast forest stretching for miles to the level horizon. It is a pleasing landscape, verdant and smiling under a summer sun yet seemingly lonely in its broad stretches of little-inhabited upland. At first one has the impression that the even surface is unbroken by any "topographic accidents," as the inevitable results of immutable laws in operation are unwittingly called. But a searching of the horizon reveals a low east-west ridge in the dim distance, rising a few hundred feet higher and promising better views of the plateau surface if one could reach it. A short walk brings one to the margin of a valley, sharply trenched several hundred feet into the plateau rocks, its bottom bright green with meadows and the lower slopes of its walls dotted with picturesque villages. Turning back to the crest of the escarpment and looking eastward over the lowland, how striking the contrast! In the foreground, the lower and gentler slopes of the scarp are completely covered with the world-famous vineyards of the Champagne (Figs. 59 and 60). Trees are rare, and villages are the only noticeable objects to break a rather pleasing monotony of countless small patches of vines, planted so close together and so much to the exclusion of all other growth as to give the impression of one vast vineyard. Beyond, and reaching to the horizon, is seen the low faintly undulating chalk plain (Fig. 60), often arid and bare between the winding green ribbons of stream valleys or the straight white ribbons of chalk roads, but, unlike the plain of Picardy and Artois, mottled with numerous dark patches of pine woods—a dry, desolate, repellant land, which the natives describe by a figurative expression meaning useless or wretched country: *la Champagne pouilleuse*. The traveler finds between these two divisions of the Paris Basin almost as great a difference of aspect as that between the humid plain of Flanders and the arid plain of the Somme.

The Paris Basin is not perfectly symmetrical, for Nature seldom uses perfect geometrical patterns. But in the region we are to study the departures from the ideal scheme of a circular plateau and rimming lowland are so slight as to be inconsiderable. The basin, moreover, is not so small as we have thus far pictured it; there are still other layers of hard and soft rocks underneath the chalk which come to the surface farther northeast, east, and southeast to form additional circular escarpments bordered by rimming lowlands (Fig. 61). They are of interest as affecting the strategic position of the Marne battlefield and are briefly described on a later page (p. 224). The Battlefield of the Marne itself properly embraces parts of the central plateau and of the gently inclined chalk plain of the Champagne. Napoleon's brilliant but hopeless campaign in 1814 showed how difficult it is to separate these two regions when military operations are under discussion, and the campaign one hundred years later reinforced this truth. The rivers of the Champagne flow into the Marne plateau; the natural defensive lines of the plateau often find their logical continuation eastward in the Champagne. An invader's position on the plateau can most easily be outflanked by an advance northwestward around the plateau margin, following the chalk lowland. In 1914 it was on the plain of the Champagne that the Germans made one of their chief efforts to win the battle raging on the plateau; and the Allied victory on the plateau made a German retreat in the Champagne inevitable. The Battle of the Marne spread eastward beyond the limits of the Champagne; but the most vital operations took place on the terrain included within the limits of what has here been called "the Battlefield of the Marne" (Fig. 56), which extends from east of Vitry at the eastern border of the chalk plain to west of Meaux in the heart of the central plateau.

Strategic Position of the Marne Battlefield

Whether or not all roads lead to Rome, one is tempted to say for northern France that all rivers lead to Paris. From the north, east, and south a remarkable series of valleys (Fig. 56) converge on the city which is in point of wealth, industrial development, political importance, and sentiment "the heart of France." From whatever direction an invader may come, valley routes point the way to the most vital point to attack. On the south the Loire appears to have had its former continuation northward into the Seine, and the old gravels of its bed are found along the low depression which seems to mark its abandoned course. Even on the west the valley of the Seine offers a natural pathway from the sea upstream to the capital. Lying near the point of convergence of so many rivers, and on that side of Paris which faces toward most of the other countries of Europe, the Battlefield of the Marne was destined to play an important part in the military history of France. From the earliest days hostile hordes have again and again poured along the valleys leading to the center of the Paris Basin, while the level plain of the Champagne tempted the Romans to build their roads across it in every direction and served as a meeting place and a grand route of travel for the international merchants of a later period.

If the rivers point the way to Paris, the plateau scarps bar that way. Each resistant rock layer coming to the surface slopes gradually upward to the northeast, east, or southeast, to fall off in a steep erosion scarp facing away from Paris or toward an approaching enemy (Fig. 61). Imagine yourself standing on the crest of the scarp rimming the central Marne plateau, near Rheims or Épernay, looking eastward over the Champagne lowland spread out before you like a gigantic map, and you will not find it difficult to realize the military significance of such a terrain. For miles in every direction all important movements on the lowland lie open to your observation. No enemy could successfully maneuver for a surprise attack so long as you were watching him from the crest. The roads stand out like white ribbons against the dull greens or browns of the plain and hence are easily registered for accurate artillery fire. Enemy movements, being largely confined to the roads, can then be broken up before assaults on the heights are organized. Artillery of the defenders finds ample cover in the ravines and valleys of the plateau; but on the exposed plain the attacking army finds little concealment for the batteries necessary to the proper support of infantry charges. Uphill assaults against entrenched positions on the slopes and crest of the plateau would be made with every advantage in favor of the defending troops.

If you descend into the plain of Champagne and move eastward over its surface, your course will lie for 35 or 40 miles upon the thirsty chalk soil of this unfruitful "Dry Champagne," where only scattered pine woods and infertile fields relieve the semi-desert aspect. Soon you note that the plain is sloping gradually upward as you advance, and at last you suddenly find yourself standing on the brink of a second escarpment, lower than that of the central plateau and composed of the chalk instead of the higher limestone formation, but high enough to give you good views over the next lowland to the east. Here you see a landscape clothed with verdure, watered by numerous streams, and dotted with occasional marshes and ponds. This is the "Wet Champagne," where impervious marls and clays, coming up to the surface from under the chalk, hold more water on the surface (Fig. 56).

If you continued the eastward journey, as indeed we shall do later (Chapters VII and IX), before reaching the Vosges Mountains you would cross not less than six of these east-facing escarpments, arranged in concentric lines around the north-eastern, eastern, and southeastern sides of the Paris Basin. Their value as massive walls of defense upon which no artillery could make an impression cannot escape even the most casual observer. The converging rivers do offer easy passage through the obstacles, carving natural gateways toward which roads, railways, and canals converge; but those having the highest strategic value, like Laon and Rheims at the innermost scarp, are protected by permanent fortifications; while the others, like Épernay at the inner scarp, and Rethel and Vitry-le-François at the second or chalk scarp, can quickly be fortified by field works in case of war. The advantages enjoyed by the forces of defense guarding the strategic gateways, holding the crests of the uplands, and maneuvering on the gentle backslopes out of sight of the enemy, as against a foe compelled to advance across the flat open lowlands fully exposed to observation and artillery fire from the heights, are so obvious that the concentric escarpments have long been called "the natural defenses of Paris."

The Battlefield of the Marne is the core of this defensive system, the central keep of the many-walled castle of France. Whatever victories advancing hosts may win on the outer walls, a final struggle awaits them on the central plateau. Attila and his Huns were conquered on the level plain of the Champagne in the eastern part of the field, just in front of the innermost wall. After defeating his enemies on many a distant field, Napoleon was finally driven to bay on the inner wall, where his supreme genius enabled him to dash from one part of the natural fortress to another and for a time to hurl back successively the enemy columns seeking to breach it simultaneously from the northeast, east, and southeast. When the Prussians in 1870 had crushed the military power of France on the outer walls, there still remained a months-long struggle in the center of the fortress before Paris was in their hands. In 1914 the fundamental purpose of German grand strategy was to avoid delaying actions on the outer walls by striking swiftly at the core of the defensive system, thus carrying the fight into the Battlefield of the Marne during the very first weeks of the war.

Surface Features of the Battlefield of the Marne

It will be profitable to examine more carefully the terrain of each of the two main subdivisions of the Marne battlefield, the Marne plateau and the plain of Champagne (Fig. 62).

THE MARNE PLATEAU

A very hasty survey is sufficient to show that the hard cap rock of the central plateau is not a single layer of one kind of limestone but rather a series of resistant beds lying one upon the other. Some layers are of very pure limestone; others are calcareous marls, fine-grained enough to detain downward-seeping rainfall for a time and thus to form fairly distinct water horizons; still others contain gypsum or sandstone. Calcareous rocks predominate, and to speak of the upland as a limestone plateau is not far wrong.

One interesting feature of the plateau, and one which has distinct military significance, is that these several layers of the resistant cap are not only tilted in toward the center of the basin but have been beveled across by erosion so that different layers come to the surface over different parts of the plateau, as an east-west cross-section south of Épernay will show (Fig. 58). It so happens that in the Marne battlefield the dip of the beds from north to south is more constant than that from east to west, while at the same time the beveling by erosion at the present surface seems to have been more complete. As a result the belted structure due to the outcropping of different layers of the cap is unusually distinct (Fig. 57). We thus have the curious result that, whereas the general structure of the Paris Basin gives concentric belts of different types of country with the rocks dipping inward toward the center, the central plateau itself shows nearly east-west belts of country with the rocks dipping prevailingly southward. This contrast in direction of the topographic belts, which has, as we shall see, very important military consequences, may be expressed diagramatically by Figure 63, in which the arrows show the directions of rock dip and the names are those given to the different belts by the inhabitants.

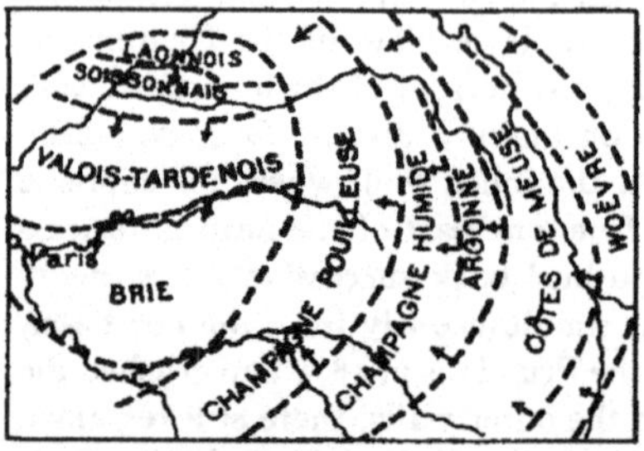

FIG. 63—Diagrammatic representation of the contrast between the east-west topographic belts of the central plateau, where the rocks have a prevailing southward dip, and the concentric (roughly north-south) belts, where the rocks dip prevailingly inward toward the center of the Paris Basin.

The Brie Region

If one starts at the southern side of the Marne battlefield and advances northward over the plateau, the topographic contrasts of the successive belts will readily become apparent. In the belt between the valleys of the Seine and Marne Rivers, known as "the Brie" region (Fig. 56), the layer of cap rock exposed at the surface is an unusually resistant "millstone" formation, in which are found quarries over four centuries old. Only the larger streams and their principal branches are able to cut through it. As the soft sands and unconsolidated clays are here far below the surface, owing to their southward dip (Fig. 57), even the deepest valleys do not reach them and hence have no opportunity to undermine the cap rock and dissect the plateau into fragments. The most striking features of the Brie, therefore, are the vast expanses of level plateau upland, relatively unbroken by minor stream valleys and only trenched by a limited number of widely spaced major valleys. While the top layer of rock in much of the Brie is more or less cavernous, owing to solution of its calcareous material, and so permits surface water to escape readily underground, the marl layers next below are somewhat impervious. They check the water's downward progress and cause it to come out on the valley sides in the form of springs, some of them notable for their abundant flow. Hence the pleasing verdure of the Brie valleys, among which the Grand Morin and the Petit Morin are the most important.

Where impervious beds are close to the surface, not only is the vegetation more vigorous but ponds and small lakes abound on the plateau. The dip of the rocks shown in Figure 58, and a lateral change in the composition of certain layers, combine to bring about this condition in eastern Brie, where we find great forests interspersed with lakes southwest of Épernay, making a peculiarly difficult country in which to conduct military operations. Here Blücher's fleeing troops were caught in the bogs or drowned in the marshes and lakes a hundred years before the Kaiser's hordes battled on the plateau against d'Esperey and Foch. Western Brie, on the contrary, has its better-drained surface layers covered with a fertile loam which gives excellent wheat land. The upland is here much cultivated, grainfields alternating with extensive pastures and only occasional forest patches. The straight roads across the cultivated upland are often bordered on either side by a single row of trees, which give excellent ranging marks for artillery fire. Extensive forests like those of Crécy and Armainvilliers occur where the rich loam covering seems to fail, while others like Tartre Wood grow on erosion remnants of a higher sand formation (Fig. 57). This is the same sandstone on which grows the Forest of Fontainebleau south of the Seine, and it forms a number of isolated hills and ridges scattered over the plateau, some of which, like the butte of Doue near Coulommiers, have been cleared and afford excellent views in all directions.

Valois and Tardenois

The Valois-Tardenois belt next north (Figs. 56 and 57), extending from the Marne valley to the east-west ridge known as the "crest of Villers-Cotterets," shows a different type of terrain. Here the layers of cap rock forming the surface of the Brie occur only in scattered patches, and the next underlying beds come up to influence the topography. In the Valois portion, lying west of the Ourcq valley, these beds are less resistant than the layer of millstone forming so much of the Brie surface; and hence this part of the plateau is more dissected by small valleys. The beds are also more pervious than the marls which kept so much moisture near the surface of the Brie, and lie directly upon a massive limestone formation through which water easily passes downward by means of fissures and solution cavities. Consequently the smaller valleys are apt to be dry, and those which cut into the massive limestone have the narrow gorge forms so characteristic of limestone regions. The ponds and marshes so abundant in parts of Brie are rare in Valois. As the surface of the plateau is covered with a fertile loam retaining just enough moisture to make it productive, much of the Valois upland is cultivated. Only in the more infertile tracts, where a sandy layer protrudes through the loam, are there large forests, such as those of Halatte, Ermenonville, and Villers-Cotterets. Isolated erosion remnants capped by the Fontainebleau sandstone are rather frequent, especially in the west, where they form lines of hills and ridges protecting Paris on the north and east. One ridge in particular, the Dammartin line of hills which dominates the plateau northwest of Meaux (Fig. 56), is of vital interest in connection with the First Battle of the Marne.

The loam which covers much of the upland in Brie and Valois, like the loam on the chalk plain of the Somme, becomes a serious impediment to military operations when soaked with heavy rains or melting snows. In combination with the water-soaked soils of the more impervious beds forming other parts of the plateau surface, it may then make unmetaled roads extremely difficult to traverse. The Allies' campaign against Napoleon in 1814 was greatly impeded by the bad condition of the main highways; Blücher's columns advanced slowly, and Yorck arrived late at the battle of Montmirail because his artillery could make poor headway through the mud. During Napoleon's pursuit of Blücher late in February "the artillery stuck fast in the mud and delayed him much," and Napoleon himself on another occasion wrote: "I am a little annoyed by the roads. They are horrible. There are six feet of mud."[1] But neither in Brie nor in Valois, nor in any other part of the Marne plateau, does the loam seem to have been sufficiently thick or impervious, or sufficiently reinforced by impermeable underlying deposits, to make it the awful curse it proved on the Battlefield of the Somme. Under normal conditions the roads of the upland are good.

Just as eastern Brie differs from western Brie, owing to the westward dip of the beveled rocks, bringing different layers to the surface, and to lateral changes in the composition of the same bed, so for the same reasons Tardenois differs from Valois. The dip carries the rocks upward to the east (Fig. 58), and, as beveling by erosion has not been complete, the plateau in Tardenois is

[1] T. A. Dodge: Napoleon: A History of the Art of War, 4 vols., Boston, 1907; reference in Vol. 4, pp. 366, 407.

higher and the streams cut deeper than in Valois. In places they even cut through the massive limestone into the soft sands and clays below, thus giving opportunity for undermining and a more extensive dissection of the plateau. The fairly resistant cap rock of Valois is not present above the limestone in Tardenois; it is replaced by soft, unconsolidated marls, which erode easily themselves and permit the soluble limestone below to be more easily eroded. Tardenois is, therefore, much more dissected than Valois. The valleys are more open and have gentler slopes; and, since the marls are sufficiently impervious to cause considerable water to move over their surface and flow down the valley walls, and since the valley bottoms often trench the water horizon in the soft sands overlying the dense clays at the bottom of the plateau series, the plateau of Tardenois is frequently forested and the valleys are rich in verdure. Tilted up to the highest elevation, projecting like a bastion from the general margin of the plateau out into the Champagne, and clothed with a dense forest, that part of Tardenois called the Mountain of Rheims is one of the most imposing military obstacles in this difficult terrain (Fig. 62).

The Soissonnais

North of Villers-Cotterets ridge, itself a long east-west erosion remnant capped by the Fontainebleau sandstone, the Soissonnais stretches northward beyond the River Aisne (Figs. 56 and 57). Here the massive limestone, the *calcaire grossier* of the French, has reached the surface, while the underlying soft sands have risen so high that the valleys cut clear through them to the underlying clays. This means that the rivers, abundantly supplied by water filtering through the limestone and sandstone but held up by the clays, have every opportunity to sweep away the soft sands and undermine the limestone cap rock. Accordingly, in the Soissonnais the plateau is more dissected, especially in its northern part, than any area we have described farther south. The remaining upland consists of flat-topped strips on the principal divides between the main streams, the margins of which have been frayed into irregular promontories, peninsulas, and bays by the gnawing action of small tributaries working headward into the divides. The long east-west upland strip between the parallel valleys of the Aisne and Ailette Rivers east of Fort Malmaison, known as the Chemin des Dames (Fig. 64), is more dissected than the rest of the Soissonnais and belongs rather with the Laonnois farther north, although its unbroken continuity with the less dissected plateau to the west makes it desirable to describe it here. It shows the contrast between the shorter tributaries on the north side, and the longer ones on the south, which we should expect on southward-dipping rocks. With its comparatively straight and steep northern scarp, defended by the natural moat of the Ailette valley, and the parallel north-south ridges and ravines of its southern side affording protection against flank attacks and concealment for artillery, munitions, and men, the Chemin des Dames offered the Allies a very strong position, to gain which they made heavy sacrifices. While held by the Germans, facing southward with the valley of the Aisne as the protecting moat, it proved strong enough to defeat repeated Allied offensives. Whether attacked from the north or the south it was to prove a formidable obstacle, while the north-south ridges and ravines would make progress along it from the east or the west almost equally difficult. The Chemin des Dames is in fact one of the most important natural defenses of Paris.

It has been noted above that the fissured limestone cap permits surface water to escape rapidly downward, just as does the fissured chalk of the Somme battlefield. Like the Somme plain, therefore, the Soissonnais is arid. It is even more arid than the chalk plain, and, while a covering of loamy soil produces good crops of beets and wheat, the general appearance is dreary in the extreme. There is not a village to add a touch of life to the landscape, for they all nestle in the valleys to escape the winds sweeping unhindered across the barren upland and to profit by abundant supplies of water flowing out below the limestone cap. Trees are rare, and fields of grain and beets, growing on a thin but fertile loam soil covering, succeed each other in monotonous procession. The unobstructed view sweeps the level surface to the horizon. Departures from the horizontal are all downward, never upward, and pass unnoticed unless one happens to stand on the brink of a valley. Then a smiling landscape greets the eye. The depressions are spacious, even when the streams which drain them are tiny brooks. Below the steep upper cliff formed by the undermining of the limestone cap, the slopes are gentle and pleasing, except where the slumping of landslides has roughened

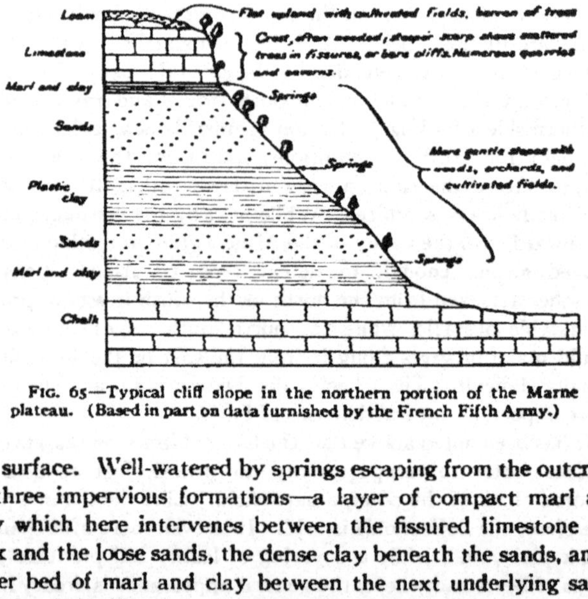

FIG. 65—Typical cliff slope in the northern portion of the Marne plateau. (Based in part on data furnished by the French Fifth Army.)

the surface. Well-watered by springs escaping from the outcrops of three impervious formations—a layer of compact marl and clay which here intervenes between the fissured limestone cap rock and the loose sands, the dense clay beneath the sands, and a lower bed of marl and clay between the next underlying sands and the chalk (Fig. 65)—the slopes are clothed with green, partly owing to vegetable gardens flourishing on the fine-grained sands enriched by lime falling down from above, and partly to orchards, vineyards, and unreclaimed woodland. The broad valley floors are apt to be marshy, for the rivers flow close to or on the clay; here meadows or open grasslands are the rule, except where a marsh remains unimproved.

The massive limestone cap is exceptionally easy to quarry and furnishes a durable building stone. Two interesting consequences of this fact are quickly noted in the Soissonnais. The villages are numerous, solidly built, and of great antiquity; for man "found in this district, together with a fertile soil and abundant waters, an inexhaustible source of excellent building materials. The smallest habitations are built of dressed stone from the *calcaire grossier;* everywhere are old churches with sculptured towers, sometimes even the remains of Roman or Merovingian walls, testifying to the facilities which construction work has at all times found in this region."[1] This, in its turn, had a consequence of some military importance. Not only were the buildings more formidable as fortified redoubts than those of less solid construction in the Somme plain, but—far more important —the extraction of such quantities of stone throughout long centuries had left the plateau honeycombed with quarries, caverns, and vast subterranean galleries in which whole armies could find secure shelter from the heaviest artillery bombardments. The traveler from Soissons to Laon passes, near Fort Malmaison and within a few feet of the road, an inconspicuous opening in a side ravine, which is typical of countless others like it. Enter and explore the cavern, unwinding a string to mark the route behind you if you are new to its labyrinths, and you will discover abundant traces of German occupation—the remains of an electric lighting system and of telephone lines, signs to guide the former inhabitants, remnants of the comfortable furnishings of officers' quarters, and the miscellaneous débris left by a horde of soldiers. In this single cavern several thousand Germans are said to have been quartered. What must have been the military value of a strong natural position like the Chemin des Dames, furnished with numberless ready-made subterranean shelters on such a scale?

The Laonnois

The Laonnois is the Soissonnais carried one step further. Here the underlying formations rise so high that the streams

[1] Albert de Lapparent: Description géologique du Bassin Parisien et des régions adjacentes, Paris, 1888, p. 61.

frequently cut through the clays and into the chalk (Figs. 56 and 57). Valleys are wider, and the dissection of the plateau has gone far enough to break it up into skeleton ridges or more or less isolated mesas and buttes, of which latter the fortified hill of Laon (Fig. 64), towering 350 feet above the plain below, is the type; while the St. Gobain, Noyon, and Lassigny massifs, with their broader, forest-clad uplands, belong to the class of larger mesas. Those remnants lying northwest of the Oise valley, including the Noyon and Lassigny massifs, are usually treated separately as the Noyonnais; but they may here be grouped with the Laonnois, with which they have a common origin and history. It should be noted, however, that on the Noyon and Lassigny uplands, and even more on the St. Gobain upland, the removal of the limestone cap from all but the central area has exposed broad expanses of the underlying sands, while sandy and clayey phases of overlying beds here still surmount what is left of the limestone. It is these relatively infertile and relatively inaccessible uplands of sand and clay which bear the extensive forests covering by far the greater portion of the massifs. We have found, in our review of military operations on the Somme battlefield (Chapter IV), what an important rôle these wooded plateau remnants played throughout the war. Again and again the battle line on the plain, imperiled by major operations in the open country, anchored itself fast at the southern end to the unshakable buttress formed by one or another of the wooded tablelands.

The valleys between the mesa remnants are well watered, although the short streams are not large. So humid have they become in some cases that canals are necessary to drain the marshy land. It was the Russians' skillful defense of a defile between the marshes southwest of Laon which in 1814 stopped repeated attacks by Ney and destroyed the French hope of getting possession of the strategic stronghold of Laon without a battle. When Napoleon with his 35,000 men finally attacked the formidable plateau of Laon, defended by Blücher with nearly 100,000, it was with these marshes and the narrow defile in his rear. The whole operation was so dangerous that Blücher in his astonishment could see in it only the indication of some hidden maneuver by the great master of strategy.[3] That Napoleon's impossible undertaking was defeated gives less surprise than does his utter disregard of the topographic difficulties against which he hurled his little army. It was but one sign among many of the decadence of his remarkable powers.

A survey of the several belts of topography characterizing the central plateau (Fig. 56) has revealed to us a transition from the little dissected Brie plateau on the south, through a moderately dissected Valois-Tardenois belt in the center, to the much dissected upland of the Soissonnais next north, and finally to the extreme fragmentation of the northernmost margin in the Laonnois. From the military point of view the terrain becomes increasingly difficult from south to north. An army capable of beating an enemy in the southern area might find the task beyond its power if that enemy stood on the northern terrain. On the other hand, a blow heavy enough to drive an army southward from the difficult terrain in the north might go far before its energy became exhausted in the more favorable country encountered in the advance. We shall see that both of these conditions were realized in the military operations of 1914–1918 on the Battlefield of the Marne.

Parallel Lineaments of the Marne Plateau

The southward dip of the rocks capping the plateau is not the only cause of an east-west trend of topographic lineaments in the Battlefield of the Marne. In examining the Somme battlefield we found that a series of shallow northwest-southeast trending folds affected the orientation of valleys and upland crests. This system of disturbances invaded the central plateau of the Paris Basin and, while too faint to bend the rocks into folds noticeable to the eye, was sufficient to affect so delicate a thing as the direction taken by running water, and hence the lines of greatest and least erosion. As a result, long ridges, or lines of hills, left where erosion was least effective, rise above

[3] Dodge, Napoleon, Vol. 4, p. 430.

the general level of the plateau. Because the folds swing more to the eastward in the central part of the basin, these ridges run more nearly east-west than do the topographic elements of the Somme area; but the inclination toward the northwest is still very noticeable, especially in the region about Paris.

Transport yourself for the moment to some point on the high ridge west of Paris which is clothed with the enchanting Forest of Marly and which there forms the southern wall of the broad valley of the winding Seine. From some point of vantage on the upland crest, perhaps from the ancient aqueduct towering over the valley above the little village of Louveciennes, you can read in the surface of the land unmistakable evidences of that hidden structure which determined the architecture of the inner defenses of the Parisian plateau. First note that the ridge on which you stand trends northwest-southeast and that the valley below you, however much its great river may swing in serpentine curves, pursues a general course northwestward to the sea. Behind you lies the open, smiling valley in which so much of French history was written at Versailles; beyond it another ridge, and beyond that the valley of Chevreuse—all keeping parallel northwest-southeast courses. Far to your left, where the Oise flows into the Seine, you see an upland crowned with trees rising above the valley floor; this is the height of Hautil, and it stretches far away to the northwest. In front, where the Seine swings rudely in against the vineyard-clad slopes, you see a zigzag road mounting past the towers of Cormeilles to the level crest above. It is the northwest-southeast ridge of Cormeilles-Sannois, which almost hides the triple-crested massif of Montmorency, whose forested upland rigidly preserves the same northwest-southeast trend. Off to the right, the heights of Vaujours run northwestward from the Marne but then swing suddenly around toward Paris, seeming thus a rude interruption of Nature's otherwise symmetrical pattern. But if you will go to those heights and again look northeast, you may descry in the distance the long line of hills running from Montmélian past Dammartin ridge and Montgé to Monthyon and its neighbors, the latter of which overlook the Marne valley near Meaux. Here again the northwest-southeast parallelism of form is beautifully shown.

There is nothing haphazard in Nature's workings. Immutable laws operate to produce inevitable results. Here near the center of the Paris Basin the rock layers appear horizontal; but water running on horizontal rocks could never produce a parallel system of heights such as surrounds Paris. A faint warping of the beds into parallel undulations, so faint as to be imperceptible to the closest scrutiny and determinable only by instrumental methods, would produce the results observed. Rock structure, however inconspicuous, is the key to Nature's architecture in the Paris region and hence to an understanding of the defenses of the city. Obsolete forts, like those of Liège, can be destroyed in a few hours by 16-inch howitzers; but, contrary to a widespread misconception, the better types of fortification, like Verdun or Belfort, have as much importance today[4] as had any forts in the days of Marlborough, Condé, and Vauban. The true modern stronghold is designed with due regard to the indestructible barriers erected by Nature and with a fine understanding of their architecture, in order that the army of defense may have confidence in the success of its operations.

It is to the faint folding, therefore, rather than to the more pronounced southward dip of the rocks, that we must ascribe some of the linear features of the Battlefield of the Marne. The Dammartin line of heights ending near Meaux has already been mentioned (Fig. 56). Farther north, near Senlis, another ridge runs slightly south of east from the mound of Montépilloy, the main height being crowned by the village of Rozières. Even more commanding and continuous is the long ridge of Villers-Cotterets, broken only at one point, near its center, by the Savières headwater of the Ourcq and trending nearly due east from the Compiègne Forest to a point north of Fère-en-Tardenois, a distance of some 25 miles. Parallel valleys as well as parallel ridges must result from the folding, and undoubtedly the pro-

[4] H. F. P. J. von Freytag-Loringhoven: Deductions from the World War, New York, 1918, pp. 76–78, 94.

nounced east-west (or slightly southeast-northwest) trend noticeable in the valleys of the plateau is in part to be ascribed to this cause, but in part also, perhaps, to the exposure of east-west belts of weaker rock due to the more pronounced and continuous southward dip of the cap-rock formation. Detailed examination and careful measurements of the rock layers appear to have demonstrated that the Petit Morin, Grand Morin, Marne, Aisne, and other rivers are in part located in shallow east-west synclines or parallel downwarped basins, above which the anticlinal arches usually rose less than one or two hundred feet. Elsewhere erosion has quite commonly attacked the exposed arches most vigorously, with the result that ridges or lines of hills remain along the downwarped synclines. In either case the shallow folding brought about a parallelism of topographic lines.[5]

Add the fact that the concentric courses of the rivers flowing from the rim of the Paris Basin toward its center must be expected to produce, in the eastern sector, one or more lines of valley and divide with an approximate east-west trend, and we have three factors involved in giving to the central Marne plateau a topographic trend at right angles to that of the succcession of cuestas and uplands farther east. Among the valleys which show this trend are the Ailette, Aisne, the upper Ourcq and many tributaries of the lower Ourcq, the larger tributaries of the Oise next south of the Aisne, the Petit Morin and the Grand Morin, and others of less importance. A map emphasizing the prevailing east-west trend of the lineaments of the plateau is shown in Figure 66.

Military Significance of the Parallel Lineaments

Pause for a moment and consider the military significance of this discordance in direction of topographic elements between the outer rims of the Paris Basin and its central core. From whatever direction an enemy approaches Paris he must run counter to the grain of the country at some stage of his advance.

[5] G. F. Dollfus: Recherches sur les ondulations des couches tertiaires dans le Bassin Parisien, *Bull. Serv. de la Carte géol.*, Vol. 2, 1890, pp. 17–35; *idem:* Relations entre la structure géologique du Bassin de Paris et son hydrographie, *Ann. de Géogr.*, Vol. 9, 1900, pp. 413–433.

If he elects to fight the defenders on the outer walls and is successful in a direct attack against the east-facing escarpments, then he may flank the barriers of the central plateau and advance parallel with them. But if he seeks to flank the outer barriers by swinging far around to the north, where the scarps decrease in height and almost disappear owing to an imperfection in the

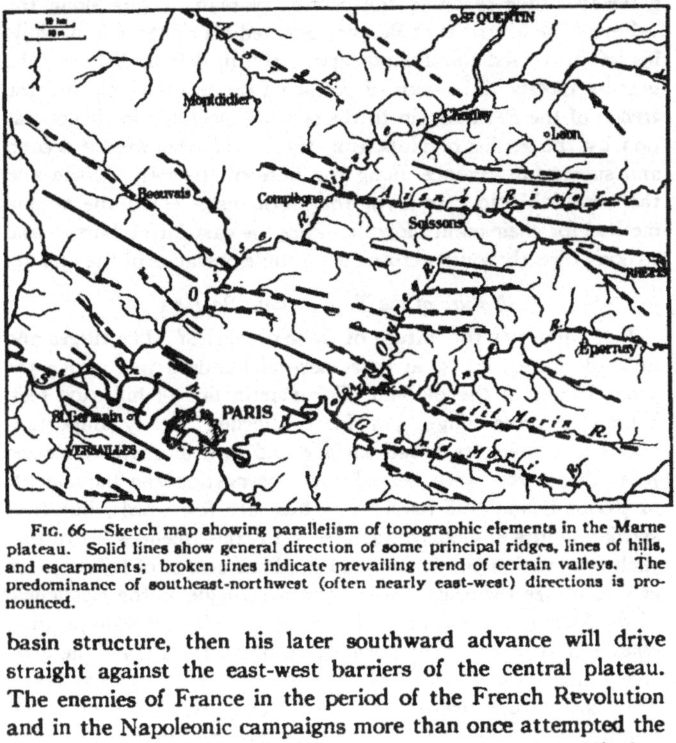

FIG. 66—Sketch map showing parallelism of topographic elements in the Marne plateau. Solid lines show general direction of some principal ridges, lines of hills, and escarpments; broken lines indicate prevailing trend of certain valleys. The predominance of southeast-northwest (often nearly east-west) directions is pronounced.

basin structure, then his later southward advance will drive straight against the east-west barriers of the central plateau. The enemies of France in the period of the French Revolution and in the Napoleonic campaigns more than once attempted the former method, first suffering defeat on the outer walls but later reaching the core of the system and moving on Paris parallel with its ridge and valley system. The first attempt at outside interference with the French Revolution was defeated on the eastern face of the second or chalk escarpment at Valmy. When the Allies invaded France in 1813–1814, throwing Ney back from the seventh line of escarpment at Épinal, forcing Mortier from the fifth scarp at Langres, defeating him in several actions on the fourth in the region Chaumont–Bar-sur-Aube,* and forcing the line of the second scarp at the Vitry and Troyes gateways, thus reaching the central plateau, it was along the valley of the Marne that Blücher directed his right column, while his left traversed the upland farther south, defiling between the marshy forests and ponds of eastern Brie on his north and the trench of the Petit Morin to the south. Defeated in this operation by the genius of Napoleon, the Allies attempted a second and successful advance along the lines of the Petit Morin and the Grand Morin. In 1914 the Germans selected the second method for their main blow, flanking the east-facing barrier and striking directly south across the multiple obstacles of the upland.

Nature of the Topographic Barriers

Just what was the nature of these obstacles? The ridges and isolated buttes, rising in cases several hundred feet above the general level of the plateau (Dammartin line of hills 300 feet, Villers-Cotterets ridge 325 feet) afforded vitally important observation over its broad surface and in a measure concealed from those who were denied this observation the movements of troops behind the barriers. These are few words, but they speak volumes in modern warfare. The consequences of utilizing such observation points to the best advantage, or of failing properly to utilize them, in the four years' struggle on the Battlefield of the Marne will appear in the sequel. As for the valleys, their character rendered them formidable obstacles despite the relatively small size of most of the streams which traverse them. They are natural trenches or moats of giant size, from a few hundred to as much as three or four thousand yards across and several hundred feet deep (Figs. 67 and 68). Furthermore, the type of valley cross-profile most frequently found in the plateau favors excellent observation of enemy operations. Where the

* Chaumont is 50 miles east-southeast, Bar-sur-Aube 30 miles east, of Troyes.

surface formations of a region are soft and unconsolidated and the underlying rock is resistant, valley walls and hill slopes are apt to be convex upward (Fig. 69, A), with the steepest slopes near the bottom. On such slopes there is no good military crest, and an enemy in the valley bottom may be quite invisible to an observer posted but a few feet above him. On the other hand, where there is a hard cap rock, with softer beds below, the profile is

FIG. 67—Flooded floor of the Petit Morin valley, a tributary of the Marne which played an important rôle in preventing the British from effectively pursuing the Germans withdrawing northward during the 1914 Battle of the Marne. (French official photograph.)

more likely to be concave upward, with the steepest slope at the top (Fig. 69, B). In this case the whole of the valley lies open to observation from the crest of either wall, and the slopes are exceptionally difficult of ascent. The Marne plateau, with its resistant calcareous cap rock and its underlying weak sands, usually gives valleys of the second type. Profiles of the type shown in Figure 69, B, are particularly well developed throughout the Soissonnais and Laonnois; and the Aisne and Ailette valleys, with the intervening Chemin des Dames, in any case formidable barriers as we have already seen (p. 231), derive much additional value as military obstacles from this circum-

FIG. 68—The flat-floored valley of the Marne River in the vicinity of Meaux. It was the exposure to enemy fire of troops crossing the valley floor, rather than the passage of the river itself, which made the Marne such a formidable obstacle. (French official photograph.)

stance. Here the army which holds the crest enjoys a tremendous advantage over an enemy which must assault the fully exposed slopes.

The valley bottoms usually offer little forest cover for troops or artillery. Open meadows or cultivated fields leave both exposed to accurately directed fire from above. To descend one valley wall, cross the open valley floor, and assault the steep wall beyond, exposed the while to artillery, machine-gun, and rifle fire, would be a difficult task even were no further obstacle encountered. When there is added the necessity of bridging an unfordable river, or of constructing causeways across a treacherous morass under such fire, the difficulties imposed by Nature are further magnified. Many of the rivers of the plateau are impracticable except where bridged (Figs. 70 and 71), and the passage of the Aisne by the Allied armies after the Battle of the Marne proved a task of tremendous difficulty. Marshes are sufficiently numerous to add to the strength of the barriers, although not nearly so dangerous as in the Somme plain.

Rivers subject to floods become yet more formidable during their periods of high water. Napoleon sometimes definitely counted on such obstacles in planning his battles, as when he wrote to Joseph in February, 1814: "The Yères [Yerres] is an overflowed river which is not fordable. It can cover the army at least three days. On the 17th I shall be ready to attack."[6] We have observed that most of the rivers of the Somme battlefield are little subject to flooding, because they rise and flow on the pervious chalk formation, whereas those on the impervious clay plain of the Flanders battlefield repeatedly inundate their valleys. The rivers of the Marne plateau in part originate on its composite formations, in part on the pervious chalk of the Dry Champagne, and in part on the impervious clays of the Wet Champagne, or even farther east. As a result their behavior is variable, but flooding is common. The Marne, subject to heavy floods in the Wet Champagne, has its flood waters decreased in crossing the broad belt of the pervious chalk plain of the Dry Champagne, but suffers to some extent in the plateau. The Aisne undergoes a similar moderation of its régime in the chalk plain but transmits enough of the flood waters to become a source of danger farther west. When the steep slopes of the plateau valleys shed rainfall or melting snows, particularly if the ground be already saturated by previous downpours, the local floods may become dangerous or combine with floods from a more distant source to cause disastrous inundations.

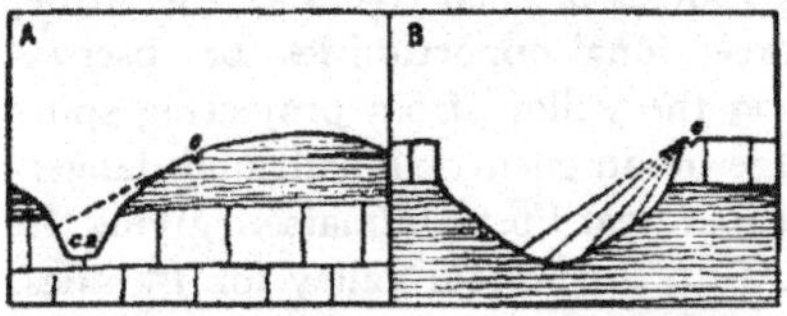

FIG. 69—Two types of valley profiles, in one of which (A) there is no good "military crest," and large concealed areas (c. a.) remain hidden from an observer (o) even when he is part way down the slope; whereas in the other type (B) all parts of the valley floor and walls are easily visible from the crest.

[6] Dodge, Napoleon, Vol. 4, p. 378.

It was a flood in the Aisne during the winter of 1914–1915 which, by destroying the military bridges over the river, isolated the French on the northern bank and compelled their retreat to a safer position south of the barrier (p. 296). Valleys which may artificially be inundated can quickly be transformed into serious military obstacles (Fig. 72).

Observation Along the Valleys

The remarkably straight courses of such rivers as the Aisne, Ardre, and Ailette afford exceptional opportunities for observation and enfilading fire along the valleys from projecting spurs and thus render their passage by an enemy difficult and dangerous. It was the capture of a spur near Fort Malmaison, giving the French observation eastward up the Ailette valley for 15 miles, which in 1917 compelled the Germans to abandon their strong positions on the Chemin des Dames and retire north of the river barrier (p. 304). From a spur near the head of the rectilinear Ardre valley 9 miles southwest of Rheims an observer could sweep almost the entire 20 miles of its length and see any German movement on sixteen different roads and additional minor lanes leading southward to cross the river (Fig. 62). As the valley lay along the base of the great Marne salient, in which the Germans were busy during June and the first half of July, 1918, preparing their last great offensive by which they hoped to surprise and overwhelm the Allied armies, one can easily imagine the enormous value of accurate observation of their movements as they brought troops and supplies for the offensive into the salient from the north. To get an adequate idea of the character of observation possible with the aid of a good telescope, one would have to read the reports of observers prepared on blanks ruled for the recording of all information of possible military value, such as movements of trains on railways hour by hour, traffic on the roads, prosecution of enemy engineering works and the operations of artillery and other arms of the service.

Thus from his vantage point at the head of the Ardre valley one observer noted between certain hours on July 11, four days before the Germans struck their great blow, "65 camions, 25 autos, 78 wagons, and 113 cavalry horses led by the bridle descending the north slope of the valley toward Crugny," thirteen miles distant; "24 camions, 4 autos, 185 wagons, and 181 cavalrymen going toward Hourges or Unchair; 22 wagons en route between Vandeuil and Savigny," and so on; another observer

FIG. 70—The valley of the Marne near Dormans, showing the wooded crest of the Marne plateau, the cleared valley walls and valley bottom, and German war material lost in attempting the passage of the barrier. (French official photograph.)

noted that within the salient there was an "unusual circulation of men on foot and of cavalrymen" upon certain roads, "20 boches in Indian file" on another road, so many loaded wagons entering a certain village, "a group of boches stationed in the edge" of such a wood, men busy building camouflage to conceal certain roads, "a batallion, in sections spaced about 200 meters apart, descends the spur south of Bois le Prêtre and goes toward the south, probably toward Villars Agron; 30 cavalrymen debouch from the same spot and follow the same route." These details are summed up by the observers in general terms, as for example, in regard to movements on the roads: - "Isolated men en route are *very numerous*, one sees them everywhere and all day long going in every direction. The same is true of small groups of three or four men. Many cavalrymen come and go leading their horses by the bridle. Reconnaissances of the terrain by cavalrymen. In short, unusual activity of men on foot and cavalrymen, giving the impression of a densely occupied sector."

FIG. 71—Bridge over the Marne at Château-Thierry, destroyed to render passage of the natural barrier as difficult as possible. In the distance is seen the crest of the Marne plateau. (French official photograph.)

Leaving out of account the importance of observation in controlling artillery fire, it will be evident that such reports as those quoted above, coming in from a large number of good observatories, must give the commanding officers quite a clear picture of what the enemy is doing behind his front. It was on the basis of such observations, supplemented of course from other sources, that Foch accurately gauged the Germans' preparations for their offensive of July 15, 1918, and made his own dispositions to combat them. The straight valleys, even when held in large part by the enemy, became serious obstacles to the successful prosecution of that enemy's designs. Hence the violent combats for possession of points giving observation along the valleys, such as the Malmaison spur referred to above, dominating the Ailette valley, and Mont de Bligny (southwest of Rheims), dominating the Ardre valley.

Napoleon's Campaign of 1814 on the Marne Plateau

Such, then, are the several characteristics of the rivers and valleys of the Marne plateau which make them notable military obstacles. That they have played an important rôle in the

FIG. 72—The valley of the Vesle near Fismes inundated by the Germans to increase its defensive strength. In the right background is seen the crest of the Marne plateau. (French official photograph.)

history of the Marne battlefield can easily be demonstrated. It is sufficient to take a single chapter from this history, such as Napoleon's brilliant campaign of 1814, to see how largely the tactics and strategy of a great military leader must be affected by these elements of the terrain. As soon as Blücher directed his advance on Paris, the retiring French forces under Macdonald sent back parties to seize and defend the vital crossings of the Marne at Château-Thierry and La Ferté-sous-Jouarre (Fig. 62), in order that, if defeated in front, they should not have an impassable barrier in their rear. Later the French crossed the river at La Ferté and destroyed the bridges, to place a serious obstacle between themselves and their pursuers. Napoleon, having meanwhile come upon the scene and overwhelmed the enemy at Champaubert and Montmirail, drove the defeated Allies northward with the hope that Macdonald would have again seized the crossings there and so enable him to pin the fleeing army against the Marne barrier. He was disappointed, however, and was himself compelled to stop and bridge the obstacle. Leaving the pursuit to one of his lieutenants, he returned to the north bank of the Petit Morin, where he defeated Blücher largely by the skillful use of cavalry on the level upland of the plateau near Vauchamps. Two days later he joined Oudinot, Victor, and Macdonald, who were standing behind the trench of the Yerres River facing Schwarzenberg's army, and, debouching from that defensive position, drove the enemy behind the Seine. This formidable barrier seriously delayed the French pursuit. The efforts of Macdonald and Oudinot to cross at Bray and Nogent failed, and the bridge at Montereau * was gained only after an all-day battle and after Napoleon himself had taken charge of the operation. The single crossing here was wholly insufficient. The Emperor was greatly annoyed and wrote to his brother: "It took us all day to pass that horrible defile of Montereau." Protected by the Seine below its bend at Méry, Schwarzenberg's army retreated eastward behind the upper Seine above the bend, and later behind the Aube. It was to the obstacle which the Seine trench interposed between him and his enemy that Napoleon attributed the defeat of his designs. "If I had had a bridge equipage of ten pontoons," he wrote, "the war would be finished and Schwarzenberg's army would no longer exist. I should have taken his army in detail; but wanting boats I could not cross the Seine, when I ought to have been able to cross at will."[7]

Blücher, who had joined Schwarzenberg on orders but who chafed at retreating before a foe who had but two men to the Allies' five or more, now obtained permission to attempt another advance on Paris in conjunction with new Allied forces coming southward from Belgium. Opposing the latter menace was

* Forty-five miles southeast of Paris, at the mouth of the Yonne.
[7] Dodge, Napoleon, Vol. 4, pp. 391, 403.

a small French force holding the line of the Aisne at Soissons; while the two French marshals, Marmont and Mortier, were maneuvering, the first to oppose Blücher, who had driven him from the escarpment at Sézanne and the valley of the Grand Morin in the expectation of pinning him against the Marne, the second to control the vital crossings of the Marne barrier at Château-Thierry and La Ferté-sous-Jouarre. Both marshals were soon compelled to withdraw their very small armies behind the northeast-southwest barrier formed by the Ourcq-Marne valley, there to dispute Blücher's advance on Paris.

Now began a "Battle of the Ourcq" which in some of its phases foreshadowed another struggle that just one hundred years later was to bear the same name. Part of Blücher's forces crossed the Ourcq near its junction with the Marne at Lizy while the remainder attacked the line of the Marne at Meaux. Defeated in the latter operation, Blücher shifted more troops north to push the movement across the Ourcq. The French thereupon struck north, overwhelmed the vanguard which had crossed west of the river, and threw them back to the east bank. As Napoleon was approaching from the southeast, Blücher withdrew all his forces opposite Meaux to the north bank of the Marne and destroyed the bridges, in order to put a formidable obstacle between himself and his dangerous antagonist. He likewise destroyed the bridges over the Ourcq, so was protected on the west by that river just as on the south by the Marne. Let the forces of the two marshals stand for the French Sixth Army under Maunoury attacking toward the Ourcq, Napoleon's forces for the British army advancing against the line of the Marne, and Blücher's army of invasion for Von Kluck's army of invasion, and we have almost an exact parallel for one stage in the Battle of the Ourcq of 1914.

Blücher now shifted forces farther north with the object of turning the left of the marshals' army, just as Von Kluck a hundred years later shifted his forces north to turn the left of Maunoury's army. In both cases the operation failed, in Blücher's because he could not breach the Ourcq defenses. In both cases the Marne barrier, after holding the assailants at bay for a time, was ultimately forced; and in both cases the invaders thereupon fell back on the line of the Aisne and fought a battle on the heights of the Chemin des Dames. The same topography produced in two campaigns a century apart a similarity of tactics and strategy which cannot fail to impress the student of military geography.

Napoleon attributed Blücher's escape from the Marne position to the delays caused by the river trench. He reached the river at La Ferté-sous-Jouarre on the afternoon of March 1, but repairs on the bridge were not completed until the afternoon of March 3. It was difficult to maneuver successfully under these conditions. On March 2 he wrote his brother urging that bridge equipment be sent him from Paris, saying: "This morning I should have destroyed Blücher if I had had one."[8]

We have seen that the line of the Aisne was held by a French garrison at Soissons. Blücher's retreat was therefore embarrassed by the fact that he had an enemy on the barrier to which he was retiring. Napoleon's intention was to outmarch Blücher and, by getting east of him, prevent his junction with the new Allied forces in Rheims; then, in conjunction with the army of the two marshals farther west, to pin the invaders back against the Aisne and defeat them. This design could be accomplished only in case the French held the passages of the river at Soissons long enough for Napoleon to dispose of Blücher. Unfortunately the commanderat Soissons surrendered to some of the Allied reinforcements before making proper resistance, and Blücher withdrew his army safely behind the strong barrier. "When the Emperor learned of the fall of Soissons, he was bitterly angered by its slight resistance. Instead of being able with his whole force to fight Blücher, cut off from his colleagues by an impassable river, he would now, in order to reach him, be obliged to fight for his own passage against a force threefold his own."

Blücher now assembled his army on the strong position of the Chemin des Dames, leaving forces in the valley to defend the

[8] Dodge, Napoleon, Vol. 4, p. 413.

passages across the natural moat that protected his front and extending his left to the crossing at Berry-au-Bac (Fig. 64). The invaders thus held the line of the Aisne on a front of 30 miles. Napoleon ordered Marmont and Mortier to force a passage at Soissons, Nansouty to seize the bridge at Berry-au-Bac, while other forces would throw temporary bridges over the river at two other points. Blücher was deceived into believing Napoleon would attack at Vailly and, in concentrating his forces on the upland north of that sector of the barrier, uncovered Berry-au-Bac, where Nansouty promptly seized the bridge by a surprise attack. At the west the barrier held firm against all assaults, but at the east both it and the Chemin des Dames were being turned by French forces hurrying forward from the Berry-au-Bac crossing to seize the stronghold of the Laon mesa in Blücher's rear. By this operation Napoleon hoped to cut the invaders' line of retreat and pen them up in the angle between the Aisne and Oise Rivers. Blücher then moved part of his forces out to the eastern end of the plateau to threaten from this dominating position the flank of the French columns advancing northward on the plain below, and so compel them to fight, while with the remainder of his army he should retire on Laon. The maneuver was successful, for while Napoleon won the "Battle of Craonne," the attack on the strong Chemin des Dames position was so difficult and occupied the Emperor so fully that Blücher easily effected his intended retirement. The steep northern face of the upland, the deep ravines scouring its southern wall, the slopes fully exposed to artillery fire from above, the numberless ditches in the wet plain below, all combined to make the attacks of the French extremely arduous. Again and again they were hurled back down the slopes, and not until Blücher, assured of his safe retreat to Laon, gave orders to begin withdrawing from the Chemin des Dames, did the French secure a permanent foothold on the upland and drive off the remaining Allies. The line of the Aisne was lost to Blücher, but he assembled his troops for a new battle about the natural fortress of Laon, where every effort of Napoleon to dislodge them was murderously repulsed.

Schwarzenberg's army was once more threatening, and Napoleon now instructed his lieutenants to destroy the bridge at Berry-au-Bac if necessary and to spare no effort to hold Blücher north of the Aisne barrier as long as possible, while he set off to deal with the new danger. Blücher remained ten days behind the Aisne before moving southeast to effect a junction with Schwarzenberg, who had defeated Napoleon on the line of the Aube (p. 263). Then began a new and joint advance on Paris by way of the Grand and the Petit Morin, while the Emperor vainly tried to draw them away from the capital by attacking their communications in the east. Paris was entered, and the remarkable campaign was over. Is it too much to say that throughout its brief but brilliant history the natural trenches of the Marne plateau exercised a decisive influence upon the tactics and strategy of the opposing forces, if not, as Napoleon apparently believed, upon the final results?

The rôle played by these same obstacles in the World War will become clear as we review the military operations from 1914 to 1918 on the Battlefield of the Marne (Chapter VI).

THE CHAMPAGNE

Bounded on the west by the high escarpment of the central plateau, the low, rolling chalk plain of the Dry Champagne rises gradually eastward until it drops off abruptly in a second escarpment facing the verdure-clad lowland of the Wet Champagne. For the present our chief interest is in the chalk country, which is but the southeastward continuation of the same rock we learned to know in the Battlefield of the Somme. Let us examine first the border lands of this region and then compare the plain itself with the chalk plain of the Somme country.

When seen from a distance the escarpment of the Marne plateau appears as an even-crested slope, its steep upper portion usually covered with trees while the gentler declivities below are planted with vineyards (Fig. 73). The evenness of the crest, due to the level character of the upland surface, robs it of some of its impressiveness; yet it rises from 400 to 600 feet above the level of the plain. Because the surface cap rocks were beveled across by erosion (p. 226), different layers form the crest at different places, giving some variety of aspect; and because the rocks are warped into shallow folds and unevenly attacked by the under-mining processes of erosion, the escarpment shows deep re-entrants and projecting salients, the re-entrants often coinciding with the shallow down-folds, or synclines, the salients with the up-folds, or anticlines. This emphasizes the character of the

FIG. 73—Escarpment of the Marne plateau south of Épernay as seen from the plain of Champagne. The steep upper slopes on the calcareous cap rock are wooded, the lower slopes on the underlying sand and chalk formations are covered with vineyards.

plateau as a natural fortress, in which projecting bastions alternate with re-entrant curtains (Fig. 62). Of the bastions the Mountain of Rheims is the highest and the farthest advanced into the plain, while in the well-defined curtains on the north and south are located the cities of Rheims and Épernay. Because of its fortress form and because of the magnificent observation far out over the plain from its summit, the escarpment is a military obstacle of much significance.

Outliers of the plateau, in the form of buttes or mesas, are less numerous and less striking than on the borders of the Laonnois to the northwest. Yet they are not unknown and possess a high military value. Most important of all are the three near Rheims lying in a northwest-southeast alignment between the Vesle and

Suippe Rivers (Fig. 62)—the butte of Brimont, Berru mountain, and the Moronvilliers massif (Fig. 75), each capped by the higher plateau formations and covered with forests. They rise respectively 200, 530, and 500 feet, approximately, above the chalk plain, and from the earliest days were utilized as lookout points and defensive positions by the primitive inhabitants. Together with the Mountain of Rheims bastion between the Marne and Vesle and the smaller St. Thierry bastion between the Vesle and Aisne, these outlying buttes completely encircle the Rheims curtain with dominating heights, forming what is frequently called the "Rheims Basin" (Fig. 74). Here the chalk has a composition giving less dry and infertile soils, is mixed with wash from the plateau beds above, and, especially when covered with *limon*, proves rich enough to support a denser population than other parts of the Champagne. The chalk also furnishes a good building stone which has been so extensively quarried as to leave the region honeycombed with vast subterranean caverns, utilized as wine cellars in time of peace but capable of sheltering whole armies in time of war. Close to the Aisne, Vesle, and Marne gateways through the plateau scarp and on the historic pathway skirting the plateau on the level plain, Rheims is a center from which radiate five main and several smaller railways, a larger number of principal highways, and innumerable minor roads, not to mention canal connections with the Aisne and Marne. All these features combined to make the Rheims Basin a stronghold of the highest military and political importance, and the heights have long been fortified to transform the basin into one of the great entrenched camps of France.

The Chalk Escarpment

The second line of escarpment, terminating the chalk plain on the east, is far less imposing. Usually rising not more than 100 to 300 feet, sometimes less, above the flat plain of the Wet Champagne, the dissected cuesta hardly deserves the name "Monts de Champagne" given to the typical portion between the Aisne gateway at Rethel and the Marne gateway at Vitry-le-

François (Fig. 56). Nevertheless, the "Monts" are striking features, their white or gray slopes contrasting strongly with the rich green of the plain below, even when the white shows only in patches between the rows of a plantation of pines. Elevations of a hundred feet or more give commanding observation over the low clay plain to the east, conceal the western slope of the chalk plain, and deflect many roads and railways to the easier routes through such gateways as Rethel, Vitry, and Troyes. Hence the military significance of the second escarpment. The Battle of Valmy, on the face of the scarp northeast of Châlons, which rolled back the tide of hostile invasion in 1792 and so altered history as to make the engagement one of the decisive battles of the world; the great rôles played by the Troyes, Vitry, and Aube gateways in the Napoleonic campaign of 1814; the important fighting at the Rethel and Vitry gateways in the fall of 1914—these are but a few episodes in the military history of this important crest.

With the northern and southern limits of the Champagne we are not here primarily concerned. Suffice it to say that south of the Seine the chalk is soon masked by an overlying formation which gives a new topography, making a special geographic province. To the north, beyond the Aisne, there is a gradual transition to the type of country found in the chalk plain of Picardy, a transition so evident that the natives speak of the better soil as *terres de Picardie.* Precise limits for the true Champagne need not be sought, for our interest is centered in the typical chalk lands well within its boundaries between the Aisne on the north and the Seine on the south.

Character of the Chalk Plain

From the base of the western scarp to the crest of the eastern, the chalk plain of the Dry Champagne rises almost imperceptibly in faint undulations. Near the western border the vineyards descend the scarp slopes and spread out over the gently rolling surface (Fig. 59), as far as wash from the limestone cap rock gives a favorable soil. Farther east the white surface of the more barren chalk begins to appear, blotched here and there with the somber

FIG. 74—Looking northeast over the comparatively fertile Basin of Rheims, from a tree-top observatory in the forest on the Mountain of Rheims. Throughout most of the war the Germans held the northern and eastern sides of the Basin, the French the southern and western sides, as well as the center at Rheims (which lies to the left of the view, and beyond its limits). The excellent observation of German forces attempting to advance into the Basin which the French enjoyed from the Mountain of Rheims bastion, and the possibility of concealing their own troops and artillery in the forest of the upland and in the lateral ravines, made this one of the strongest points on the French front.

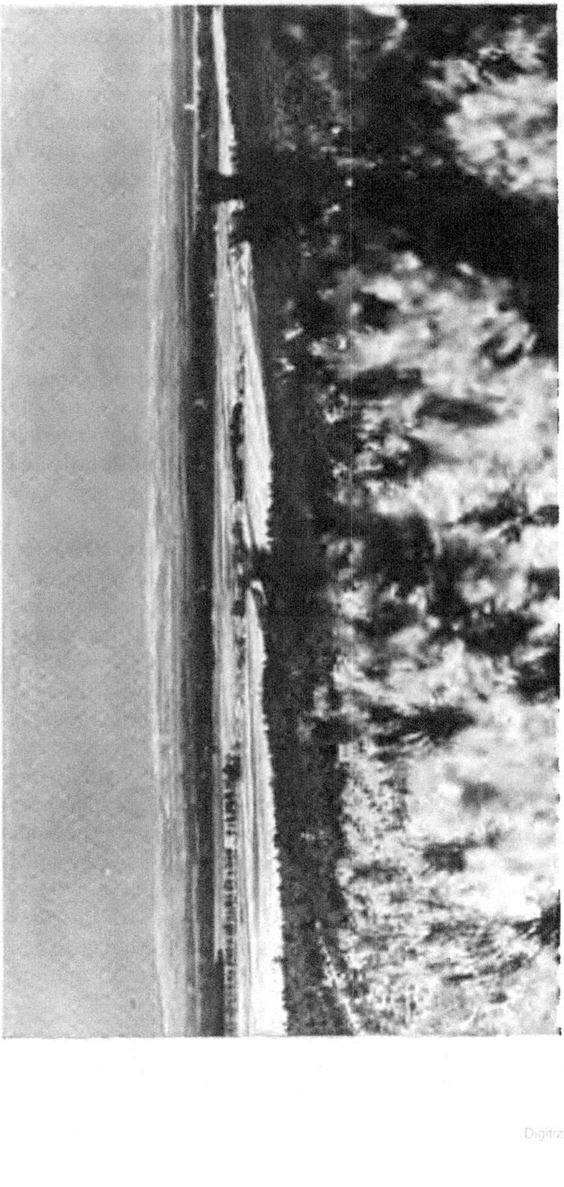

FIG. 75—The Moronvilliers massif as seen from a tree-top observatory north of the Camp of Châlons in 1918. Pine plantation in the foreground. The butte has been stripped of its forest cover by shell fire, which also powdered the chalk to give the appearance of a snow-clad summit.

hue of pine woods planted by man (Fig. 60). This is the real Champagne as one sees it today. As far as the eye can sweep the horizon there is not an object higher than a tree to arrest the vision, unless one happens to be near one of the few isolated erosion buttes which diversify the otherwise monotonous plain. The surface is more nearly level, less noticeably undulating than is the chalk plain of the Somme. Indeed, over vast areas the country is almost as flat as the Flanders plain, the valleys lie almost on the surface, and one cannot speak of an "upland" as we did in describing the Battlefield of the Somme. Only in the narrow belt along the eastern margin of the chalk, where the upward-sloping plain has attained its highest elevation and streams cut into it to the lower level of the Wet Champagne beyond, is there anything to compare with the Thiepval and other so-called "plateaus" bordering the Ancre and Somme valleys. Here also are the only ravines of any depth and importance, carved by the numerous streams descending the steeper slope. If the plain of Picardy and Artois is monotonous as to form, that of most of the Champagne is doubly so.

There is no need to describe in detail all the physical characteristics of the Champagne, the nature and behavior of the rock underlying its surface, and their effect on military operations. To do so would be in large measure to repeat what has already been said in describing the Battlefield of the Somme. For it is the same rock which underlies the terrain in both cases, the same white, fissured, pervious chalk. In both cases the structure is of the simplest: in one, faintly folded, in the other, faintly tilted toward the center of the Paris Basin; but in neither case departing greatly from horizontality. In both the erosion history has been much the same. Now, the same kind of rocks having the same general structure and subjected to the same erosive forces should produce much the same topography. It will be sufficient, therefore, if we compare the chalk plain of Champagne with the chalk plain of the Somme region, pointing out those elements of resemblance and contrast most important to the student of military geography.

Aridity of the Plain

Like the plain of the Somme, the plain of Champagne is arid because the fissured chalk permits the water to descend too far below the surface for plant roots to reach it. The Champagne is indeed much the more arid of the two, for it lacks both the widespread deposits of the clay-and-flint formation and the fertile loam which in Picardy held enough water on the surface to support a virgin forest and later an abundant agriculture. In Champagne the bare white chalk shows at the surface, not merely as white splotches on steeper slopes, but over vast stretches of the level surface. As a result the vegetation history in the two regions has been reversed. In Picardy the land was originally forest-clad, and the labors of man changed it into a treeless region in order to make the fertile soil produce rich harvests. In Champagne the land appears to have been originally a barren desert, but man has labored diligently to give it a forest cover by planting pines, the only thing he could make the sterile soil produce (Fig. 75). The monotony of form, so much more pronounced in the Champagne, is relieved in a measure by frequent and extensive patches of woodland, something practically unknown in the Somme country, where even small copses but seldom interrupt the treeless expanse.

The natural barrenness of the Champagne is due not only to its greater aridity but also to the absence of the fertilizing elements contained in the loam of Picardy and to the further fact that the chalk crumbles so badly as to afford a poor root-hold for trees. When too wet it loses all solidity, and when dry it blows away, exposing the roots to the danger of freezing if cold weather comes.[9] Broad areas of the plain still remain in the condition which justified Michelet's characterization of it as "a melancholy sea of stubble spreading over a plain of plaster." One might travel for hours seeing only the boundless white plain and the arching blue sky, with never a tree or a bush in sight. So poor was the land it gave rise to the saying that in the Champagne an acre of land, when it had a hare on it, was worth just two francs. Today the

[9] Émile Chantriot: La Champagne: Étude de géographie régionale, Paris, 1906, p. 125.

pine plantations have softened the harshness of the landscape, and near the villages the unending toil of the farmer has created ever-expanding oases where careful fertilizing enables the naturally sterile soil to yield a harvest of cereals and other produce. Still the Champagne is, over much of its extent, a country for sheep raising, pine growing, and hunting. Nothing could be more desolate than the arid wastes of the great Camp of Châlons, in the midst of the flat chalk plain.

Military Aspects of the Plain

Nor could a land be better adapted to military maneuvers. Attila chose it as best suited to the operation of his cavalry hosts, and modern military leaders have confirmed his judgment. The level surfaces, the absence of serious topographic obstacles, the dryness of the soil, the large open spaces almost devoid of vegetation, and the cheapness of the land led Napoleon III to establish the Camp of Châlons as a field for military displays and maneuvers on a grand scale. Today its level wastes constitute a modern entrenched camp of the first importance, a field for long-distance artillery practice and for the exercise of other arms of the service, and a point of concentration for an army in case of war. Thirty miles farther south is the equally vast Camp of Mailly, covering the more hilly topography of a subsidiary terrace or cuesta of the chalk, just west of the main escarpment near Vitry-le-François, and serving as a garrison camp as well as a field of maneuver. Another variant of the Champagne topography is organized into an entrenched camp at Rheims (p. 254). The city of Châlons partakes of the martial character of the land and is a military and administrative center rather than a center of commerce and industry.

Trench warfare found its best habitat in the Champagne. Not only does the water drain readily from the pervious chalk, but the clay and loam which proved such a curse when rains deluged the plain of the Somme are generally lacking. Trenches, dugouts, and tunnels were usually dry, and the subterranean cellars and caverns in the chalk offered warm and dry shelter to the combatants. The vicissitudes of the weather wreak less havoc with military plans on such a terrain. Roads are plentiful in every direction and most of the time are in fair condition; while the construction of the networks of new roads and railways made necessary by a war of position encountered a minimum of difficulty in this nearly level country. In all the broad expanse of the chalk Champagne only three tunnels for the standard railways were found necessary, and all of these are small ones piercing the crest of the eastern escarpment. On the main surface of the plain no engineering works of importance were needed.

As in the plain of the Somme, so in the Champagne the sinking of water far below the surface of the fissured chalk causes the population to congregate in compact villages around communal wells. Springs and surface streams are comparatively rare, and out of 104 communes in the arrondissement of Châlons-sur-Marne, 97 draw all their water from wells, some of which are very deep.[10] Scattered farmhouses are rare, and smaller villages (often less than 100 inhabitants) than those in the richer plains of Picardy, separated by wider expanses of uninhabited plain, are the rule. Hence positional warfare developed the same fortified village strong points, buttressing the trench systems, that became typical of the fighting on the Somme. Even in the more humid valleys the villages are usually isolated, and only exceptionally, as in the upper Vesle valley east of Châlons, do they spread along the stream to coalesce with each other and form a continuous belt of habitations several miles in length. Unlike the villages of the Somme region, however, they conceal themselves in the depressions of the undulating plain; for there is no rich upland loam to attract the peasants to the heights, and in hollows the wells do not have to pass through so much chalk before reaching the groundwater level. Hence the country, poor and sparsely populated though it be, appears less inhabited than is really the case.

Rivers of the Champagne

We found that in the chalk plain of the Somme surface streams were not numerous, forming only in the major valleys which

[10] Chantriot, La Champagne, p. 122.

reached the groundwater level lying deep in the pervious rock, while shallower ravines were dry; that the streams which did form were clear and tranquil because fed by perennial springs from the underground supply; and that in their tranquil waters extensive peat bogs formed. In the clay plain of Flanders, where the impervious soil increased surface run-off, the régime of the rivers was very different. Streams were numerous and fed by an indefinite number of branches; their waters were muddy and frequently spread in floods over the land—conditions unfavorable to the formation of peat.

The rivers of the Dry Champagne belong to two classes. First are those which rise on the chalk plain itself and therefore resemble the rivers of the chalk plain of the Somme. They are fed by springs, often a single big spring called the source, or "*somme,*" of the river, as Somme-Py, the source of the Py; Somme-Suippe, the source of the Suippe; Somme-Vesle, the source of the Vesle; and many another which catches the eye on a detailed map of the region, for about each spring a village has grown up and taken the name of the water which gave it birth. These rivers have few branches, their waters are clear and tranquil, floods are rare, and they flow through marshy peat bogs. They are, in short, the typical rivers of a chalk country and hence of some importance as military barriers. Between them lie broad bands of undulating arid plain in which dry. shallow hollows endlessly alternate with dry, rounded swells.

The second class of rivers in the Dry Champagne are those which rise on the impervious surface of the Wet Champagne or receive numerous and important tributaries from that or similar regions. They enter the chalk plain already endowed with the characteristics of the Flanders rivers: muddy waters, a volume which varies from less than normal in dry seasons to dangerous floods at other times, and a valley floored with much alluvium but comparatively little peat. In the chalk plain itself they receive few tributaries, and part of the flood waters are absorbed by the fissured rock. But the characters acquired in the upper parts of their basins cannot be wholly changed, and traversing the Dry Champagne these rivers form a class by themselves. The Seine rises on pervious, calcareous formations southeast of the Wet Champagne and is so little influenced by its short passage across the impervious belt that it is unusually free from floods in this part of its course and is bordered by extensive peat deposits. It belongs rather with the first class of streams typical of the chalk. The Marne, on the other hand, rising in part on impervious beds and receiving some important affluents in the Wet Champagne, carries floods throughout its course across the Dry Champagne, filling its bed with alluvium and forming very little peat. The Aisne, draining a far greater area of the impervious belt east of the chalk, floods its valley repeatedly and forms almost no peat. But, though they lack the treacherous bogs of the chalk rivers, the broad valleys of the Marne and Aisne, often floored with marshy meadows and traversed by an unfordable stream even when not in flood, oppose serious difficulties to the movements of an army.

Military Value of the Rivers and Valleys

Both classes of rivers may, therefore, in a region relatively free from topographic accidents, serve as valuable defensive barriers or as dangerous obstacles, as the case may be. It was flagrant disregard of the natural obstacles found in the valley of the Aube that near the close of the 1814 winter campaign led Napoleon into the most dangerous topographic trap he ever occupied. The town of Arcis-sur-Aube lies on the southern bank of the river, 33 miles south of Châlons, on a level and open plain bordered on the south by a semicircle of low chalk hills. On the north the river could be crossed by one bridge only; and just beyond lay a belt of marshes which must be traversed by a narrow defile. It was on the plain south of the Aube that Napoleon had stationed his army, with the enemy holding the semicircle of dominating heights on three sides and the almost impassable barrier in the rear. "At the view of this absolutely fatal situation, he lost not for a moment his presence of mind. He recognized his strategic error and saw that he was outmaneuvered;

he saw how ill he had reconnoitered; he understood his peril."[11] From this trap Napoleon escaped only because Schwarzenberg had made all his dispositions to receive an attack, not to deliver one; and while his enemy waited the Emperor planned his dangerous retreat. As it was, the French lost 5,000 men to the Allies' 300.

The valleys gain in military significance because they function as more than usually important highways of travel and are the centers of the densest population. From eastern France to Paris the valley of the Marne is the most vital artery, the Nancy-Paris railway running through it being paralleled by the Marne canal and by important highways. The flood waters of the river renew periodically the richness of the valley soils and encourage the growth of forage, cereals, and vegetables in quantities sufficient to support a population which in the Champagne must be considered large. The valley of the Suippe is an oasis occupied by a large industrial population. Nearness to transportation attracts the quarrying industries, with the result that the valley trenches gain strength as defensive lines from numerous excavations admirably adapted to serve as shelters from artillery fire. Strategic crossing points along the chief valleys early gained a military renown which they often preserve under conditions of modern warfare. Berry-au-Bac and Châlons are familiar names in the communiqués of the late war which conjure up memories of the Napoleonic and other historic campaigns.

Changed River Courses

It will be noticed that, while some of the rivers flow down the gentle westward slope of the Champagne to enter the plateau scarp through natural gateways, others flow westward toward the escarpment only to turn abruptly north or south to join the Marne or Aube. Thus the Soude almost reaches the scarp near Blücher's old headquarters at Vertus, but turns north to the Marne. The little Somme and the Vaure head straight for the gateway of the Petit Morin, but, just before reaching the St.

[11] Dodge, Napoleon, Vol. 4, p. 450.

Gond marshes lying in the gateway, the Somme turns north to the Marne, the Vaure south toward the Aube. The Maurienne branch of the Superbe pursues the westward course nearly to Sézanne, then turns south to flow into the Aube. Curiously enough the original direction of the Soude valley is continued westward on the plateau by the valley of the Surmelin, which heads in a notch in the crest of the scarp, although there is no stream at present to carve such a notch (Fig. 62). So also the

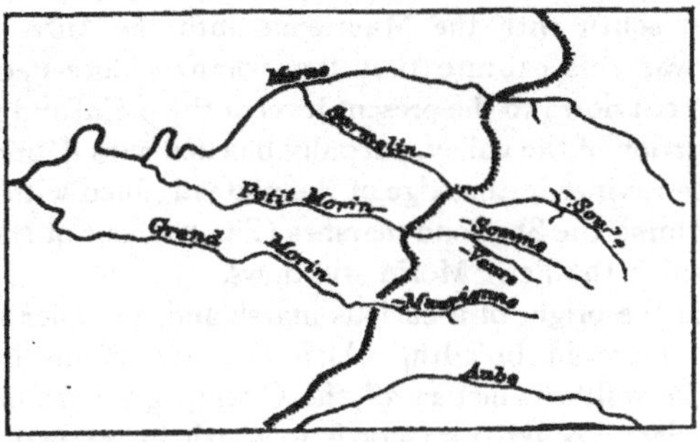

FIG. 76—Supposed former courses of the Soude, Somme, Vaure, and Maurienne. (After Davis.)

courses of the upper Somme and Vaure are continued by the valley of the Petit Morin, while that of the Maurienne is in line with a deep notch in the scarp at Sézanne continued westward as the valley of the Grand Morin. To the geographer this assemblage of features tells a remarkable story of successive river captures, which has been fully set forth in a classic essay by Davis.[12] Before the weak rocks of the Champagne were worn down by erosion so far below the plateau level, the Soude continued westward as the Surmelin (Fig. 76). A branch of the Marne, gnawing its way southward along the weaker beds, captured the upper part of the stream and turned its waters northward to the main river. As the capture took place early, before the Soude-Surmelin had

[12] W. M. Davis: The Seine, the Meuse, and the Moselle, in his "Geographical Essays," Boston, 1909, pp. 587–616.

deeply entrenched itself, the notch in the plateau upland is not very deep, although sufficiently so to cut through the cap rock. Later the Maurienne portion of the original Maurienne-Grand Morin river was captured in a similar manner by a branch of the Aube; but not until it had deeply entrenched itself. Hence the old valley notch at Sézanne is so deep that a railway takes advantage of it to enter the plateau. Finally, very recently as the geologist would count time, the Somme and Vaure were captured and the former turned north with the Soude into the Marne, the latter south with the Maurienne into the Aube. So late, indeed, was this capture that the Somme-Vaure-Petit Morin river had cut down to the present level of the plain, and the abandoned portion of the valley, partially blocked with débris brought in by side ravines in the edge of the plateau, filled with stagnant water forming the St. Gond marshes (Fig. 77), out of the western end of which the Petit Morin still flows.

Such is the origin of a curious marshland, 12 miles long by 2 miles or more in breadth, which has played an interesting rôle in the military history of the Champagne. It is an east-west barrier, crossed from north to south by several roads on causeways and bridges. To one looking out over its broad expanse of reeds and grass, broken by occasional tree clumps, the marsh presents an aspect dreary in the extreme. Drainage ditches have been constructed and the streams confined between dikes, in an effort to reclaim the clay-floored basin. But when heavy rains fall the waters rise and overflow the meadows, and the whole region becomes an impassable swamp. The causeways are low, and as the roads are not metaled they soon become mired. Even in dry weather the reclamation is not complete enough to make it safe for an army to maneuver at will over the low surface, while the roads can be held under accurate artillery fire directed from the surrounding heights.

Hills of the Champagne

The hills of the Champagne, as well as the rivers, valleys, and marshes, rank as important military obstacles. Of the formidable bounding scarps, prime elements in the defensive system against an enemy approaching from the east, no more need be said. Against a hostile advance from the north the outlying erosion remnants of the plateau formations, left isolated on the chalk plain, have a high value when properly disposed. Mont Aimé and the hill of Toulon north of the St. Gond marshes, Mont Août and the plateau spurs near Montdement on the south (Fig. 62), help

FIG. 77—The St. Gond marshes, which prolong the Petit Morin barrier eastward into the Champagne and which proved an obstacle of some tactical importance in the First Battle of the Marne. (French official photograph.)

make the passage of the marsh barrier in either direction more difficult. The butte northwest of Neufchâtel-sur-Aisne would serve to protect a bridgehead north of the Aisne barrier. But it is the line of buttes north and east of Rheims—Brimont, Berru, and Moronvilliers (p. 254)—which do most to bar invasion of the plain from the north. They form a natural prolongation of the east-west defensive line of the Chemin des Dames and Aisne, or Aisne-Vesle, and, via the hills north of the Suippe near Somme-Py and the ridges and ravines of the Monts de Champagne near Monthois,* may be connected with the Argonne barrier at the pass of Grandpré.* Brimont and

* Monthois is 33 miles, Grandpré 42 miles east of Rheims.

Berru still bear on their summits masses of the higher plateau formations; but from the great Moronvilliers massif (Fig. 75) the beds above the chalk have been almost completely removed by erosion. It rises from the forest-splotched white plain north of the Camp of Châlons as a sprawling mass with subdued profiles, formerly black with forests, today stripped of its cover and showing white, like a snow-clad mound, with the chalk plowed up by countless shells. From its summit the whole broad plain for miles in every direction lies under observation. It was inevitable, therefore, that it should form a strong point in the system of natural defenses seized by the invaders in 1914 and hotly disputed till the end of the war.

CHAPTER VI

MILITARY OPERATIONS ON THE BATTLEFIELD OF THE MARNE

The First Battle of the Marne

In the last days of August, 1914, the British and the French Fifth Armies, retreating before the onrush of the powerful German right wing, halted for a brief space behind the defensive barrier of the Oise River. This stream cuts from northeast to southwest obliquely across the northwestern margin of the Marne plateau, lopping off the western ends of the Laonnois and Soissonais; so the retreat was carrying the struggle into the Battlefield of the Marne. On August 29 the British fell back on the strong position of the Chemin des Dames plateau north of the Aisne trench between Soissons and Compiègne. The extreme left of the Allied line formed by the French Sixth Army was coming into position along the Oise from Compiègne to Verberie, whence it extended westward past Clermont to Beauvais along the edge of a westward projection of the Brie plateau.

On September 1 General Joffre gave the order for a continued retreat toward, and if necessary behind, the Seine-Aube barrier. This meant that no permanent stand was to be made on the northern barriers of the Marne plateau, however strong they might be tactically and however squarely they might oppose the southward advance of the invaders. The Allied armies had been caught unprepared, they had been severely handled at Mons and during the retreat, and were not yet ready to engage the supreme struggle. Broad considerations affecting the extent of the strategic retreat and the proper moment for resuming the offensive dictated a further retirement; and the strong line of the Seine was set as the possible limit.

An earlier order of the generalissimo, the famous "Instruction Générale" of August 25 which first directed the general retreat, had emphasized the importance of employing natural obstacles for arresting the enemy's pursuit. "The movement will be covered by rearguards left behind favorable depressions of the terrain, in such manner as to utilize every obstacle to arrest by short and violent counterattacks, in which artillery shall form the principal element, the enemy's march, or at least to retard it." In compliance with this instruction rearguard actions were fought on a number of the northern barriers. Thus in falling back from the line of the Aisne, the British temporarily stood behind the valley of the Authonne, a branch of the Oise south of the Forest of Compiègne and the first east-west trench after leaving the Aisne. The northern wall of the trench is unfortunately higher than, and commands, the southern wall; and a vigorous onslaught by the Germans soon carried the position. In retiring farther south the British held for a time the Montépilloy-Rozières ridge,* while forces of the French Sixth Army fought a delaying action along the line of the Nonette, the next east-west trench south of the Authonne, against heavy German forces seeking to debouch from Senlis.* Maunoury's Sixth Army shifted to the east side of the Oise and in the first days of September was basing on the northern defenses of Paris, its left standing on the plain north of the ridge of Montmorency,* its center advanced near the Dammartin ridge,* and its extreme right near the Marne in front of the heights of Vaujours, with headquarters on the fortified butte of Écouen,* an outlier of the Montmorency massif. The British army now lay behind the great Marne barrier, from the eastern end of the Vaujours ridge at Lagny* to near the important crossing of La Ferté-sous-Jouarre, executing General Joffre's instructions "to defend the passages of the river as long as possible, and to blow up the bridges" along its front.[1] This work of

* Montépilloy-Rozières ridge is 27 miles NE, Senlis 24 miles NNE, Montmorency ridge 10 miles N, Dammartin ridge 18 miles NE, Vaujours ridge 9 miles ENE, Écouen 8 miles N, Lagny 14 miles E, of Paris.

[1] Sir John French: Despatches: Mons; The Marne; The Aisne; Flanders, London, 1914, p. 25.

delaying the enemy accomplished, the British on September 4 fell back behind the line of the Grand Morin.

During the retreat of the extreme left wing of the Allies, the armies of the center and right (considering only the front west of Verdun) were rapidly falling back in accordance with the order of September 1, fighting rearguard actions where undue pressure by the enemy or a favorable terrain made it advisable. The direction of the main German advance—from the north—outflanked the north-south escarpments east of Paris, so there was no opportunity for a real test of the strength of those defenses. German strategy was so ordered as deliberately to avoid that test. It was for the Germans "a matter of life and death" to sweep through the Belgian plain and gain the Marne plateau with all possible speed, before Russia could become dangerous. To advance from the Franco-German border directly westward across the formidable concentric walls would mean a succession of battles, each undertaken with the terrain wholly in favor of the defending forces.[2] Paris could not possibly be reached until after a long and costly campaign; and meanwhile the Russian bear might be knocking at the doors of Berlin.

To meet and defeat the German maneuver Joffre undertook a retreat which should pivot on the solid buttress of the fortified camp of Verdun on the east, the topography of which rendered it a practically impregnable bastion (p. 387). When the exigencies of the situation had carried the retreat so far that the Allied left wing had swung back nearly to Paris without the development of conditions favorable for the resumption of the offensive, his orders envisaged the further retreat of the center to the natural defensive line of the Seine-Aube valley, the right still pivoting on Verdun and the left on the entrenched camp of Paris with its fortress-crowned parallel ridges. As a result, between the terminal buttresses the center would bend far south until the Germans were caught in a great *strategic salient* and subjected to simultaneous attacks from the west, south, and east.

[2] D. W. Johnson: Topography and Strategy in the War, New York, 1917, pp. 20–25.

It was because of this plan of campaign that the various armies had made no determined stand on the northern defenses of the Marne plateau; and it was because the maneuver was still in process of execution when orders were suddenly issued to stop the retreat, that the Allied front showed little adjustment to the topographic features of the battlefield. The British extreme left was still north of the Grand Morin, while its center and right had abandoned the defense of that trench and were falling back toward the Seine. The French Fifth Army, next east of the British and now under command of General d'Esperey, was falling back with its front obliquely across the grain of the country, its left lying successively behind different branches of the Yerres and Grand Morin, its right near the crest of the plateau scarp west of Sézanne.* Foch's Ninth Army, earlier assembled north of the Aisne and put into line between the Fifth and Fourth, retreated over the Champagne plain until it stood behind the St. Gond marshes, with its right extending eastward to rest on the entrenched Camp of Mailly.* De Langle de Cary's Fourth Army paused with its left holding the hills of the chalk cuesta just south of the Vitry gateway,* its center and right protected by the marshy valleys of the Saulx and Ornain, parallel branches of the Marne, and the Marne Canal, in the difficult terrain of the Wet Champagne. Sarrail's Third Army, including the mobile garrison of Verdun, continued the front northeastward across the Wet Champagne to connect with the great Verdun bastion. Joffre's strategic salient was established, but was not as deep as contemplated; the terminal buttresses held firm, but between them the front sometimes lay obliquely astride serious obstacles, at other times in secure positions behind them. Back of the German front, however, lay a series of east-west topographic barriers in a hostile country, across which their over-extended lines of communication had to be maintained at a great expense in men and energy. Conditions were on the whole favorable for an Allied counterblow.

In the meantime Von Kluck's First Army, now separated from

* Sézanne is 60 miles, Camp de Mailly 80 miles, and Vitry-le-François 100 miles, east of Paris.

the rest of the German armies by a gap only imperfectly filled by cavalry, executed its remarkable change of front. Bending abruptly southeastward, it crossed the line of the Ourcq (thus closing in on the rest of the German front), then turned south to cross the Marne. This flank march across the front of the French Sixth Army was made over the level plateau upland, north of the Dammartin line of heights (p. 236), which stretches from the Forest of Chantilly on the northwest as a straight though somewhat broken wall 20 miles long southeastward to the Marne at Meaux. From this ridge, rising 200 to 300 feet above the adjacent plateau, commanding observation is possible in all directions, as the fertile surface of this part of Valois is most completely cleared of forest cover. Elementary common sense would seem to dictate the seizure of this vitally important member of the northwest-southeast ridge system by the invaders. To do so would place the French left wing, Maunoury's Sixth Army, at a tremendous disadvantage, penning it up within the natural defensive wall while the flank march was carried out in security, or compelling it to fight a battle with the critical element of observation in favor of the Germans before it could debouch from behind the obstacle. In any case the opportunity for Von Kluck to complete his delicate and dangerous maneuver unhindered would have been gained. At the very least, possession of the ridge by Von Kluck would have made a surprise attack on his exposed right flank impossible. He would have received earlier warning of the blow which soon involved the whole German front from Paris to Verdun in defeat. Von Kluck's great blunder was not in effecting his change of front toward the southeast. His own position with reference to the rest of the German armies, the direction of retreat of the main Allied forces, the impossibility of overcoming the defenses of Paris until defeat of the field armies of the enemy released more German troops for that large undertaking, and the necessity of bringing all possible strength to bear in the supreme task of compassing that defeat—all dictated the maneuver which the commander of the German First Army attempted. His guilt lay in contemptuously underestimating the strength of his enemy and in calmly neglecting the most obvious military precautions.[3]

Instead of firmly seizing the Dammartin ridge as a protection for his flank, Von Kluck seems, if not to have ignored its existence, at least to have left totally inadequate forces to occupy parts of it only. With the critical defense line thus weakly garrisoned, the German First Army marched blithely along in front of it, in fancied security. Seldom has a general's stupid neglect of the topographic advantages offered him by Nature been more swiftly punished. Joffre grasped quickly the opportunity presented by the error of his adversary, stopped the retreat to the Seine, hurled Maunoury's army against Von Kluck's unprotected flank, and ordered the rest of the Allied front as far eastward as Verdun to strike north and northwest in one great concentric blow against the imperiled German line. The decisive Battle of the Marne was engaged.

Maunoury's Sixth Army, debouching with little difficulty from behind the Dammartin wall on September 5 and 6, swung around toward the east under orders to seize and cross the line of the Ourcq, pivoting its right on the Marne barrier near Meaux. Near the pivot German rearguards were soon encountered about Monthyon;* but the French left wing had swept 25 miles across the plateau, and far beyond the Dammartin wall, before the surprised enemy could take effective counter measures.[4] Von Kluck acted promptly to correct his blunder as soon as he was apprised of its consequences. No doubt he saw at once the peculiar peril involved in the French drive at the line of the Ourcq. This valley, nearly 300 feet deep below the plateau in places, with the steep sides characteristic of stream trenches cut well into the massive limestone formation, its open floor exposed to artillery fire and traversed by both river and canal, is an obstacle of serious

[3] For a discussion of the possible consequences had Von Kluck seized the Dammartin line of heights, cf. Général Berthaut: De la Marne à la Mer du Nord, Paris, 1919, pp. 34–38. For Von Kluck's explanation of the change of front, citations of orders, and defense of his conduct, including his account of measures taken by him to protect his right flank, cf. Alexander von Kluck: The March on Paris, 1914, London, 1920, especially p. 97.

* Four miles northeast of Meaux.

[4] Berthaut, p. 33.

magnitude. Undefended by important German forces, it could of course be passed with ease. But even wholly to prevent the French from crossing would avail nothing to the Germans. If Maunoury's army could reach this excellent defensive position and so firmly establish itself along the western bank that it could not be dislodged, a German retreat would become necessary. For the Ourcq trends northeast-southwest, making an acute angle with the Marne, just as the Oise trends northeast-southwest to make an acute angle with the Aisne (p. 139). And just as surely as De Castelnau's attack, a few days later, to press the Germans back on the Oise barrier would, if successful, have made that sharp salient untenable, so surely would Maunoury's reaching the Ourcq have cleared the Ourcq-Marne (or Ourcq-Morin) salient. In 1814 Blücher could for a time think of maintaining himself in so precarious a position with the French on the river in his rear. But a century later the improved range and power of artillery and better methods of observation rendered such a thought impossible. Von Kluck's lines of communication, moreover, were cut by the prolongation of the oblique line of the Ourcq barrier almost to Soissons. There was no choice left the German general. He must, at all hazards, withdraw his forces from south of the Marne and hurl them against the French Sixth Army in a desperate effort to prevent it from reaching the line of the Ourcq. The battle which resulted is properly called the "Battle of the Ourcq," even though fought on the plateau west of that river trench. It was the battle *for* the Ourcq, a prize of the first importance.

The task of Von Kluck was extremely difficult. To withdraw a great army across two, in some places three, river barriers with the speed necessary to throw them into battle west of the critical line of the Ourcq, demanded not only quick decisions and able staff work but also the most skillful use of the topographic advantages offered by the terrain. Leaving rearguard forces to defend first the trench of the Grand Morin and, when that was forced, the trench of the Marne-Petit Morin, and finally the trench of the Marne alone, the commander of the German First Army soon had the bulk of his forces, with additions borrowed from other parts of the line, facing west against Maunoury, 3 to 12 miles west of the Ourcq. The transfer was not accomplished without difficulty, however. The French Eighth Division, established on the wooded spur which is surrounded by the big meander of the Marne east of Meaux and which dominates the crossings from Changis past Lizy-sur-Ourcq and Varreddes to Trilport,* harassed and delayed all movements across the trench in that vicinity.

Von Kluck's weakening of his own front south of the Grand Morin, as well as that of the German Second Army on his left, in order to meet effectively the threat to his right flank west of the Ourcq, opened the way to an advance by the British and French Fifth Armies. On September 6 they observed that the German advance had halted and was being followed by a retrograde movement, back to the Grand Morin. The British reached that barrier at Coulommiers by evening. On their right the Fifth Army met stiff resistance but seized the crossing of the Aubetin branch of the Grand Morin at Courtaçon and strong positions at Châtillon and Esternay on the upper waters of the main river. In the dissected ridges and ravines of the edge of the escarpment above Sézanne, where the deep notch of the former course of the Maurienne is cut through, the fighting was particularly severe, as the wooded crests afforded excellent natural protection for the invaders.

Next day the British fought a battle for the crossings of the Grand Morin east and west of Coulommiers. The object of the Germans was to hold the British and French armies at bay as long as possible until the Battle of the Ourcq should be won; but for this purpose they could spare only the cavalry under Von Marwitz, as all infantry was needed to extricate Von Kluck's flank from its perilous situation. The parallel east-west trenches were therefore relied upon to strengthen the defense against an advance from the south. Three cavalry divisions were distributed along the Grand Morin trench for a number of miles east and west of Coulommiers, to fight a delaying action against the

* Unless otherwise stated, the places named throughout the rest of this chapter may readily be located on Pl. III in the pocket, on or near the river or other topographic feature with which the names are associated in the text.

British. The latter finally succeeded in forcing a passage, but only at the cost of serious casualties. Thereupon the German cavalry fell back to the line of the Petit Morin, and the British seized control of the Grand Morin trench as far upstream as the vicinity of La Ferté-Gaucher. Meanwhile the Fifth Army had reached, and in some places forced passages over, the Grand Morin barrier between La Ferté-Gaucher and Esternay. Farther east the forests of the less fertile eastern Brie were beginning to make trouble, the right of the Fifth Army being held up by a strong defense of the Forest of Gault northeast of Esternay, which was only overcome by outflanking the obstacle. The natural barriers of the plateau were standing the Germans in good stead. Against them the Allies, and particularly the British who were located at the critical point, seemed unable to press the Germans northward with sufficient haste to disarrange Von Kluck's movements, or to prevent his main army from breaking contact with the enemy in his front and shifting its bulk, without serious confusion, to his imperiled right flank. Aided by the temporary protection of the valley trenches Von Kluck was escaping from the dangerous trap into which he had blundered.

On September 8 the British pressed on to the line of the Petit Morin, a formidable trench cut nearly 500 feet deep in places, with steep sides frequently forested. The story of the struggle for this defensive barrier is well told in Major Whitton's account of the Marne campaign.

On reaching that river it was soon realized that the German rearguard would not yield their line without a struggle, especially as the steep valley, covered with small but thick woods, distinctly favored the defense. On the British right two battalions of the 1st corps were sharply engaged about Sablonnières and suffered a number of casualties before they succeeded in clearing the Germans out of the village in conjunction with the 1st Cavalry Brigade. A couple of miles to the west the left of the same corps met with considerable opposition, the 2nd Division being held up for some time at La Trétoire; but two Guards battalions of the division and some cavalry managed to get across the river higher up and turned the flank of the German rearguard, which was dislodged with considerable loss. . . . Lower down the river the 2nd corps was engaged between Orly and Jouarre, the 8th Brigade experiencing some trouble near the former village; but after a horse artillery battery had bombarded the position for some time two infantry battalions rushed it, driving off the Germans and capturing several machine guns. On the left the 3rd corps . . attacked in the direction of La Ferté-sous-Jouarre, supported by some French guns, while the British howitzers shelled the bridges of that place across which Germans were streaming northward. The attack succeeded except for the fact that the Germans destroyed the bridges at La Ferté-sous-Jouarre and, by holding the town on the farther bank, prevented the British from crossing. By evening the British had made good the Petit Morin.

Like the British Army, the French Fifth Army does not appear to have experienced any very considerable difficulty until the Petit Morin was in sight, but, that river once reached, the Germans showed that they were not prepared to abandon it without a struggle. On the left General Maud'huy's 18th corps forced the passage of the river near La Celle and, pushing on towards his objective at Fontenelle, gained some ground on the right. But the opposition was too severe to enable him to arrive farther than a line east and west, roughly, of Marchais-en-Brie. The French 3rd corps had a difficult task at Montmirail, for the heights on the right bank were thick with German troops, and the Germans brought many guns into action, so that the fighting lasted for more than eight hours. . . On the right of this French gain the 1st corps had succeeded in reaching Vauchamps. . . . As for the 10th corps on the extreme right of the Fifth Army, it managed to reach, and apparently to cross, the Petit Morin north of Charleville. . . . The actual objectives assigned by the commander had not been reached by the Fifth Army, but . . . the capture of Montmirail and the river line east and west of it represented a good day's work.[1]

A whole day had been lost, however, in passing the obstacle, at a time when the complete success of Joffre's offensive required driving the Germans swiftly northward beyond the Marne.

On the following day the difficult task of crossing the winding trench of the Marne was undertaken. At La Ferté-sous-Jouarre, the vital point of passage which figured so prominently in the Napoleonic campaign a century earlier, the bridges had been effectively destroyed, and the north bank of the valley was held by a German rearguard supplied with numerous machine guns. The British engineers found it impossible to throw a bridge over the obstacle under the galling fire from the northern bank, until

[1] F. E. Whitton: The Marne Campaign, London, 1917, pp. 186–188.

darkness came to protect them. Hence it was not until after nightfall that a crossing was effected.[6] Farther downstream a footing was secured on the northern side at Changis. The center crossed above La Ferté after overcoming the opposition of a rearguard supported by artillery handled with great skill and bravery. Farther east, between Charly-sur-Marne and Château-Thierry, the latter like La Ferté-sous-Jouarre a critical point in Napoleon's 1814 maneuvers, the British right was held up at the barrier for some time by heavy machine-gun fire. As the bridge at Château-Thierry had been destroyed, the construction of a pontoon bridge under fire was necessary before the passage could be accomplished. The negotiation of the barrier would have been even more difficult but for the poor behavior of German troops at some of the crossings. At Charly-sur-Marne, for example, the bridge was left intact and the natural amphitheater commanding it on the north undefended, because the forces assigned to this duty got hopelessly drunk.[7] The French Fifth Army crossed with the British in the Château-Thierry sector; but farther east, owing to the longer march required to reach the Marne barrier after crossing the Grand Morin, the passage was not effected until later.

From the south bank of the Grand Morin east and west of Coulommiers, to the north bank of the Marne east and west of La Ferté-sous-Jouarre, is a distance of from 12 to 15 miles. It took three full days to traverse this distance with nothing but limited rearguard forces to combat. Nothing, that is, except the obstacles interposed by Nature. To these must a large share of the delay be accredited. He who would measure the influence of the trenches of the Grand and Petit Morin and the Marne Rivers, must try to imagine the consequences had the German First Army been so vigorously pressed on a level plateau devoid of obstacles that it could not readily break contact with the enemy in its front, leaving Maunoury's Sixth Army, almost unopposed, to reach and cross the line of the Ourcq.

[6] Sir John French's Despatches, p. 31.
[7] Whitton, p. 203.

With its communications cut, with a relentless foe holding it in front and a victorious host assailing it in the rear, Von Kluck's First Army could hardly have escaped overwhelming disaster. As it was, the story runs somewhat differently.

While the successive river trenches were holding back the enemy on the south, permitting Von Kluck to break contact there with most of his troops and to shift them north of the Marne and west of the Ourcq, Maunoury's Sixth Army was finding its advance opposed by constantly increasing enemy forces, supported by artillery concealed in the ravines cut by minor branches of the west-to-east trenches tributary to the Ourcq. Batteries of heavy guns on the reverse slopes of the plateau spur at Trocy 7 miles northeast of Meaux, on the very terrain where Blücher's vanguard, debouching from behind the Ourcq barrier, was overwhelmed by the French a century before, caused very heavy losses to the French under Maunoury. On all the broad surface of the plateau, and in the trenches which near the Ourcq cut it into parallel east-west strips, the fighting was extremely severe. By the morning of the 8th the French front was approximately 5 or 6 miles west of the Ourcq.

This day the effects of the Morin trenches in holding back the enemy at the south began to tell heavily against Maunoury. The growing pressure of German forces brought north to oppose him was forcing back his left flank. Reinforcements were hurriedly sent out from Paris by railway and commandeered motor vehicles of every description; but meanwhile the outflanking of the French left, attempted in vain on the Ourcq battlefield a hundred years before, was succeeding under Von Kluck to an alarming extent. The northern end of the French line, unprotected by any natural obstacle on the exceptionally level upland west of the heads of the Ourcq tributaries, had by the 9th been bent back 8 or 9 miles and was facing northward. The situation was critical. But it was on the 9th that British and French forces at last crossed the Marne barrier and found themselves in the rear of the German First Army west of the Ourcq. Von Kluck saw that his position, supportable so long as the Morin and Marne trenches lay between him and his foes, was now absolutely untenable. That evening he ordered the retreat up the valley of the Ourcq toward the line of the Aisne.

In the meantime a battle only second in importance to that of the Ourcq had been raging along the front of Foch's Ninth Army. Its retreat had brought it behind the St. Gond marshes, which "all felt was the last barrier providentially placed on the route of the invasion."[8] But Foch was carrying out the orders for a retreat to the Seine-Aube line, and nothing more serious than a rearguard action was contemplated along the marsh barrier. Even the plateau spur at Mondement, "the strategic key to the marshes," was but weakly held against a strong German attack. His army was still pushing southward when, on the morning of the 5th of September, Foch received Joffre's order of the night before. Immediately he faced back to the north and hastened to secure the protection of the marshes. At this time his left rested against the edge of the great escarpment just south of the deep notch or gateway cut by the Petit Morin River where it drains from the marshes into the plateau; while his right reached across the chalk plain to the entrenched Camp of Mailly.

General Joffre had not failed to see the advantages offered by this historic natural barrier for assuring the safety of the armies resuming the offensive on the plateau against a flank attack, quite possible in case the Germans pushed southward over the level chalk plain below until they were in the rear of the plateau forces. His famous order of September 4 directed that "The Ninth Army will cover the right of the Fifth Army by holding the southern outlets of the St. Gond marshes and by placing a part of its forces on the tableland north of Sézanne." The tableland referred to is one of the projecting bastions of the plateau (p. 253), and its northern wall dominates the marshes as well as the trench of the Petit Morin where it leaves them for its course through the upland (Fig. 62). Because the village and château of Mondement occupy a commanding position on the plateau spur, we may call it the Mondement bastion. On the following day, in

[8] Charles Le Goffic: Les Marais de Saint Gond, Paris, 1916, p. 23.

preparation for the battle, Foch issued the detailed orders required to execute the broad instructions of the generalissimo. The forces constituting his left wing were directed to cross the Petit Morin trench and push northwestward over the Marne plateau, thus keeping in touch with the proposed northward advance of the Fifth Army, operating just west of the Ninth. The center was to take up a defensive position along the southern edge of the marshes of St. Gond, maintaining strong advanced guards north of the marshes, where the Germans had failed to seize the hill at Toulon (p. 267) "a sort of great natural *rédan* of which the high wooded scarp and lower glacis-like slopes"[9] commanded both marsh and plain. The right was likewise to take position behind the trench of the small Somme River, "an excellent line of defense," from the eastern end of the marshes to near the head of that little stream. Thus the Ninth Army utilized the whole upper portion of the combined Somme-Petit Morin valley trench, just as if there had been no geologically recent river capture separating it into two valleys. Indeed, we have already seen (p. 266) that the capture had made the barrier more formidable by converting part of it into the St. Gond marshes. However, between the point where the Somme bent abruptly northeast to its new course at Écury-le-Repos and the point where the old valley began to get marshy at Morains, there was now a small unprotected gap in the natural barrier about 2 miles wide.

On September 6 the Ninth Army found the enemy pressure too heavy to make the full execution of Foch's orders possible, although the broader objects outlined by Joffre were in general attained. Instead of advancing north of the Petit Morin trench, the left flank was actually pushed back to the south; but it held a strong position on the tableland. All the advanced guards north of the marshes were driven back to the south side, after unsuccessfully attempting an attack against the plateau scarp at Congy, held by the invaders. Advanced French forces holding the butte of Toulon found themselves being crushed by artillery fire concentrated upon them by German

[9] Le Goffic, p. 33.

batteries along the rim of the escarpment to the northwest and were forced to retire. An attempt to recapture the important height, which Foch had ordered to be held *à tout prix*, was a failure. "The retreat through the marshes was terrible. For more than 3 kilometers extended the forest of reeds. It was impossible to use the road, where one would be immediately under accurately ranged artillery fire; and among the reeds one would sink in sometimes to the waist."[10] Part of the retiring forces entrenched themselves in strong positions near the quarries on Mont Août. The enemy followed along the defiles through the marshes and repeatedly made desperate efforts to debouch. But "his wave will seek in vain to reach the slopes of Mont Août, grand point of articulation, solitary like an island, which the 52nd Reserve Division put in a state of defense and which dominates from its height of 221 meters the vast marshy expanse. To the end of the battle, except during a few hours on the afternoon of the 9th, Mont Août will remain securely ours, and the IX Corps will find in it the most solid of supports."[11] The main body of the French IXth Corps "by holding the exits was able to maintain its position without difficulty."[12] At the gap in the barrier at Écury-le-Repos, however, the enemy had pushed southward, an operation which would, if carried much farther, outflank both the Somme River and St. Gond marsh parts of the barrier on either side. A portion of the Somme valley for several miles above Écury was already lost.

Next day the Germans attacked the Ninth Army with a fury evidently born of the desire to help extricate the Imperial armies from the difficulties into which Von Kluck had plunged them. The center, "protected as it was by the formidable marshes to its front, was able to maintain its line without difficulty."[13] A large part of the German effort in this sector was now concentrated against the plateau bastion of Mondement, projecting eastward some 5 or 6 miles between the western end of the

[10] Le Goffic, p. 70.
[11] *Ibid*, p. 78.
[12] Whitton, p. 153.
[13] *Ibid*, p. 169.

marshes on the north and Sézanne on the the south. So effectively did its heights command the marsh barrier and its passages, that General Humbert is reported to have said: "The Germans are bottled up. Mondement is the cork of the bottle. At any price it must hold." Nor is the whole value of the bastion measured on the north alone. From its projecting point the scarp retreats rapidly to the southwest, leaving the whole expanse of the chalk plain to the south and east exposed to view, with its roads, railways and villages spread out below like a giant map. On a clear day the towers of Troyes, on the opposite side of the plain, are visible. With such commanding heights in German control an extended retreat of the Allies under most painful conditions would become necessary. Small wonder, then, that German efforts were for several days centered on capturing the Mondement plateau. Little ground seems to have been lost here on the 7th, despite desperate German assaults, but on the extreme right the less formidable obstacle of the small Somme valley was already proving insufficient protection against the enemy's heavy attacks.

On September 8 picked German troops, including the Guard Corps, again launched "repeated onslaughts conducted with the most reckless violence" in an effort to smash through the French center. The Somme position at the right was lost, and its defenders, pushed backward southwestward some 8 miles to the line of the upper Superbe (the Maurienne), were now endeavoring to re-form under the protection of that obstacle. This retreat exposed the flank and rear of the troops defending the St. Gond marsh barrier. Some of these forces had to fall back southward and face east against the new danger. Taking advantage of the confused situation the Germans began to filter through the marshes under protection of heavy artillery fire and to hasten the French withdrawal from the southern margin of the barrier, a withdrawal made inevitable by the outflanking movement farther east. On the plateau at the extreme left the French line held firm and even gained ground; but all along the western border of the marshes it was being pushed back against the steep northeastern wall of the Mondement bastion, and the bastion itself was with difficulty resisting extremely heavy attacks. Believing, however, that the formidable positions on the bastion could be held, with the aid of the Fifth Army's right wing, by a part only of the forces stationed there, Foch now withdrew a division from his left and placed it in a convenient position to support his right as needed. The latter, reorganized, had debouched from behind the Maurienne and was gaining ground in an attack northward toward its former position on the Somme.

The following day was heavy with fate. The Germans, leaving holding forces at the northern exits of the St. Gond marshes, issued in great numbers, the Guards included, past the eastern end of the barrier and pushed southward beyond Fère Champenoise in a desperate endeavor to complete their success against the Ninth Army by a break-through. Assailed by the concentrated weight of three German corps, Foch's right fell back to the line of the Maurienne, then 3 or 4 miles farther back, to the vicinity of Salon. The marshes were effectively outflanked and the center was compelled to fall farther back, completely uncovering the southern exits. Mont Août was abandoned to the enemy, and the wall of the bastion at the château of Mondement was scaled by hostile troops. It was one of the most critical moments of the battle. But Foch's "calculated tenacity" was equal to the emergency. "The situation is excellent. Attack!" continued to be his order of the day. Flinging his reserve division into the breach on the afternoon of the 9th and directing an advance along the whole line, he had measurably relieved the situation before nightfall. Mondement was retaken, and the left improved its hold on the difficult terrain of the plateau by a further northward advance, made possible through the excellent progress of the Fifth Army's right wing. Plans were perfected for launching Foch's left eastward, north of the marshes, on the morrow to attack the Germans in the rear.

That same day, however, the issue of the Battle of the Marne had already been decided on the plateau to the west, and as night came on orders for the great German retreat were given. When the morning of the 10th dawned, Foch found in front of him only German rearguards and quickly advanced to his old position along the Somme valley-St. Gond marsh barrier. By night his right had pushed on to the northeast and stood on the line of the Soude, facing toward Châlons. This rapid advance over the level plain of the Champagne, while on the east the German rearguards were dislodged but slowly from the difficult terrain of the dissected chalk scarp about the Vitry gateway and on the west equally slowly from the difficult plateau terrain with its forests and lakes, formed a wedge projecting into the German center which at first led many, among them the present author, to attribute the German retreat to Foch's offensive. The French official communiqué of September 10, reviewing the battle, lent color to this erroneous view. On the basis of the fuller information later available, supplemented by personal statements of some of those directing the operations, it can safely be said that the German retreat had been rendered inevitable, and was indeed under way, some hours before Foch's wedge was formed. To acknowledge this is to detract nothing from the "beautiful maneuvers"[14] which must themselves in time have brought about the retreat of the enemy.

There seems to be little support for the legend that the Prussian Guard was overwhelmed in its retreat through the St. Gond marshes, becoming mired and losing thousands of its numbers in the morass. On the contrary, the retreat of the night of September 9 was orderly and conducted secretly under cover of darkness. According to Louis Madelin, "no one stuck in the marshes, for during those months no one could really stick there. After the hot summer and in spite of light rains, they were merely a sort of basin in which reeds and grasses grew out of the gray, cracked earth."[15] Others state that after rains during the battle "the clay pocket of St. Gond became a quagmire" and

[14] Joseph Mangin: Comment finit la guerre, *Rev. des Deux Mondes*, Vol. 56, 1920, pp. 481–520, 721–762; Vol. 57, 1920, pp. 241–285, 481–537, 774–815; Vol. 58, 1920, pp. 74–101; reference in Vol. 56, p. 511.

[15] Louis Madelin: La victoire de la Marne, *Rev. des Deux Mondes*, Vol. 35, 1916, pp. 241–287; reference on p. 268.

describe a fearful slaughter of the picked German troops while "gunners up to their knees" and "gun carriages up to the axles" struggled in mud and water. The truth seems to lie between the two extreme opinions. The writer has seen the St. Gond marshes late in July when they showed broad expanses of reeds rising out of water instead of dry, cracked clay. Charles Le Goffic, whose special study of this sector of the battle front is entitled to credence, describes the three roads which were "the best, the most solid, and alone capable of bearing heavy artillery" across the "quaking meadows," states specifically that some *arabas*, light Moroccan carts, "became mired in crossing the marshes," and quotes a participant in the French retirement through the marshes on the first day of the battle who tells of "men sinking sometimes up to the waist."[16] Accounts written by soldiers engaged on the borders of the marshes refer repeatedly on different days to "pouring rain," "a light thundershower which cleared the air," "the rain began falling again, a good downpour which wet the men to the skin," "the following morning a fine rain which ceased about ten o'clock." Apparently the marshes, partially drained and much less of an obstruction in summer than in winter, were nevertheless, especially when showers filled the depressions along the causeways and the drainage ditches with water and made the clay sticky and slippery, enough of a barrier to render progress off the roads difficult and hence to give a solid meaning to Joffre's order, to hold "the southern outlets (*débouchés*) of the marshes of St. Gond," an order which would have no real significance if the whole region was in the summer months always a dry clay plain which could be crossed anywhere. On the other hand, the great bogs which made the marshes an object of fear in earlier days, and engulfed the remnants of Pachod's troops when they sought refuge among the reeds from the pursuing Allies a century before, no longer existed in 1914. That some Germans, and Frenchmen as well, may have found the remaining undrained areas and the rain-soaked clay a trap, is quite possible. But no masses of the Guard were swallowed

[16] Le Goffic, pp. 36, 52, 67, 69.

alive, no reapers cut with their blades "corpses projecting half-way out of the mud." [17]

The chief rôle of the St. Gond marshes in the Battle of the Marne was to serve as a barrier behind which Joffre could expect the center of his great battle front to hold firm while the flanking blows were achieving their purpose. That center driven in or broken would mean the failure of his offensive operation, might mean irreparable disaster. But the center standing unshaken, an impassable barrier to the enemy, was all that was needed to make victory sure. The marshes contributed greatly to this end, and the fact that they were ultimately outflanked by an advance elsewhere does not lessen the significance of their rôle as a barrier during the first days of the fighting, or their value in the subsequent maneuvers. Even when the southern outlets fell into their hands, the Germans were seriously embarrassed in supporting an offensive across an obstacle pierced by few practicable roads, all under accurate Allied fire.

The plateau bastion of Mondement was, however, the strong point which saved the situation at the Allied center when the marsh barrier was lost. When Foch's right was overwhelmed and his center threatened, his left rested unshaken on the massive buttress of the plateau spur. It was the natural strength of that position which, in full battle, enabled him to withdraw a portion of its defenders and throw them into line farther east where the danger was greater. It was from the plateau that Foch determined to launch on September 10 the flank attack against the German communications north of the marshes. The only part of the Ninth Army front which did not lose considerable ground was the part posted on the Mondement bastion. In the words of the French official communiqué of September 10: "By a succession of violent attacks the Germans tried to break our center, but in vain; our success on the high ground north of Sézanne has enabled us in turn to pass to the offensive."

During these critical days the French Third and Fourth Armies at the eastern end of the far-flung battle front found great dif-

[17] Florian Parmentier: *Visions de guerre*.

ficulty not merely in making progress but even in holding their own. The Fourth Army, standing behind the marshy barrier of the Saulx-Ornain valley and resting its right on the hill country of the dissected chalk scarp, successfully resisted violent attempts to pierce its front. A break-through at this point would isolate Verdun, and was especially dangerous because an effort to achieve the same result was being made simultaneously from the eastern side of the sharp Verdun salient (p. 385). This latter attempt kept Sarrail's Third Army in a measure immobilized, for the threat in its rear was decidedly serious. For several days the Fourth Army beat off terrific enemy attacks, sometimes giving ground, sometimes gaining a little; while the Third Army, less actively engaged, changed front but little. On the 10th the left of the Fourth Army advanced slightly, pressing the Germans, who were now becoming involved in the general retreat. It is interesting to note that the German Third Army in retiring, at first did so only on its right wing, its left being solidly buttressed in the rough hill country of the dissected chalk scarp near Vitry-le-François. This natural stronghold resisted the efforts directed against it by De Langle de Cary's Fourth Army and materially handicapped the French advance. But on the 11th and 12th the German armies were in full retreat, jeopardized by the defeat of Von Kluck far to the west. One strong position after another was surrendered without a battle, and only in front of Sarrail's army did the Germans await serious attacks before yielding.

The Battle of the Aisne

When compelled to retreat from the banks of the Ourcq and the marshes of St. Gond, the German high command sought some natural topographic barrier which would aid in checking the Allied pursuit and behind which their badly shaken military machine could be reorganized. The Marne itself above Château-Thierry, prolonged northwestward past Bouresches and Belleau Woods by the east-west trenches of the Clignon and Grivelle branches of the Ourcq, was unavailable because it was too close behind the defeated armies. There would be no opportunity to redistribute forces while retiring so as to strengthen the weak sector of the line held only by Von Marwitz's cavalry, which, as we have seen, was unable to keep the Allies from crossing the formidable barriers of the two Morins and the Marne below Château-Thierry. On its right wing this position would be poorly defended against flank attacks, and neither branch of the Ourcq was as formidable an obstacle as the situation demanded. Besides this, the barrier in question was already being broken at its weakest point, between the Marne and the upper branches of the Clignon, by the advance of the British and French near Château-Thierry and Bouresches.

Still farther back was the truly formidable line of the Villers-Cotterets ridge, prolonged westward to the Oise by the east-west trench of the Authonne with its higher northern wall backed by the vast Forest of Compiègne, which offered admirable concealment for the defending forces, their artillery and supplies, and their movements; and prolonged eastward by the bastion of Rheims Mountain with its forest cover. But this line required time for its proper organization. The western half of Villers-Cotterets ridge is wooded, and in front of a large section of it stretches the great Villers-Cotterets forest, offering concealment for an enemy maneuvering to storm the ridge. "To have a great forest mass in front of one is one of the worst conditions for a battle line, because under the shelter of the mass one never knows what is happening. A forest is a bandage over the eyes."[18] Not until after openings had been cut in the woods on the heights to give good observation over the plain, and a zone had been cleared between the forest on the plain and the ridge, would the line be free from a serious threat to its center. From the eastern end of the ridge to the Rheims bastion there was a gap of some miles to be blocked by artificial works alone. Needless to say, the time for performing such labors had not been available.

Behind the Villers-Cotterets ridge lay the great natural moat of the Aisne, 2 miles wide and 350 to 425 feet deep, dominated on the north by the strong wall of the Chemin des Dames plateau

[18] Général Berthaut, De la Marne à la Mer du Nord. p. 55.

strip (Fig. 64) with its steep upper scarp so difficult to scale, its exposed lower slopes swept by fire from the crest, and its numberless quarries and caverns providing ready-made shelter for artillery and supplies and for whole armies of men. On the west it was protected by large forests, the trench of the Oise, and the outlying natural fortresses of the Noyon-Lassigny massifs. To the east it could be continued by the Vesle-Ardre trench and the Rheims Mountain bastion; or if this last formidable massif was too close to the scene of defeat to permit its orderly occupation and organization, then by the outlying heights of Brimont, Berru, and Moronvilliers, with advanced defenses along the Ardre and Vesle trenches and the Rheims Mountain and St. Thierry bastions (Fig. 62). The River Aisne, flowing through the flat, open bottom of its east-west trench, is from 150 to 200 feet wide and some 10 or 15 feet deep, and hence can be crossed only where bridged. All bridges could be brought under the direct fire of artillery concealed on wooded portions of the valley slopes and plateau spurs or in the ravines cut back into the crest (p. 231). Movements on the open plateau surface were largely screened from direct observation, thanks to the top of the tree fringe left growing on the steep upper slopes. The heights of Brimont, Berru, and Moronvilliers dominate every avenue of approach across the low plain of Champagne. Here was one of the most formidable defensive positions in France, 85 miles long from the Oise valley to the eastern spur of Moronvilliers, and far enough in the rear of the Ourcq and St. Gond battlefields to serve the needs of the German high command.

By the afternoon of September 12 the French Sixth Army and the British army were encountering such determined resistance that the British Commander-in-Chief "formed the opinion that the enemy had, for the moment at any rate, arrested his retreat and was preparing to dispute the passage of the Aisne with some vigor."[19] General Joffre had already foreseen that the Germans would make a stand on the line of the Aisne and that this "immense fortress prepared by Nature" would prove too strong to be

[19] Sir John French's Despatches, p. 37.

taken by frontal attack. On September 11 he wrote to Maunoury: "It must be expected that, the enemy facing about on the Aisne, it will be very difficult for you to attack from in front, and it seems necessary for you to have as large forces as possible ascending the right bank of the Oise in order to outflank the enemy's right wing."[20]

Without delay the British army was ordered to assault the barrier from Soissons eastward and secure the river crossings. Fortunately some of the bridges were not completely demolished. At Pont Arcy troops crossed in single file on a girder not completely submerged, at Vailly on a single plank connecting the ruins of the former bridge. A little farther east a passage was effected by the aqueduct carrying the Oise-Aisne Canal over the river, which was poorly defended by the Germans. The bridge at Condé was left intact but was so dominated by the heights above that it was more of a trap than a passage and remained in the control of the enemy till the end of the battle. Bridges at Venizel, Missy, and Vailly were not damaged beyond repair. Elsewhere men were rafted or ferried over in boats. Pontoon bridges were thrown across at eight points, and the construction of more substantial structures was begun by the Royal Engineers. All of this work was carried on under heavy artillery fire and at a serious cost in casualties. Approaches to the crossings were swept by enemy guns, bridges were damaged and operations interrupted, and the completion of pontoons at some points was rendered impossible. British troops established on the north bank, with a powerful enemy strongly entrenched in front and a dangerous obstacle imperfectly bridged at their backs, were in a hazardous situation.

The assault on the plateau scarp was immediately pressed by the troops which had negotiated the barrier, in the hope of winning the crest and pushing the enemy farther back from the vital crossings. It was difficult to lend the infantry proper artillery support in this operation, both because of the steepness of the slopes and because of the absence of concealed gun positions on

[20] Victor Giraud: Histoire de la Grande Guerre, Paris, 1919, p. 178.

the exposed valley floor. Enemy artillery, on the other hand, enjoyed the advantages of excellent positions and dominating observation. These advantages soon began to tell. The attack, although continued with heroic determination against appalling odds, was a failure. The British pushed up the ravines toward the plateau top and attempted to seize the southward projecting spurs which formed successive barriers against a westward advance parallel with the crest. But they were halted on an oblique front across the southern face of the plateau, their right clinging to a precarious position on the margin of the upland west of Craonne, their left down in the valley from east of Vailly to Soissons.

Of all the plateau spurs the one terminating near Chivres and Missy projects farthest south, forming a strong bastion from which long stretches of the valley to the east and west may be brought under enfilading fire. It was this spur which effectively dominated the Condé bridge at its base and made the passage of the river farther east so difficult. To relieve this situation an attempt was made to capture the Chivres bastion, "as a necessary preliminary to a further advance northward;" but the position was too strongly defended. Forces which scaled the southern point of the bastion were compelled to fall back into the valley floor. Sir John French later discussed with his commanding officers the possibility of seizing the Condé bridge but abandoned the project because the operation, conducted under fire from the bastion, would prove more costly than the general situation warranted. Cavalry which had crossed the river for a hoped-for pursuit of the enemy had to return single file over a narrow bridge near Vailly under heavy fire from the heights.

But, while the British were unable to press forward and capture the plateau, the Germans found the task of dislodging them equally beyond their strength. The north-south ridges and ravines, once seized by a foe, afforded good defensive positions. Hence the Battle of the Aisne developed into a series of heavy attacks and counterattacks lasting until September 28 without much change in the battle front. The little British army alone had sacrificed more than 13,500 men in a vain effort to breach the Aisne barrier.

At the right of the British the end of the Chemin des Dames plateau had been seized by the French after a bitter struggle in which Craonne changed hands several times. Napoleon had considered the position an extremely difficult one to carry by assault, and the French of a century later could amply confirm his judgment. A part of the Fifth Army advanced over the chalk plain east of the plateau as far as Corbeny at the head of the Ailette valley, threatening to penetrate the valley and take the Chemin des Dames plateau in the rear while the British attacked it in front and thus to repeat Napoleon's maneuver on the same terrain. But after a fierce struggle the French were forced back from both Corbeny and Craonne.

Meanwhile, Maunoury's Sixth Army was delivering frontal attacks along the Aisne barrier west of Soissons, but seeking more particularly to outflank the whole mass of the formidable obstacle by an advance up the valley of the Oise. At Carlepont, 5 miles south of Noyon, the northward drive of the French seemed in the way of success; but the natural protection the enemy enjoyed on this flank was too strong to be overcome. The Forests of Laigue, Carlepont, and others, the wooded heights of the Lassigny-Noyon massif, and the wooded western spurs of the Soissonnais plateau made a complex of obstacles through which the French could not push their way indefinitely. German advanced lines of defense, profiting by these advantages even where no attempt was made to hold them permanently, broke the force of the French blow. The western spur ends of the plateau strip were ascended, but Maunoury had shot his bolt. Fierce local struggles, about Tracy-le-Mont near the northern crest of the westernmost spur and Tracy-le-Val in the valley below, marked the end of the abortive attempt at outflanking the Aisne barrier. The Germans could not in any case hold the narrow salient lying within the acute angle formed by the Aisne and Oise. Their line must cross the plateau obliquely, from the Aisne above the confluence to the Oise. Such a line was established when the French were brought to bay along the cross ridges of the plateau between Attichy* and Tracy-le-Val.

Maunoury's frontal attacks against the southern scarp of the plateau west of Soissons at first gave promise of an important success. Crossings of the river were effected under great difficulties at Vic-sur-Aisne and Fontenoy with the aid of pontoon bridges. North of Vic a valley, unusually large for this plateau strip, heads northward and eastward into the upland (Fig. 62). Along this trench the French pushed boldly forward, hoping to outflank the defenders on the heights above; but the position of the enemy was too strong, and the French fell back to the main river. Assaults on the plateau at other points were ultimately beaten off, and efforts to build a pontoon bridge at Soissons were thwarted by enemy artillery fire accurately directed from the higher ground. The natural fortress of the plateau wall and Aisne trench had saved the German armies from a retreat into the low plain north of the Marne plateau, and assured a long and bitter war in place of a speedy expulsion of the invaders.

In the Champagne plain, where the Germans had fallen back on the northeastern heights of the Rheims Basin—the buttes of Brimont, Berru, and Moronvilliers—all attempts of the French to dislodge the enemy from his strong position failed. Brimont, violently contested, changed hands several times, finally remaining to the Germans. Berru was solidly German, and the enemy's line lay well in front of the supporting mass of Moronvilliers. Eastward the line traversed the open chalk plain north of the Camp of Châlons to cross the Monts de Champagne of the dissected chalk scarp about seven miles north of the battlefield of Valmy.*

For some months the operations on the Aisne and Champagne fronts were restricted to minor projects designed to effect local improvements in the battle front. The British sector was taken over by the French in October. About the middle of January, 1915, the French launched an attack against the plateau north of Soissons with the object of pushing the enemy from his observation posts on the crest and thus disengaging the town from the

* Attichy is 12 miles W of Soissons, Valmy 20 miles NE of Châlons.

accurate bombardment to which it had long been subjected. Crouy, at the entrance to one of the north-south ravines two miles northeast of Soissons and at the base of one of the plateau spurs dominating the town, was captured, and the limited offensive gave promise of success, when a sudden flood in the Aisne swept away the bridges at Missy and Venizel upon which the attackers largely depended for their connection with the south bank of the river. The situation was decidedly serious, as Von Kluck was rapidly bringing up reinforcements. The bridges were re-established but again destroyed by the flood. Isolated on the northern bank of the river, with a powerful enemy launching attacks from an impregnable natural stronghold in front, and an angry flood sweeping away the only means of securing support from the rear, the French found themselves in an intolerable position. There was no choice but to retreat under the most painful and dangerous conditions. Deluged by a storm of projectiles the French fell back to the south side of the river with heavy losses, leaving only sufficient forces on the north bank to maintain a small bridgehead at Soissons.

A local operation of some importance was directed early in the following June against the western flank of the plateau position. Both Tracy-le-Mont and Tracy-le-Val, as well as part of the wood on the spur just east of them, had been seized in minor attacks, when an advance eastward across the north-south ridges and ravines defending the German flank was attempted. The suddenness of the onslaught took the enemy by surprise, and two of his lines of defense were captured before he began to react with vigor. Then progress across the grain of the country was quickly halted. A succession of strong positions on the cross ridges was always available to the enemy, and his flank was practically impregnable against local attacks.

The First Battle of the Champagne

Turning to the extension of the Aisne position across the chalk plain of the Champagne, we should note that in December, 1914, the French had launched a series of attacks on this front, selecting for their operations the only sector which promised any probability of success. Between the Moronvilliers massif on the west and the Monts de Champagne formed by the dissected crest of the chalk escarpment on the east, there is a stretch of some 15 miles of gently sloping chalk plain on which topographic inequalities of any magnitude are lacking. Farther west the line of buttes bars the way toward the north, while eastward the Monts de Champagne are soon succeeded by the formidable terrain of the Argonne. As Somme-Py* is one of the most important points in the 15-mile gap between regions of more difficult country, we may for convenience call this the Somme-Py gateway. An advance through it and into the northern Champagne beyond would outflank the Brimont-Berru-Moronvilliers barrier, threaten the important gateway of Rethel, and, if pushed far enough, turn the enemy's formidable Aisne position. We must be prepared, therefore, to see repeated attempts against this significant sector of the German front.

The first efforts by De Langle de Cary's Fourth Army gained a few kilometers of ground in the neighborhood of Hurlus. Realizing the danger of an advance at this point, the Germans reacted vigorously, delivering more than twenty counterattacks at a heavy cost in casualties. The Allies had not yet learned how to deal effectively with successive trench systems, and small advances over the exposed plain were dearly purchased.

About the middle of February, 1915, the French began a series of violent attacks in this same gateway which lasted for five or six weeks and which may be known as the first great Battle of the Champagne. The front of operations was 8 miles long, from Souain to Beauséjour northeast of Hurlus, and one of the objectives was to aid the Russians by preventing the transport of troops to the eastern front. Whatever hopes of a possible break-through may have been entertained were destined to disappointment. Slight advances could be effected at the cost of proper artillery preparation, which wrought havoc with the enemy trenches and caused the defenders heavy losses. Ten thousand

* The points involved in this battle lie from 17 to 22 miles NNE of Châlons.

German corpses are said to have been found on the field of battle.[21] But the German army still blocked the gateway.

The Second Battle of the Champagne

When in September, 1915, two simultaneous offensives were launched against the great Noyon salient, one in Artois (the Second Battle of Vimy Ridge) and one in Champagne, it was again the Somme-Py gateway which was selected for attack. This time not only the whole of the gateway, from Aubérive in the Suippe valley at the southeastern base of the Moronvilliers massif to Massiges at the base of the Monts de Champagne, was included in the field of operations, but also a portion of the Wet Champagne about Ville-sur-Tourbe, east of the chalk scarp. The undertaking was planned by General de Castelnau, and the troops were exhorted "not only to take the first enemy trenches, but to push on without pause, day and night, past the positions of the first and second lines, into the open country beyond."[22] A three days' bombardment prepared the way.

The assault was terrific. In a few hours the French had smashed through the first system of defenses, on the whole front of 20 miles, to a depth of more than 2 miles in places. But just as on the plains of the north, so on those of the Champagne, the Allies were to learn that a long bombardment gives the enemy time to mass his reserves for the shock, without destroying his rear defenses; and that a continued advance across an exposed plain under these conditions is practically impossible. Heavy rain softened the ground at the critical moment and, while proving less of an obstacle than on the loam-covered chalk of the Somme region, retarded the bringing up of reinforcements and the advance of the artillery to new positions. The offensive ended almost as suddenly as it had begun. The Somme-Py gateway was still in enemy hands.

The Hindenburg Retreat

The extensive enemy retirement on the western front made necessary by the 1916 Battle of the Somme (p. 158), and which

[21] Giraud, Histoire de la Grande Guerre, p. 285.
[22] *Ibid*, p. 311.

took place early in the spring of 1917, involved a readjustment of the German front across the northern end of the Marne plateau. The enemy had decided to retire from the Lassigny-Nöyon massif and to pivot his new line on the St. Gobain massif. This automatically freed all those portions of the Aisne valley and plateau strip west of the Soissons region. The new front, after crossing the western margin of the St. Gobain tableland, reached the Chemin des Dames plateau strip north of Soissons and traversed it obliquely toward the southeast to reach the Aisne valley near the apex of the strong Chivres bastion, 5 or 6 miles east of Soissons. The important crossings of the Aisne trench at Soissons and all of the western half of the trench were now securely in Allied control. Ultimate conquest of the great natural fortress appeared more feasible than ever before. On the east the French still clung to the positions gained by the British on the margin of the Chemin des Dames upland west of Craonne; their new front on the upland northeast of Soissons threatened the flank of the defenders; and only for the 10 miles east of Missy were the attackers still held in the bottom of the Aisne trench.

The First Battle of the Chemin des Dames

The Allied plan of campaign in the spring of 1917 had as one of its chief objects the breaking of the new German front across the plain of the Somme by capture of the two great buttresses upon which that front was anchored. On the north the Third Battle of Vimy Ridge was to deprive the enemy of one of the buttresses (p. 162). On the south the St. Gobain buttress was too strong to take by direct assault, so it was decided to try, simultaneously with the attack on the Vimy buttress, an outflanking operation on a large scale which would, if successful, compel the evacuation of the St. Gobain region. For this purpose an offensive against the remaining portion of the Chemin des Dames stronghold and against the line of buttes supporting the Champagne front was undertaken by Joffre's successor, Nivelle, apparently with inadequate appreciation of the immense natural strength of the obstacles to be attacked.

After a ten days' heavy bombardment of the plateau defenses, the Sixth Army under Mangin and the Fifth under Mazel on April 16 assaulted from in front the escarpment dominating the Aisne valley. Poor weather conditions, by preventing proper aërial observation on the part of the French, enormously increased the advantages of direct observation enjoyed by the Germans from the heights above. Under a terrific fire the magnificent armies of France were hurled up the slopes, against some of the most formidable positions which ever defied armed force. Across the lower declivities, swept by enemy machine guns, up the steep cornice above, through fortified orchards and woods, over quarries and caverns innumerable, the assaulting wave rose higher and higher, faltered, neared the crest, but broke at last against the grim masonry of Nature. For three weeks the struggle was continued, degenerating into local combats for improvement of particular points in the line. The Chivres bastion was largely cleared of the enemy. From Craonne to Cerny the narrow upland was at last French; but from Cerny on west to Laffaux (northeast of Soissons) the disheartened Frenchmen clung desperately to the margin of the plateau, their backs to the brink of the deep Aisne trench, their great offensive stopped dead despite a terrible sacrifice in human lives. That the Germans had suffered heavily was poor consolation for those whose hopes of a decisive and overwhelming victory had run high. A crisis in the military affairs of France supervened, and the morale of the French army was for the moment shaken to its foundations. The strength of the Aisne defenses was indirectly threatening the very life of the great Republic.

The Third Battle of the Champagne

Two attempts at the Somme-Py gateway having failed to gain important results, it was now determined, in conjunction with the attack on the Chemin des Dames, to assault the chief buttresses of the Champagne front. If Brimont butte and the Moronvilliers massif could be seized and securely held, a converging fire could be directed on Mont Berru, between the two, and its capture rendered comparatively simple. The operations against Brimont were undertaken in connection with the assault on the Chemin des Dames, the Fifth Army attacking from the western part of the plateau, past the Berry-au-Bac crossing of the Aisne, to the Rheims sector. Its objective was not only the capture of the butte of Brimont but also the conquest of two smaller hills near Berry-au-Bac which blocked the line of advance into the plain behind the Moronvilliers massif. The smaller hills were seized after a fierce struggle, but the important height of Brimont, although closely threatened and even reported captured in some accounts,[23] resisted every effort to take it.

The assault against the all-important Moronvilliers massif was delivered along a 10-mile front on April 17, the day following the initiation of the Chemin des Dames attack. The dark forest patches cloaking the slopes had been mowed down by prolonged bombardment, and the fresh chalk, plowed up by the shells, gave to the stripped surfaces the appearance of snow-clad hills (Fig. 75). Sweeping up the slopes with irresistible *élan* the French Fourth Army reached the crest of the heights. The advance, swift as it was, was not accomplished without heavy losses. Enjoying the incalculable advantage of direct observation from the heights, the Germans were able to direct an accurate artillery fire upon the attacking columns. Tunnels through the chalk hills connected advanced positions with the rear, enabling the defenders to fight viciously to the last moment and then to withdraw underground to the more protected northern slopes. As in the chalk and limestone formations of other regions, however, the numerous tunnels, subways, deep dugouts and caverns were, with the increasing power of heavy artillery and the greater swiftness of enemy advances following the destruction of defensive trench systems, becoming fatal man traps. It is reported that some of the tunnels through the Moronvilliers massif were so blocked by débris from the bombardment that hundreds of Germans in them were suffocated,[24] while many more

[23] Giraud, p. 500.
[24] W. M. Davis: A Handbook of Northern France, Cambridge, Mass., 1918, p. 107.

were captured in the labyrinth of excavations in the chalk. But the difficulty of the task exhausted the force of the drive. When the enemy's positions on the reverse slopes were encountered, the weakened powers of the offensive were inadequate to the task of overcoming defenses which had not been destroyed by the bombardment because not visible. Again the natural obstacle had blocked the path to victory.

It might appear that possession of a dominating crest would alone be sufficient to assure the holder effective observation over the plain beyond and to render his control of the heights secure. But such is by no means the case. When the enemy is close to the crest he is constantly destroying observation posts or rendering them untenable, with the result that it is extremely difficult to watch his movements. There is constant danger that he may mass forces without his plans being detected and launch a surprise attack strong enough to sweep the defenders of the crest back into the lowland. The only way to avoid such a calamity is to keep large forces always at hand, ready for any emergency. But this is costly, for it means a big casualty list during all the time the men are concentrated so close to the front; and their numbers, being immobilized, are unavailable for service elsewhere. Thus the holding of a ridge when the crest alone is gained is both expensive and uncertain, at the same time that the value of the ridge for observation is enormously diminished. Only when the heights are securely held and the enemy pushed so far back from the observation posts as to be able to interfere with them but little, can the operation against a dominating crest be regarded as successful. This success had not been achieved at Moronvilliers. The enemy had, however, been deprived of his view to the south, and he was anxious lest the French might maintain sufficient observation northward to cause him serious embarrassment. "The loss" of the heights, writes Ludendorff, "was a severe blow, as they afforded a view to the north, right over the whole country."[25] The enemy's hold on the vital line of hills

[25] Erich von Ludendorff: Ludendorff's Own Story, August, 1914–November, 1918: The Great War from the Siege of Liège to the Signing of the Armistice As Viewed from the Grand Headquarters of the German Army, 2 vols., New York, 1919; reference in Vol. 2, p. 26.

buttressing his front and barring the advance of the Allied armies was shaken but not broken.

A great offensive intended to pave the way to a speedy ending of the war had been directed against the strongest natural obstacles on the Aisne-Champagne front, and had stopped short when it struck the stone wall. Nivelle was replaced, and a newly organized French high command set about solving the problem of breaking the German defense.

Seeking to profit by the grave crisis in the morale of the French army, and aware of the incalculable value of the topographic positions which had been partially wrested from them, the Germans during the space of several months launched repeated counterattacks against the Chemin des Dames and the Champagne positions. It has been stated that no less than 49 separate enemy divisions were employed in attacks against the Chemin des Dames; while 16 assaults by shock troops were launched against the Moronvilliers massif within ten days. But the obstacles which resisted the supreme efforts of the French were not to fall to German troops, and the enemy's heavy sacrifices netted him but slight gains.

The Second Battle of the Chemin des Dames

A brilliant operation carried through in the best French style on October 22, 1917, although small as to the territory directly affected, was large in its consequences. It has been noted (p. 244) that northeast of Soissons the plateau projects northward where the Ailette valley makes its double bend near Chavignon, in such manner as to give commanding observation up the valley of the Ailette for a distance of 15 miles. This vital point had long before been crowned by Fort Malmaison. As the French front was now only about 2 miles south of the Fort Malmaison spur, an advance of a little more than that distance would bring under directly controlled artillery fire all the lines of communication leading from the German positions on the Chemin des Dames northward across the Ailette valley. Caught in the rear by an enfilading fire, and with his communications cut, the enemy would be compelled to evacuate the Chemin des Dames and fall back north of the Ailette to the last remnant of the Marne plateau. The peculiar form of the terrain offered the opportunity of capturing 18 miles of the strongest part of the whole German line, at the price of advancing two or three miles on a five-mile front.

A violent bombardment lasting several days, followed by the use of tanks on the level upland, smashed the enemy's surface defenses across the plateau. Preceded by airplanes flying low and firing on the defenders, the French swept forward, seizing fortified quarries, caverns, trench systems, Fort Malmaison—a whole vast complex of defensive works—in a single day. On the morrow the advance was pushed down to the foot of the slopes, and French observers, secure on the plateau spur, looked away up the valley behind the German lines. The sequel was not long delayed. A few days later the enemy evacuated the Chemin des Dames plateau and withdrew north of the Ailette. At last the whole of the formidable Aisne-Chemin des Dames barrier was in Allied hands. The enemy had, after three years of bitter struggles, been pushed back upon the last rampart of the Marne plateau.

The Third Battle of the Chemin des Dames

After Ludendorff had driven his great wedge across the plain of the Somme in March, 1918 (p. 170), and followed with a second drive across the plain of Flanders (p. 75) without wresting victory from the Allies, he sought to obtain a decision by a third gigantic effort in the last days of May. Surprise was the fundamental element of his program, and to obtain it he spared no sacrifice. He selected as his third field of operations one of the strongest natural positions on the whole Allied front, the Chemin des Dames, which the Germans had lost the preceding autumn. Ludendorff was under no misapprehensions as to the difficulties of the terrain, although he considered it as less formidable than that of the Italian Alps which had been successfully overcome at Caporetto. If he could surprise the Allies by an attack on what was by nature their strongest front, at a time when they had stripped it of men to defend weaker and apparently more seriously threatened parts, success might be achieved.

Ludendorff's calculations were only too well founded, his preparations for the surprise only too skillful. The drive across the Somme plain had already come perilously near to separating the French and British armies, and a resumption of the offensive there might have disastrous results. Men must be secured to reinforce the danger spots. So formidable a position as the Chemin des Dames, with the natural moat of the Ailette in front of it, could be left to weak holding forces, and exhausted units from active sectors of the front sent there to recuperate. French forces were accordingly withdrawn and ordered west to Amiens, while from the British front five worn-out divisions were moved to the Chemin des Dames region during the month of May. Ludendorff must have smiled grimly when he received intelligence of this fact.

There remained only the necessity of maintaining the utmost secrecy as to the plans of attack. Because he enjoyed the advantages of interior lines, Ludendorff could concentrate his forces within a reasonable distance of the front without betraying any designs upon the Chemin des Dames; for the point of concentration was equally adapted to a supposed renewal of the Somme offensive. At the last moment the assault troops were moved to the front under cover of darkness and concealed during the day in the woods, caverns, and houses. No unusual movement on the roads was permitted when it was light, and it has been said that the Germans purposely avoided any effort to prevent Allied airplanes from surveying their back areas, in order to make still stronger the impression that here there was nothing to be concealed. Everything depended on the element of surprise, for Ludendorff well knew that if the Allies suspected an attack and were prepared for it, he could not hope to progress far against the imposing series of barriers which lay before him—the Ailette trench, the great wall of the Chemin des Dames, the deep trench of the Aisne with its unfordable river, the Villers-

Cotterets ridge, and, beyond that, other parallel trenches and ridges.

On May 27 the storm broke with a bombardment said by veterans to be one of the heaviest of the war. The whole region was soaked in gas, and masks had to be worn miles behind the front.[28] On the plateau the surface was blown into a maze of great craters which intersected each other, making a picture of wild confusion impossible to describe. The surprise was complete, and the weakened defense simply melted before the German onslaught. No definite information concerning the impending storm had reached the Allies until the 26th, only a few hours before the attack and too late to be of any great service. The enemy in overwhelming force swept everything before him. There was no time to rally on the successive lines of defense. In a few hours the Chemin des Dames, most formidable of barriers, was overrun, the Aisne trench passed and left behind, the Vesle crossed. By the 30th the Germans had reached the Marne.

During such a sweep, when adequate forces for the defensive are lacking and those available are thrown into indescribable confusion, little resistance can be expected on even the strongest natural defense lines. It is only when the blow has spent itself, when the retiring forces are in a measure reconstituted, and after fresh reserves have taken their place in line, that the defensive advantages offered by Nature can be turned to account with any great measure of success. Yet even as the German flood rushed on we find it suffering checks here and there, where an island of resistance skillfully organized along some valley or on some dominating crest delayed its progress. In crossing the Aisne, along the south bank of which a temporary stand was made by the French and British, great numbers of Germans were killed. The French held the enemy at bay for a time on the Chivres bastion. Southwest of Soissons they checked him again along the deep ravine of the Crise, which prolongs to Soissons the southwest-northeast line of the Ourcq. The Germans en-

[28] A. Conan Doyle: The British Campaign in France and Flanders: 1918, Vol. 5, London, 1919.

deavored in vain to open out their new wedge to the west, to filter down the Ourcq valley, and to seize the Forest of Villers-Cotterets, of such inestimable value for shielding concentrations of troops from enemy eyes; and Ludendorff in his memoirs complains bitterly of the incapacity of the army headquarters responsible for the operations in this sector. Fierce fighting occurred at Longpont, where the wall of Villers-Cotterets ridge is trenched by the Savières branch of the Ourcq.

As the British fell back to the south and east, yielding the lower valley of the Ardre to the enemy, the mountain of Bligny, commanding a magnificent view of the whole length of the valley and of the roads crossing it from north to south, became the scene of a bitter struggle. Held by the British, the "mountain," really a plateau spur between two branches of the Ardre (Fig. 62), brought German communications and battery positions under heavy fire. One German assault after another failed to dislodge the British from the crest. Realizing the critical value of the height and that his further operations must suffer seriously in case his back areas were observed and harassed from its summit, the enemy returned to the attack in overwhelming force and carried the hill by storm. With even greater determination the British in their turn stormed the coveted height, only to lose much of it again when their line was pressed farther back toward the Rheims bastion. As it was Allied observation along the length of the Ardre valley which grievously handicapped the German plans for their final offensive of July 15 and helped to turn it into a great Allied victory (p. 309), one can appreciate the willingness of both sides to pay dearly for possession of the dominating height.

Early in June two American divisions were thrown into line to relieve the weary Frenchmen who had checked the enemy along the impressive trench of the Marne from Château-Thierry eastward to Dormans and along the smaller trench formed by a branch of the Clignon northwest of Château-Thierry. At Bouresches and Belleau Wood, on the southwestern slope of the smaller trench, the Americans later defeated and threw back the enemy in brilliant local engagements. Ludendorff remarks that the Americans "were unskillfully led, attacked in dense masses, and failed."[27] Similar "failures" were soon to drive the hosts of militarism from the terrible country of the Argonne Forest, albeit at heavy cost.

When the German drive was finally halted, it had broken against the Ourcq Valley-Villers-Cotterets Forest barrier on the west, the Marne barrier from Château-Thierry to Dormans on the south, and the forested marsh and lake country of eastern Brie and the Mountain of Rheims on the east. Violent efforts to widen the salient by piercing the western and eastern barriers were unavailing, and an attempt to smash the Rheims stronghold, now projecting like a bastion into the German front but guarded by the plateau bastion towering above it, was a complete failure.

The Battle of the Oise

Early in June, 1918, Ludendorff made a determined effort to capture the Allied wedge projecting into his line between the Montdidier salient produced by the drive of March 21 (p. 170) and the Marne salient produced by the drive of May 27, just described. The apex of this wedge (Fig. 55) was protected by the western continuation of the Chemin des Dames plateau strip, the Lassigny massif, the Forests of Laigue, Compiègne, and Villers-Cotterets, and the trench of the Aisne continued westward by the trench of the Aronde; and was bisected by the valley of the Oise. This time the Allies were not taken by surprise, so it was no occasion for wonder that the drive of June 9, down the Oise and to the west of it, unlike its predecessors soon encountered a vigorous resistance. The energy required to break through the "very strong positions on the high ground just west of the Oise," as Ludendorff calls the Lassigny massif, so weakened the offensive that it was halted along the trench of the Aronde and thrown back some distance to the north. On the spurs of the plateau at Tracy-le-Mont and westward along the northern margin of the Forest of Laigue, where the Oise valley turns

[27] Ludendorff, Vol. 2, p. 269.

eastward to reinforce the obstacle, even less progress was made. In a few days the German high command gave up the task as hopeless and ordered the attacks to cease.

The Second Battle of the Marne

There could be no doubt that the invader would soon launch another drive in the desperate attempt to snatch victory from the stubborn Allies. Ludendorff would make one more gigantic effort to overthrow Germany's adversaries and thus let victory and world power, instead of downfall, be the portion of the German people. He must act quickly, for the American army was landing on the shores of France at the rate of from 200,000 to 300,000 men a month, and the "failure" near Château-Thierry had taught him respect for the new antagonist. In feverish anxiety to get his great military machine in motion with the least possible delay, he grew careless of those rigid precautions of secrecy which had contributed so greatly to the success of his earlier drives. Foch soon knew that the last throw of the great gambler's dice would be made on the tableland of the Marne.

The day of July 14 passed quietly. Late that evening people in a château 10 miles west of Paris, on the northeastern slope of the ridge crowned by the Forest of Marly, heard the dull rumble of distant artillery, and, looking forth from windows facing northeast, saw the whole horizon illumined with a ruddy glow which constantly flickered and flashed with yellower hues. It was the heavy artillery along the Ourcq and Marne, more than 50 miles away. Ludendorff was blasting the way for his last great offensive with a hurricane of steel never before equaled in intensity. After a preliminary feint on the western border of the Marne salient, he would strike with all the available might of the German army against the barriers on the eastern side—the lake and forest country of eastern Brie and the Mountain of Rheims massif; while simultaneously a drive would be launched southward over the level chalk plain of the Champagne farther east. When the two jaws of the giant pincers met at Épernay or Châlons, the impregnable bastion of Rheims with its entrenched camp, subterranean fortresses, and overshadowing mountain stronghold would be swallowed up, a wide breach would be opened in the Allied lines, and the ultimate defeat of the Allied armies made possible.

Foch was accurately advised as to the enemy plans. Excellent observation from critical points on the terrain (p. 246) and from the air, the increasing frequency of ammunition dump explosions in the enemy's back areas provoked without increase of bombardment, data elicited from captured prisoners, and other sources of information, enabled him to prepare for the onslaught. Pershing had already urged the employment of American troops in an attack on the exposed Marne salient,[28] and conditions were now propitious for putting such an operation into execution. The American 1st and 2nd Divisions and a famous French Moroccan division were selected to form the spearhead of the attack. Massing men and tanks at the last possible moment under the concealment of Villers-Cotterets Forest, Foch took the steps necessary to drive eastward, parallel with the ridge and valley barriers, as soon as the mass of the German reserves was fully engaged elsewhere, thus at one stroke to put a great army behind the enemy forces on the Marne and those attacking the eastern side of the salient. If the Allies could push rapidly eastward along the Aisne trench at the base of the salient, to meet their comrades advancing down the Ardre and Vesle, a worse than Sedan would end all German hopes.

To meet the blow in the Champagne, the Fourth Army under Gouraud prepared to withdraw from the Moronvilliers massif, where their precarious position on the crest without control of the northern slopes and without the security necessary for the most effective observation, threatened to involve the whole front in trouble. The magnificent observation from the Rheims Mountain bastion, by means of which every road leading southward from many miles of the German front could be brought under accurately controlled artillery fire, assured a powerful reaction

[28] Final Report of General John J. Pershing, Commander-in-Chief, American Expeditionary Forces, *Annual Repl. War Dept. for 1919*, Vol. 1, Part I, pp. 547-642, Washington, D. C., 1920, pp. 580, 581.

against the enemy's advance over the western part of the plain. Farther east tree-top observatories on low mounds or hills would be used wherever available, while forward posts on the Moronvilliers massif, maintaining their positions as long as possible and communicating with the real front by means of wireless telegraph, signals, carrier pigeons, and volunteer runners, would give valuable information as to the movements of enemy columns. Advanced units were to make a show of covering a hasty retreat, while the army in reality had taken its stand on the level plain farther south, there to surprise the onrushing enemy with a perfectly organized resistance.

The success of the defense shattered the whole German plan of campaign. Ludendorff's shock troops hurled themselves in vain against the marshy woods of eastern Brie, the badly dissected plateau of Tardenois, the impregnable bastion of Rheims Mountain. French, British, and Italian troops gave ground slowly under the initial blow; then, solidly based on the difficult terrain, brought the attack to a standstill. Along the Marne barrier on the south the right wing of the German attack collapsed with equal promptness. Between Château-Thierry and the bend of the river at Jaulgonne the American 3rd Division bore the brunt of the shock. "Day was just breaking; and through the mist, fog, and smoke one could see the boats and rafts loaded to the gunwales with enemy infantrymen and machine gunners set out for the southern bank. That was about 3.30 o'clock. Yet not one crossed that day in the center of the sector. . . . Scores of those boats were shattered and sunk or else disabled and sent drifting harmlessly down the river."[29] In the Jaulgonne bend and farther east the enemy succeeded in throwing a number of pontoon bridges over the barrier, and spread out upon the edge of the plateau to the south. But here the French, employing their "yielding defense" tactics and aided by the American 28th Division, soon halted the advance. At the western terminus of the Jaulgonne bend one regiment of the American 3rd Division,

[29] Lieutenant Lovejoy, quoted by Shipley Thomas in: The History of the A.E.F. New York, 1920, p. 118.

although outflanked on both sides by enemy forces which had crossed the river, successfully defended the barrier at its most critical point. "To this regiment had been entrusted the front where the Surmelin River, flowing due north down a narrow valley, emptied into the Marne, and along both sides of this river lay two splendid roads leading to Montmirail, upon which the Germans had planned to transport their artillery and supply trains for the exploitation of the first success. The Surmelin River valley, then, with its two roads, was the crucial point in the whole attack; and the 38th Infantry was holding this front with the object of preventing any pontoon bridges from being thrown across the [Marne] river to connect with these roads. . . Time and time again the Germans in boats and pontoons tried vainly to land, but each time the boats were sunk in midstream either by rifle fire or by hand grenades."[30]

The enemy troops which had passed the Marne barrier to the east and west of this point found themselves in an impossible position. Checked on the slopes of side ravines which they could not traverse under the heavy fire from above, and with the crossings over the river behind them being shelled, they hung on desperately, waiting for relief. Ludendorff emphasizes the "remarkable achievement" of crossing the Marne barrier and the tremendous difficulty of extricating the troops which performed the feat when it appeared they were in a trap the exits from which were under heavy enemy fire. No one who saw their twisted and torn bodies, crumpled in weird postures and scattered thick as flies over the exposed upland and valley walls, will begrudge them Ludendorff's tribute: "The troops on the Marne had lived through days of severe trial and behaved like heroes."[31]

East of Rheims the smashing of the great German attack was even more abrupt and decisive. Gouraud's Fourth Army, including the American 42nd Division, brilliantly executed the defensive measures previously prepared to break the force of the enemy's supreme blow. Swarming down the southern slopes of

[30] Thomas, The History of the A.E.F., pp. 120–122.
[31] Ludendorff, Vol. 2, p. 31.

Moronvilliers and over the plain, in pursuit of an apparently defeated foe, the Germans in the Champagne suddenly staggered under a hurricane of shells from unsuspected batteries which slaughtered them like rabbits driven into a pen. Looking north over the plain from a swaying tree-top observatory next morning one could see here and there a monster German tank, silent and motionless in front of the French line, mute symbol of a giant offensive that had halted short of its objective. The shells were still screaming overhead, but the greatest enemy attack of the war had been stopped dead in its tracks.

Then Foch launched his counterblow from the shadows of Villers-Cotterets Forest. The inadequate screen of holding troops on the western side of the Marne salient swayed under the impact and yielded ground. Soon the forces attacking the eastern side heard the roar of guns growing louder in their rear. Halted on the Marne and in the Champagne, checked in the eastward drive, and threatened with disaster in the rear, the much-heralded German "Friedensturm" collapsed utterly. Orders to abandon the offensive and fall back to the strong defensive position of the Aisne-Vesle trench were issued.

No nearer line was available to the defeated Germans. French and American forces already stood astride the Villers-Cotterets ridge ready to take in the flank an enemy who should try to halt along the eastern half of the crest. French and British commanded the whole length of the Ardre from their observation posts and were astride its upper branches. The Mountain of Rheims had remained securely in Allied hands. The first strong natural barrier not flanked by the Allies was the Aisne trench prolonged by the Vesle; and to it the Germans now beat a retreat, closely pressed by their French, British, Italian, and American pursuers.

The Allied forces driving eastward south of the Aisne unfortunately experienced considerable delay in overcoming the obstacle formed by the Crise valley and its side ravines, the Missy ravine next west, and the exposed plateau spur between these two natural trenches. The difficult terrain was well defended, and failure to cross it promptly deprived the Allies of the chance to cut off a great part of the German armies caught in the salient. The British advanced down the Ardre far enough to reconquer the mountain of Bligny with its valuable command of the valley and so to harass severely the enemy's retreat in that section; but they could not go far toward closing the Marne salient before the Germans escaped. By August 1 the salient was extinguished, and the badly shattered armies of Ludendorff were reorganizing under the protection of the Aisne-Vesle trench.

A few weeks later one of the mighty blows that were now smashing in first one part, then another of the German front, struck the sector crossing the plateau obliquely from the Aisne west of Soissons to the Oise at Ribécourt. Here the French in two days drove the enemy behind the protection of the Oise-Ailette trench from Noyon eastward to near Coucy-le-Château. German forces pinned back against the Oise barrier experienced great difficulty in extricating themselves under fire and escaping to the northern bank. After the barrier of the Vesle had delayed the advance of the Allies in that sector for some time, it was breached by the Americans near its junction with the Aisne. Two weeks more of hard fighting were required to dislodge the enemy from this strong position and to force him back behind the Aisne trench east from Soissons. Under the never-ceasing pressure of the Allies the Aisne trench east of Soissons was soon given up and a further retirement effected to the line of the Ailette as far east as the point where the Oise-Aisne Canal turns south to pierce the plateau divide; east of that, to the crest of the Chemin des Dames. When the difficult task of dislodging the enemy from the Chemin des Dames plateau strip, which he could no longer defend with his old-time energy, was finally achieved, the victorious Allies stood before the last of the plateau strongholds, the northernmost upland strip of the Laonnois, defended by the natural moat of the Ailette. Its capture from the defeated and disheartened German armies was now possible, but only at a heavy sacrifice. It could more easily be secured as the fruit of offensives pushed across less formidable terrains.

By early October attacks across the level plain of Champagne had proved so difficult to support that an extensive retreat to the Hunding-Brunhilde Line was decided upon. Beginning September 26, in unison with the American offensive in the Argonne, the French Fourth Army under Gouraud had struck a series of sledge-hammer blows for five days, scoring daily advances until caught in the left flank by German fire from the Moronvilliers massif, a position naturally so formidable that Gouraud sought to turn it instead of attacking it. This checked his advance, and along the trench of the Py River, dominated on the north by the chalk ridge of Notre Dame des Champs and Blanc Mont, the enemy fought furiously to stop the turning movement which would flank him out of the higher Moronvilliers massif. Only at heavy cost was the Py finally crossed, the chalk ridge taken by French and American troops in a combined frontal and flank attack, and the advance continued. Moronvilliers, rendered untenable, was abandoned, and the whole German front began to crack. It was then that the great retreat was ordered. First to the line of the Suippe valley, then to that of the Retourne, the retrograde movement proceeded by definite stages. Finally across the whole sweep of the Battlefield of the Marne the enemy fell back to his strong northern defenses behind the marshy valley which traverses the chalk plain from Sissonne to the Serre at Froidmont, and the trench of the Aisne from Château-Porcien upstream past the gateway of Rethel to the Argonne Forest barrier. On October 13 the Germans abandoned Laon. The Marne plateau was now wholly free of the enemy, and the Champagne soon saw the last of his rearguards.

CHAPTER VII

THE BATTLEFIELD OF VERDUN: THE CUESTA-AND-LOWLAND BATTLEFIELD

A short time ago you stood on the crest of the chalk escarpment far east of the central Marne plateau, the dreary wastes of the Dry Champagne behind you, the fairer prospect of the Wet Champagne in front. As you gazed down upon the verdant plains some two hundred feet below, watered by countless pleasant streams and sheltered on the east by the dark slopes of a forest-clad mountain barrier, the scene was one inviting exploration. Thirty miles eastward would bring you to immortal Verdun; forty more to the fortress of Metz. Straight to these distant objectives, across the grain of the country keep your unchanging course, and you may easily read in the succession of sloping upland, steep scarp, and level lowland, the secret of the part which the Argonne, Verdun, and St. Mihiel played in the World War.

Leave, then, your viewpoint on the upland of the Dry Champagne at the west (Fig. 79), descend the white slopes of the chalk scarp, and cross eastward over the Wet Champagne. Your shoes, heretofore powdered white with the dry chalk dust, may soon receive a baptism of mud, for you walk on the belt of soft clays and marls which were quickly worn down to form the depression, and water stands long on the low, impervious surface. Shallow lakes and ponds, the *étangs* of the French, compel occasional détours, and many streams flow eastward from the chalk scarp to meet the Aisne running in a deeper trench along the eastern edge of the lowland. Forest patches occur, but for the most part you pass cleared lands and culti-

NOTE. For Chapters VII and VIII the reader should constantly consult the detailed map of the battlefield in the pocket (Pl. IV) and the block diagrams (Figs. 84 and 94).

vated fields stretching far away to the north and south, like an open pathway five or six miles broad, between the pine-planted chalk upland on the west and the massive forests of the Argonne on the east—a curved pathway, since its northern and southern extensions bend strongly back toward the northwest and southwest, conforming to the curved outcrops of the Paris Basin rocks.

Beyond the Aisne, perhaps even before you reach it (Fig. 78), you begin to ascend what was once the gentle backslope of a thick sandstone formation, now deeply dissected by stream erosion. At once you plunge into the dark shadows of an almost unbroken forest, for the steep valley walls, the isolated upland surfaces, and the sandy, infertile soil are neither easy nor profitable to cultivate. This is the savage country of the Argonne Forest. Penetrating the woodland maze you climb one precipitous valley wall after another only to cross dry, stony uplands where cooling breezes murmur among the trees, and to descend abruptly into other ravines in whose damp depths silence reigns unbroken. The way is exceedingly arduous, and when you have pushed painfully on to the high eastern crest, you are only too glad to see open country again, spread out below you in the lowland drained by the River Aire. The eastern slope of the highlands on which you stand is straight and steep, in strong contrast with the western slope you have just ascended; for while both are badly cut by valleys and ravines, you have traveled eight miles in a direct line from the Aisne valley to reach the crest, whereas the Aire River, seemingly just at your feet, is in reality little more than a mile away. Clearly the westward-dipping sandstone formation has been eroded to form one of those asymmetrical plateaus, having one gentle and one steep slope, which are so common in this and other regions and are known to geographers and geologists as "cuestas."

As in the case of the eastern scarps of the Marne plateau and the Dry Champagne, the steeper slope of the Argonne cuesta faces toward Germany. To the northwest its forested

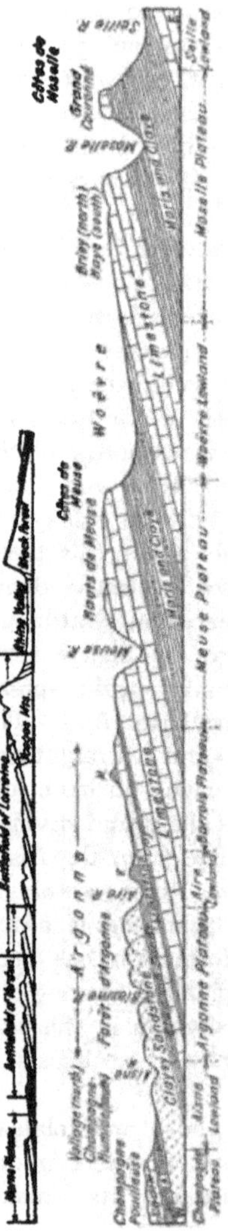

FIG. 78—Ideal east-west section across the Battlefield of Verdun, showing the relation of the rock layers to the asymmetrical plateaus (cuestas) and lowlands. The successive plateaus and lowlands are named below the diagram, while the French names for different parts of the terrain are given above. V = butte of Vauquois; M = butte of Montfaucon. The small cross-section above affords a key to the relation of Figs. 58, 78, and 99 and forms a continuous section from the Marne plateau to the Black Forest.

mass stretches for many miles, occasionally cut through from east to west by a river gateway; but southward, contrary to the rule in the Paris Basin, the upland suddenly drops down to the level of the plain and abruptly vanishes—a mystery easily understood when it is found that in this direction the resistant sandstone formation thins out until it disappears. The weak sands and clays upon which the Aire lowland was eroded are then in direct contact with the overlying weak marls and clays of the Aisne lowland; and the result is a very broad depression giving a much expanded "Wet Champagne." The forest barrier of the Argonne is not entirely lacking in the southern region, however. An infertile belt of the Wet Champagne, continuing the curved axis of the Argonne, is covered with great forest patches and innumerable lakes, constituting a zone of wild land ten miles wide lying between two belts of predominatingly cleared land on the east and west.

Plunging down through the forest of the eastern scarp of the Argonne upland, you cross the open fields of the narrow Aire lowland, and again begin to ascend the more gentle western slope

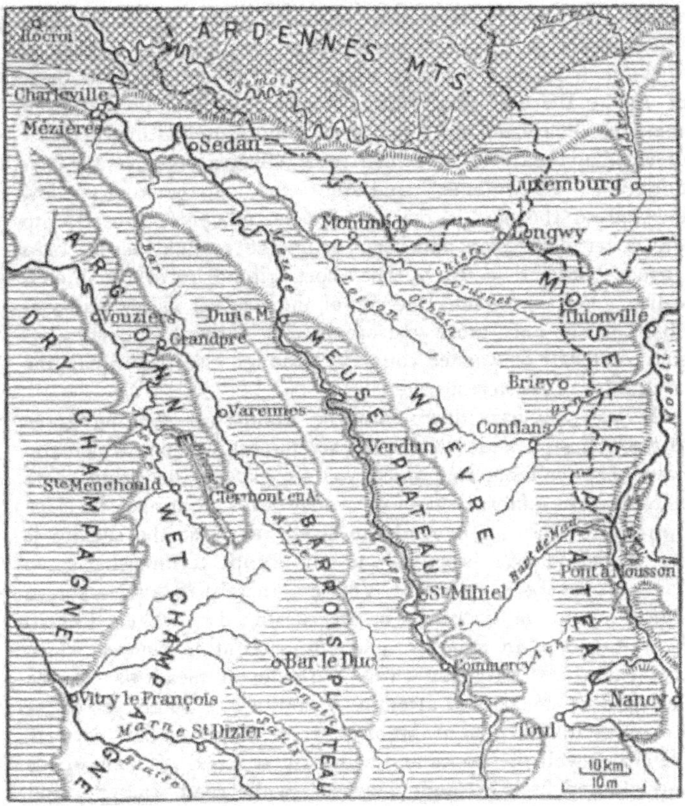

FIG. 79—Generalized sketch map of the Battlefield of Verdun. White areas are lowlands, ruled areas uplands, cross-ruled areas mountains. For topographic details and place names referred to in the text, see Pl. IV and Figs. 84 and 94.

of another asymmetrical plateau, or cuesta. The country is less rugged than in the Argonne, and woodland areas are less continuous, although extensive forests do occur. When you reach the crest, some distance west of Verdun, the eastern face is again strikingly abrupt, but the lowland beyond is shallow and poorly developed. Evidently you are on a minor limestone layer forming a subordinate cuesta on the backslope of a much grander example which you can see rising in front of you to the east (Fig. 78). Pausing only to note that even the subordinate upland constitutes a more or less distinct belt of woodland patches, in which some villages preserve the termination "-en-Argonne," and that the narrow and shallow lowland to the east is prevailingly cleared, you descend into the latter depression and begin again the gradual ascent of a surface badly dissected by the short tributaries of the Meuse valley. Before reaching the rim of the western valley wall you observe that once more you are passing through a northwest-southeast belt of wooded country, although clearings break the continuity of the forest cover.

At last you gaze down into the wonderful, winding gorge of the Meuse. Evidently the massive limestone upon which you have been walking, and which you see exposed here and there in cliffs or in quarries cut in the valley walls, continues to slope upward to the east beyond the river; at least the surface is still higher in that direction, and the abrupt terminating scarp has not yet been reached. You are merely looking upon a trench cut in the backslope of the cuesta. But what a trench! How it sweeps in majestic, graceful, serpentine curves, which the little river wriggling through the open meadows on the valley floor makes no attempt to imitate! Surely a larger stream than the present Meuse cut the larger valley pattern. On either wall the slopes and spurs are stripped of their woods and planted in fields and vineyards, while road, railway, and canal show that the northwest-southeast valley clearing is an important highway lying between the forested uplands on either side. Guarding this highway, just where the great east-west route from Rheims to Metz crosses it, the picturesque old town of Verdun sits silent in the center of its ring of fortresses, a sleepy sentinel which, roused by danger, will bellow defiance at the enemy who would pass by "the great white way."

Verdun must again be visited for a more careful study; but now you press onward, up the eastern wall of the Meuse trench, and over the upland beyond, where you traverse the somber shades of another forest belt. Some eight miles of wilderness, perhaps, and then in the twinkling of an eye there bursts upon the delighted vision a panorama of green fields, shining lakes, and woodland patches (Figs. 80 and 81) to which no pen can render justice. You stand on the crest of the limestone cuesta, and more than five hundred feet below you, a gigantic map in natural colors spread out at your feet, is the clay plain of the Woëvre. Picturesque villages nestle at the base of the steep scarp, where springs burst forth above the impervious clay: Ornes,* Eix, Vigneulles, Heudicourt, Apremont. The white strips of road stretch straight across the level plain, cutting narrow trenches through the woodland areas and passing other villages whose names awaken a thousand memories. Here and there an outlying erosion remnant or butte, like Mont Sec, reminds one of the forces which are constantly sapping the face of the scarp.

Down into the plain of the Woëvre and eastward over its muddy, marshy, lake-dotted surface you pick your way, wondering how men could ever live in the low morass while enemy guns poured fire on them from above. Slowly you begin to rise, as once more the land ascends gently toward another east-facing scarp. After some miles you find firmer footing on another massive limestone formation coming up from beneath the clays (Fig. 78). But the lowland is unusually broad, and you may travel ten or fifteen miles, in some places much more, before the forests become closely spaced and the streams flowing eastward to join the Moselle cut deeply enough into the rising land to dissect it into broad-topped plateau remnants. Suddenly the great trench of the Moselle yawns at your feet. But you press on across it, till you stand once again on the crest of a steep escarpment which dominates a vast lowland of green fields and forest

* Unless otherwise stated, places named in Chapters VII and VIII may readily be located on Pl. IV, along the plateau scarp, river, or other topographic feature with which the names are associated in the text.

FIG. 80—The level surface of the Woëvre lowland as seen from the crest of the Meuse plateau scarp, looking southeast from just above the village of Heudicourt (in the right foreground), seven miles northeast of St. Mihiel. In the middle distance, just beyond the forest, is the village of Essey, this side of which may be dimly seen the smoke of an American battery, concealed in the forest, shelling the retreating Germans during the last Battle of St. Mihiel Salient.

FIG. 81—The Woëvre lowland, looking northeast from the crest of the Meuse plateau scarp at Fort Gironville, four miles northeast of Commercy. In the left distance is the detached butte of Mont Sec, organized into a German stronghold. The first "American front" stretched from left to right across the plain just beyond the forest patches near the center of the view. Note the effective "command" of the roads on the plain enjoyed by observers on the plateau crest. The wire entanglements in the foreground form part of the defensive position of this front, based on the plateau crest.

FIG. 82.—View northeast from the lower slopes of Mont d'Amance, eight miles northeast of Nancy, across the Seille lowland to the Côte de Delme, which forms the sky line. After the Battle of the Grand Couronné the Germans retired beyond the marshy barrier of the Seille River in the center of the lowland, where they were supported by the strong Côte de Delme position, taken from the French in 1871 because of its tactical value.

patches (Fig. 82). Over the level expanse the small river Seille slowly takes its uncertain way. Just to your left rise the towers of Metz, while far away to the south Nancy lies at the base of the scarp, concealed from view by a protruding spur of the upland. On the eastern horizon the land is evidently sloping upward to the crest of yet another cuesta scarp. But you have reached the limit of the terrain which we shall wish to include in the Battlefield of Verdun and have journeyed far enough to fix firmly in mind the fundamental elements of its topography.

It is clear that the rock layers of the Paris Basin, alternately resistant and weak, and dipping gently a little south of west, have caused the forces of erosion to produce a systematic alternation of plateaus and lowlands, parallel to each other and trending northwest-southeast. All of the uplands are of the asymmetrical, or cuesta, type with the steep scarp directed toward Germany, the gentle backslope toward Paris (Figs. 78 and 79). Repeatedly a narrow valley trench is cut into the backslope of the upland roughly parallel to the crest. As a rule the uplands are forested, the lowlands and valley trenches cleared. Hence there is also a systematic alternation of parallel cleared and forested belts trending northwest-southeast.

It does not require a military expert to discern the defensive value of such a terrain in time of war. An enemy advancing on Paris from the east would find his way barred by a succession of formidable barriers: muddy and marshy lowlands, precipitous scarps, rugged zones of dissected plateau, belts of forest, and deep river trenches. Maneuvering in the lowland is particularly difficult when, in wet weather, the wet clay soil restricts movements to the roads and even renders many of these unfit for travel; especially since the roads, easily visible on the exposed plain, are subjected to accurate artillery fire directed by observers stationed on the commanding heights of the adjacent plateau rim. The military value of the escarpments lies not merely in the better observation from their crests and the difficulty of attacking their steep slopes, but also from the fact that they force roads and railways to converge and pass through the relatively small number of natural gateways which rivers have carved through them. By fortifying these gateways and concentrating mobile forces near them, the defense may be secured even if there are not sufficient troops to garrison heavily all the intermediate stretches of country. There the natural strength of the escarpments will offset the weakness in men; and not until the gateways are captured by the enemy can he move forward troops and supplies to press any advantage he might gain in the intervening areas. Hence the importance of the Grandpré gateway through the Argonne upland (Fig. 79), the Toul gateway through the Meuse upland, and the Nancy gateway through the Moselle upland. The opportunities for defense offered by the upland remnants of a dissected plateau, by fortified forests, and by the natural moat of a deep river trench have been dwelt upon in earlier pages and need no new emphasis here. Few regions offer a terrain where progress across the grain of the country would be more difficult in the face of a determined defense.

A hostile advance directed from the north or south, parallel with the axis of the barrier system, while avoiding some of the difficulties described above, would nevertheless find the task one of tremendous difficulty. The narrow cleared lowlands and valley trenches are corridors easy to traverse in time of peace; but in time of war no advance along them would be possible except as the wooded uplands dominating them on either side were cleared of the enemy. But this is a difficult and costly undertaking, given the power of resistance readily developed in a forested hill country. The narrow corridor of the Meuse, with its cross spurs projecting into meander loops and its forest belts frowning down from the eastern and western uplands, was to prove impregnable against one of the most terrific offensives launched by the Germans during the entire war—their six months' struggle to advance southward along the corridor to the fortress of Verdun. When the Allies sought to push their "victory offensive" northward along the Aisne and Aire corridors, the Americans were compelled to clear the Argonne ridge at a frightful cost in human lives. Verdun thus lies in the center of a terrain which is peculiarly formidable to an enemy, from whatever direction he may choose to approach.

The parallel belts of the Verdun terrain, caused by the erosion of southwest-dipping rock layers of different degrees of resistance, will repay a closer examination, after we have looked for a moment at the strategic location of the battlefield.

Strategic Position of the Battlefield of Verdun

The Meuse cuesta was the last of the asymmetrical plateaus which guarded Paris on the latitude of Verdun. After 1871 the crest of the next scarp to the east lay within German territory, crowned by the fortifications of Metz. The shortest road from German territory to Paris ran from the walls of Metz almost due westward, straight through the heart of the Verdun battlefield. Thus Verdun and its natural barriers stood at the danger point on the eastern frontier, a strong border march of France.

Northwest of Verdun the plateau scarps curve more and more toward the west, gradually converging and finally dying out west of Mézières. Until that point is reached they form successive walls bordering on the southwest a lowland route, which, passing through southern Luxemburg and Belgium into France, lies between the plateaus and the Ardennes Mountains (Fig. 79). Through this natural corridor runs a railway of the very highest strategic importance, and in or near it lie Sedan of fateful memory and the fortress towns of Longwy, Montmédy, and Mézières. But far more valuable for its defense is the formidable terrain of the Verdun battlefield flanking the corridor on the south. Entrenched among the fastnesses of the parallel plateaus, lowlands, valley trenches, and forest belts, an army of defense would so threaten the flank of an enemy desiring to advance by the corridor as to compel him to seek victory on a difficult terrain of the defenders' own choosing, or to leave enormous containing forces to protect his rear, before the route would become available. As Ludendorff tells us: "From a strategical point of view Verdun as the point of attack [by the Germans in their great 1916 offensive] was well chosen. This fortress had always served as a particularly dangerous sally-port, which very seriously threatened our rear communications, as the autumn of 1918 disastrously proved."[1]

South of Metz the German frontier in 1914 left the Moselle cuesta and turned southeast, cutting across other cuestas and lowlands to reach the massive buttress of the Vosges Mountains. Here Germany might seek to launch an invasion southwestward into France, following along the broader lowlands typical of this section. Rather than fortify their entire eastern frontier with permanent military works on a big scale, the French elected to place opposite Germany two stretches of fortified wall, leaving between them an open gateway through which the invasion would have to be directed. Limited to an opening thirty or forty miles wide and flanked on either side by impregnable buttresses heavily garrisoned, the enemy advance would lose much of its danger and could the more readily be controlled. For one of these flanking buttresses the mountainous terrain of the Vosges was selected, and an unbroken system of fortresses constructed from the Swiss border near Belfort to the southern crest of the Vosges and down the canyon of the Moselle River to Épinal. For the other, the difficult country of the Verdun battlefield was chosen, the chain of fortresses extending from Verdun along the Meuse valley to Toul on the Moselle. There is thus no Verdun-Toul-Épinal-Belfort line of fortresses, properly speaking; there are only a Verdun-Toul line, and an Épinal-Belfort line (Fig. 83). Between Toul and Épinal is the unfortified gateway in the center of which is the town of Charmes, and which is therefore known in French military circles as the "trouée de Charmes." In contrast to the unfortified zone along the Meuse north of Verdun including

[1] Erich von Ludendorff: Ludendorff's Own Story, August, 1914–November, 1918: The Great War from the Siege of Liège to the Signing of the Armistice As Viewed from the Grand Headquarters of the German Army, 2 vols., New York, 1919; reference in Vol. 1, p. 244.

the Sedan lowland, sometimes called the "Meuse gateway," the Charmes gap is also known as the "Moselle gateway."

The Verdun terrain is thus much more than an obstacle barring the direct route to Paris and a stronghold guarding the Sedan corridor. It is half of that formidable system of defense by which France sought to deter a powerful enemy from violating her eastern frontier. It was inevitable that its conquest must form a principal objective in any campaign directed from the east.

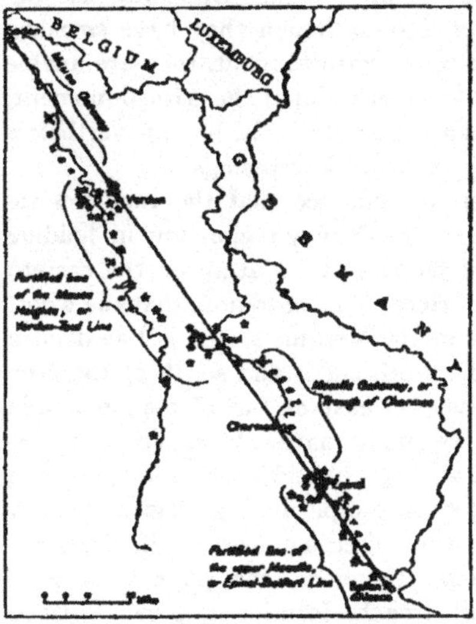

FIG. 83—The Meuse and Moselle defensive lines parallel to the German border. A straight line from the Ballon d'Alsace in the southern Vosges to Sedan is nowhere a dozen miles from one of the river barriers. The stars show the principal forts.

Both the Meuse and the Moselle Rivers flow approximately parallel to the northeastern frontier of France for great distances. If a straight line be drawn from the Ballon d'Alsace at the southern end of the Vosges Mountains to Sedan near the Belgian border, more than one hundred and fifty miles away, no part of the line will be as much as a dozen miles from one or the other of these two rivers (Fig. 83). Their trenches are by position and direction, as well as by form, well adapted to serve as defensive barriers against a German invasion. But it is not in the nature of the French, neither is it military wisdom, to condemn an army to a purely defensive rôle. Since the best defensive may be an offensive, it is vital that the defensive forces of a nation should be in position to launch an attack the moment a favorable opportunity presents itself. Hence an army, when falling back behind a river barrier, fights hard to preserve one or more "bridgeheads" on the enemy's side of the obstacle, areas large enough to keep that enemy from seizing or shelling the crossings over which the defenders may later wish to advance to an attack. Otherwise the defensive forces may discover that, although they have smashed the enemy's attacks all along the natural moat, they are unable to complete the victory by attacking him at the critical moment, because, weakened and disorganized as he is, he is nevertheless strong enough to hold the practicable crossings.

In a subsequent chapter we shall see that the very essence of De Castelnau's defense of the Nancy region lay in holding a great plateau bridgehead prepared by Nature on the eastern side of the Moselle barrier. Here we have to note that the whole system of defensive works in the Verdun sector is based on a similar natural bridgehead. Both north and south of the area in question the Meuse River lies east of the plateau scarp, in the broad lowland where we might naturally expect it to be. But between Dun-sur-Meuse (Fig. 79) and a point southwest of Toul the river has, owing to a complicated geological history which we cannot now review, acquired a course cut in the backslope of the plateau. Here, therefore, the plateau projects forward east of the valley trench, forming a great fortified wall with steep eastern face and forested summit defending the river crossings behind it. The erosion gaps at Toul and near St. Mihiel may be considered to divide the wall into several bridgeheads, of which the northernmost, opposite Verdun, is the largest and most continuous. Should the French armies be driven back to the natural trench of the Meuse, they would still be in position to pass to the offensive so long as they held any one of the several bridgeheads. Thus the strategy of defending France's eastern frontier is based in part on a striking natural peculiarity of the Verdun battlefield.

After what has been said above it will not appear surprising that the Verdun region, favored with a terrain admirably adapted to defense and possessing both in its general position and in its topographic form natural advantages of the first magnitude, should have been made one of the greatest entrenched camps of France. The ancient economic value of the Meuse valley is gone, the former political power of Verdun has vanished, and the frowning citadel of the town is today the true emblem of the region's importance as a vital frontier stronghold.

Let us now examine more carefully the successive belts of terrain which we have found systematically developed on the Battlefield of Verdun.

Surface Features of the Battlefield of Verdun

THE AISNE LOWLAND

We have already had occasion to emphasize the salient features of the verdure-clad plain eroded on the non-resistant but impervious marls and plains coming up from under the chalk cuesta. This "Champagne Humide," as the French call it, forms a very broad plain on the south, drained westward by the Ornain and Saulx, tributaries of the Marne. Here the impervious soils hold on the surface innumerable shallow ponds and lakes, and large tracts remain abandoned to forest growth. Not all of the terrain is level, or even approximately so, for the less resistant beds continuing southward the Argonne plateau-forming sandstone are sufficiently hard to give a broad zone of low hills which might be called the "Little Argonnes" after the Forest of the Argonnelles with grows upon them. Other beautiful forests, such as those of Belval, Belnoue, and Laheycourt, and a labyrinth of lakes both large and small help to make of this such a continuation of the Argonne barrier as justifies the termination "-en-Argonne" found in the names of several villages of the region.

Farther north, between the Champagne and Argonne cuestas (Fig. 79), the western part of the Wet Champagne is continued as a narrower lowland drained from west to east by the numerous tributaries of the Aisne, born in springs at the base of the chalk scarp and nourished by surface waters which cannot sink into the impervious beds of the plain. This Aisne lowland as we may properly call it, or "Vallage" as it is known to the French, is less humid than its counterpart to the south. Lakes and forests are fewer, and the broad, gently undulating surface, where covered with a fertile loam, is famous for its rich harvests, while natural prairies cover the alluvium of the valley floors. Abundant grain and hay of exceptional quality not only supply the regiments of cavalry stationed at several points in the valley and at the Camp of Châlons a few miles to the west, but encourage the raising of horses, including a breed prized in the artillery service.[1] Villages and farms are scattered over the fruitful surface, and an important railroad and highway run through the lowland from end to end. An army might traverse this natural corridor and live well as it passed.

River Barriers

But the story is not yet complete. In addition to the difficulties which a hostile advance would encounter by reason of the rugged hill country of the Monts de Champagne on the west and the Argonne on the east, from neither of which could defending forces easily be dislodged, the terrain of the lowland itself offers obstacles of some importance. The impervious and clayey soils make difficult fields for maneuvering in wet seasons and transform unmetaled roads into a succession of ruts and mud holes. The rivers tributary to the Aisne flow transversely across the plain, constituting thus an endless succession of barriers at which defending forces might dispute passage. Some among them, like the Auve, Bionne, Tourbe (French for "peat"), and Dormoise, wend their ways through marshes and peat bogs bordered by willows. Farther north the Aisne River itself and its lateral canal, crossing the lowland from Attigny on the east to Rethel on the west, constitute an obstacle of sufficient importance to have led the Germans to select it as the basis in this region of their famous Hunding-

[1] Émile Chantriot: La Champagne: Étude de géographie régionale, Paris, 1906, pp. 96–97.

Brunhilde defensive system (Fig. 55). To the south, where the Aisne lowland expands into the broader Wet Champagne, the mazes of ponds, lakes, and woods become an added obstacle. Where the interlacing channels of the Chée, Ornain, and Saulx have cut a single broad lowland, threaded by a canal and innumerable watercourses, and dominated on the south by the upland of Sermaize covered by the massive Forest of Trois-Fontaines, we have a truly formidable barrier. It was here that the German invasion of 1914 was finally halted and repeated violent offensives abruptly checked.

As on the clay plain of Flanders, so on the impervious soils of the Aisne lowland, rainfall sinks into the ground with difficulty. Hence much of it escapes as surface run-off, flooding the permanent streams and forming countless temporary brooks which disappear in dry weather. The lowland, with its southward continuation in the broader Wet Champagne, thus forms one of the gathering grounds for the floods which periodically inundate such rivers as the Aisne and Marne in their courses across the arid chalk, where tributary streams are few and surface waters scarce. The Aisne, receiving the floods brought in by its transverse tributaries in the lowland and by the streams draining the steep slopes of the Argonne upland, is particularly dangerous. Its level at Vouziers may quickly rise ten or more feet, and all efforts to control the river and confine its waters in a limited channel have failed. Hence we usually find the valley floor deserted, especially below Challerange, where the stream has acquired a goodly size, the towns and villages being perched in places of safety on the spurs of the lowland hills. So difficult, indeed, has been the problem of controlling the turbulent stream that efforts to make it navigable by canalization were finally abandoned. From Vouziers only is water traffic possible, and then not by the river but by a lateral canal parallel to its course.

River Gateways

There are certain points in the Aisne lowland and in its continuation southward which acquire considerable military importance, not so much from any features of the lowland itself as from the relation of these points to military geographical elements in the terrain of adjacent belts. Among these are the towns guarding the entrances to or exits from river gateways through neighboring highlands. Rethel and Vitry, at the entrances to the Aisne and Marne gateways through the chalk cuesta, have already been mentioned in other connections. A number of exits from deep defiles through the Argonne cuesta played a significant rôle in the World War: Attigny near the mouth of the Le Chesne gateway, Vouziers commanding the Fournelle exit, Challerange opposite the broad opening carved by the Aire, Vienne-le-Château and Vienne-la-Ville where the gorge cut by the Biesme River emerges on the lowland, and Villers-en-Argonne at the mouth of the gateway of Triaucourt which offers passage around the southern end of the Argonne highland. Revigny stands at the exit of the Ornain gorge through the subordinate Barrois cuesta. Here pass the Strassburg-Paris railway and other lines of importance, as well as the Marne-Rhine Canal, giving to the town a higher military value than attaches to the larger Bar-le-Duc, situated in the gorge itself. St. Dizier, guarding the outlet of the Marne gorge, is the natural focus of a number of important railways, highways, and canals converging upon this exit into the Wet Champagne. It has an even larger strategic significance, for with Vitry-le-François across the lowland, fortified nearly four hundred years ago by Francis I, St. Dizier guarded the passage of the Marne against enemies advancing either along the lowlands or down the converging valleys to reach the open plain of the Dry Champagne. For centuries it stood fortified against the invasions which so often menaced this part of France. Ste. Menehould, for several centuries a strong point guarding the eastern frontier of France, has much military importance today as the western portal of the Les Islettes passage, partly natural and partly by tunnel across the Argonne. This is the route of the direct railway from the entrenched camps of Rheims and Châlons to Verdun and Metz, and of a national highway, both of which at Ste. Menehould intersect the railway and highway traversing the lowland longitudinally. The value of the junction is military rather than commercial, and the activity of the little town depends in no small measure on its cavalry garrison.

THE ARGONNE PLATEAU

East of the Aisne lowland a layer of resistant sandstone some three hundred feet thick rises gradually from under the marls and clays to form the asymmetrical Argonne plateau (Fig. 78). Evidently the geological conditions under which the whitish sand and intermingled clay particles which make up this rock were accumulated caused it to be restricted to a lens-shaped deposit, thinning out to the north and south. Hence the asymmetrical plateau, or cuesta, to which it gave rise, is not so continuous as the other cuestas of the Paris Basin but disappears northward and southward, the southern termination being strikingly abrupt. As far as it goes, however, the form is quite typical, with its gently inclined western slope and steep eastern scarp; and it is therefore proper to include it as one of the series of concentric cuestas constituting the natural defenses of Paris (p. 224).

The crest of the plateau, near the eastern side of the six or eight-mile wide upland, rises only a thousand feet above sea level, or less than five hundred above the bordering lowlands. Yet the topography is rugged and picturesque in the extreme, recalling the scenic features of the Vosges Mountains (p. 415). The secret of this remarkable dissection lies in the character of the porous sandstone, which is resistant enough to stand high where erosion is not vigorous, but at the same time is so deeply fissured and so poorly cemented as to be profoundly trenched wherever streams are active. Even small ravines may have depths of three hundred feet or more below the steepsided ridges which separate them, and narrow clefts or chasms, sometimes expanding into veritable caverns, are numerous. Wild and savage in aspect, these erosion forms have invited the inhabitants to bestow upon them fantastic names, like the Fairies' Hole and the Devil's Gorge. It is in the south that the terrain is most elaborately sculptured, as in the region of the Beaulieu Wood; while, north of the section drained by the Biesme River and its branches, one can see in places large areas of the gentle backslope but little eroded. Thus on the parallel of Varennes one could, from the valley of the Aisne, ascend gradually to the crest on a broad, sloping tableland covered by the Grurie Wood, then drop abruptly down the eastern scarp to the valley of the Aire, without ever crossing a ravine of major importance. Areas so well preserved from erosion are the exception, however; and the typical Argonne terrain shows a labyrinth of deep ravines or chasms between narrow ridges and plateau strips which suggest the original form of the plateau only by the gradual westward decrease of their average altitudes. It is in this ruggedness of form that the Argonne finds its initial value as a military barrier.

The Argonne Forest

Although not of an imposing elevation, the Argonne plateau does rise high enough to provoke an abundant rainfall from the winds which pass over it—hence the heavy mantle of forest which cloaks the rugged terrain. On the ridge crests and remaining strips of plateau upland the rains sink deeply into the porous rock. As a result the upland is arid, the soil stony, and vegetation, aside from the trees, comparatively meager. Clearings in the forest are few, for man finds it difficult to live where water is so hard to reach and where the soil is so unresponsive to his labors. The few villages perched on the upland, like Beaulieu in the south, La Grange-aux-Bois dominating the route over the Les Islettes pass, Florent on an old route across the upland into the Biesme valley, and La Croix-aux-Bois on a similar but even more important upland route farther north, persist on sites chosen for defensive purposes despite the difficulties imposed by the aridity of their surroundings. In the valleys the water which sinks so quickly into the porous sandstone on the uplands, but which is held in the lower levels of the same rock by the impervious clays immediately underneath, reappears in the form of numerous springs. Consequently the valleys and ravines are more humid, vegetation is more luxuriant, and the forest is there rendered more impassable by a dense growth of underbrush. Shaded by trees and smaller growth and drenched by frequent rains and by water seeping from the lower ground, the unmetaled roads of the Argonne in places become almost impassable, the more so because the sand formation contains sufficient clay to make a tenacious mud. Chains of ponds and lakes fill some of the valley floors, especially in the south, contributing their share toward making the region difficult to traverse. But it is the hundreds of square miles of almost unbroken forest mantling over the rugged terrain which, next to the shape of the land, makes the Argonne so formidable to an attacking army.

Longitudinal Valleys

The southern half of the Argonne plateau is rent in twain lengthwise by the deep chasm carved by the Biesme River. This stream flows from southeast to northwest, maintaining the remarkable parallelism of direction manifested by the Aisne, Aire, Meuse, and Moselle Rivers, until near its mouth it curves westward to join the Aisne. Since one of its headwater branches cuts nearly through to the Aire lowland at Clermont and another almost reaches the low Triaucourt gateway near Beaulieu, it is only natural that "The Valley," as the natives call it, should have become a highway of some importance across the barrier. Les Islettes, at the junction of the two branches of the Biesme just referred to, is the strategic point where cross the east-west railway and highway connecting Rheims and Châlons with Verdun and Metz, and the north-south route along the valley floor. Unfortunately that part of the Argonne upland west of Les Islettes has no transverse gorge connecting with Ste. Menehould in the Aisne lowland, corresponding to the gorge from Les Islettes to Clermont on the east. Thus it is that the railway has to tunnel through the western upland, and the highway to surmount its crest, in order to avoid a long detour northward down the Biesme valley. This fact detracts but little from the strategic importance of Les Islettes, which derives a slight additional military significance from the fact that it is the great depot for shipping excellent hay grown on the natural prairies of the Biesme valley to cavalry garrisons of the surrounding regions, including Verdun. The natural strength of the "defile of Les Islettes" is so great that in 1792 the Duke of Brunswick considered it impregnable to direct attack.

One could almost include the Aisne with the Biesme as one of the longitudinal valleys of the Argonne; for, while it drains the Aisne lowland and in part lies on the floor of that broader depression, elsewhere

Fig. 84—Block diagram of the northern Argonne a

euse plateaus, showing the salient features of the terrain.

it is cut in the backslope of the Argonne cuesta (Fig. 78), somewhat as the Meuse River is cut in the backslope of the Meuse cuesta and the Moselle River in the backslope of the Moselle cuesta. This would seem to suggest that the same geological history which had operated to give the Meuse and Moselle valleys their peculiar positions was beginning to operate in the case of the Aisne. All we need to observe here, however, is the fact that the Argonne barrier is reinforced against an attack coming from the east by this parallel trench near its western base, even if the Aisne trench has considerably less defensive value than the deeper moats of the Meuse and Moselle.

Defiles of the Argonne

Of the transverse valleys of the Argonne the most impressive by far is the broad trench of the Aire, which from Grandpré cuts squarely across the barrier to the Aisne lowland near Challerange (Fig. 84). A railroad of some military importance runs through this broad pass to connect the military stations in the Aisne valley and the great entrenched camps of the Champagne with the steep eastern face of the Argonne plateau. An important highway also takes advantage of the pass for a part of its course. It was here that Dumouriez expected the main Prussian attack in 1792, which he prepared to resist by organizing a strong natural position that "offered several lines of defense which united with one another and mutually supported each other."[3] The Prussians found the position almost as formidable as Les Islettes and refrained from a direct attack. Farther north the little Fournelle River has carved a gorge clear through the plateau to Vouziers in the Aisne lowland. This pass is traversed by a narrow-gauge railway (Fig. 84) and in part by a national highway. Still another transverse gorge has been cut a few miles farther north by a small stream rising near Le Chesne on the east and joining the Aisne above Attigny. Here the Argonne barrier is beginning to die out, but the Le Chesne defile is of some military importance and carries a canal uniting the Aisne Canal with the Meuse River below Sedan. South of the Grandpré pass there is no through-going trench except the indirect valley of the Biesme already described. But the low gateway of Triaucourt, carved by the Aisne and its branches between the southern end of the Argonne plateau and the lake-forest barrier of the Little Argonnes, has distinct military value. It is easy to defend, being dominated by the southern end of the Argonne upland and by the butte of Senard in the midst of a large clearing almost completely surrounded by the forests of Belval, Belnoue, Laheycourt, Argonelles, Four, and those of the southern Argonne.

The defile of Les Islettes is not the only one of the Argonne passes which takes advantage of a transverse gorge for a part

[3] Arthur Chuquet: Les guerres de la Révolution, 11 vols., Paris, 1886–96; reference in Vol. 2 (Valmy), p. 49.

of the way only, then crosses by a short-cut over the upland. The well-known defile of La Chalade, connecting Vienne-la-Ville and Vienne-le-Château on the Aisne lowland side of the barrier, with Varennes in the Aire lowland, utilizes the lower, transverse portion of the Biesme gorge, then turns north-eastward up a side ravine to surmount the eastern crest. North of the Grandpré pass the famous defile of La Croix-aux-Bois, forced by the Prussians in 1792, connects Vouziers in the Aisne lowland with Boult-aux-Bois and Buzancy in the Aire lowland, by following minor ravines only at either end of its passage, the greater part of its course lying on a flat-topped ridge remnant of the plateau. Even at Le Chesne pass the principal highway leaves the gorge followed by the canal, to cross the upland into the Fournelle gorge at Quatre Champs and thus to reach Vouziers.

Defensive Value of the Argonne

Despite the valley defiles which pierce the Argonne plateau, and remnants of the upland surface which offer fair crossings, the "wilderness of deep gorges and steep ridges" covered with "an immense forest-cloak of oaks, birches, mountain ash, and pines, with an impenetrable underbrush of heather, ferns, and broom"[4] remains one of the most formidable military barriers of France. For centuries it stood as a bulwark of defense marking the frontier between the Kingdom of France and the German Empire. The boundary lay for a time along the gorge of the Biesme River, but the French kings, appreciating the high military value of the "tangle of wooded heights," early pushed eastward to secure entire control of the obstacle. The natural adaptation of the longitudinal mountain gorge for the purposes of a boundary line is evidenced by the fact that in modern France the Biesme for much of its length divides the Department of the Marne from the Department of the Meuse.

Charles V sought to avoid the costly delay involved in a direct attack on the Argonne barrier by executing a turning movement

[4] Raoul Blanchard and Millicent Todd: Geography of France, New York, 1919. p. 127.

designed to outflank it from the south, a design frustrated by the heroic resistance of St. Dizier. Later Blücher was forced to pass around the obstacle, and in 1870 the invaders' Third Army was executing a similar movement to avoid the formidable Argonne terrain when it learned of the French advance toward Sedan.

But it was in the early days of the French Revolution that the Argonne best served its rôle as a protective barrier. "Behold," said Dumouriez to his adjutant-general as he pointed to a map showing the defiles of the Argonne, "Behold the Thermopylae of France." The commander of the Revolutionary forces then seized the passages through the Argonne barrier and, with the advantages of the terrain all in his favor, awaited the onslaught of the Prussians. "The most important thing is to gain time," the Minister of War had written to Dumouriez. "We have to organize our forces, augment our material of war and stocks of provisions, and encourage the people with the sight of our efforts. . . Kellermann is marching to join you and will aid you. The country is well adapted to the defensive." Before the minister's instructions reached him Dumouriez "had recognized the true strategic position which it was necessary to defend, the line of the Argonne." At Paris there was rejoicing when it was learned that the French held the defiles from Le Chesne southward, for it was believed that now the enemy could not force the formidable barrier.[5]

The Duke of Brunswick was of like opinion, in so far as a direct assault was concerned; but if he could trap Dumouriez into relaxing vigilance at one of the northern passes by feigning a heavy attack against the defile of Grandpré, he might pierce the northern part of the barrier and, turning south along the Aisne lowland, threaten the defenders of the other passes from the rear, thus forcing a precipitate retreat. So successful was this ruse that when small Austrian forces attacked the pass of La Croix-aux-Bois they found it defended by absurdly inadequate works manned by a pitiful handful of soldiers under a captain. Short work was made of the defense, and the Argonne barrier was

[5] Chuquet, Les guerres de la Révolution, Vol. 2, pp. 33, 35, 41, 116.

crossed by the invaders, not because it was weak, but because of criminal negligence of Dumouriez.

That erratic but brilliant commander skillfully repaired the consequences of his blunder. Swinging back to the southwest, and resting his right wing on the solid buttress of the southern Argonne, his left extended toward Châlons, he faced north to meet the enemy's advance. Kellermann now joined him, and, thus reinforced, he defeated the invaders in the apparently trivial but really decisive battle of Valmy. Even when inadequately defended the Argonne barrier had delayed the enemy's advance long enough for the French to concentrate sufficient forces to win one of the world's most important victories.

Sudden attacks on France in overwhelming numbers, forbidding by the very swiftness and power of the onset any adequate defense of the Argonne, may in 1870 and 1914 have created the impression that under modern conditions of warfare its military value has vanished. The increasing number of passable routes, the reduction of forest areas on the adjacent lowlands, and the draining of marshy valleys have removed some of the difficulties of traversing the rugged region; but new methods of organizing forests for defense, the enormously increased range, precision, and power of artillery fire, permitting effective control of approaches across cleared zones in front of an obstacle, and other improvements in modern warfare, have greatly increased the value of such natural features of the terrain as are found in the Argonne. The tremendous cost to the French and Americans of dislodging the enemy from the forested stronghold was sufficient to demonstrate its defensive advantages even when subjected to a flank attack. No opportunity for testing its resistance to a frontal attack occurred in the present war; but the opinion may be hazarded that under such a test it would not have been found wanting. We have only to qualify this opinion by the remark that the great size of the modern battle front more readily subjects a barrier of limited length, like that of the Argonne, to the danger of a turning movement. It was in part for this reason that the French made no effort to increase the natural strength of the obstacle with elaborate permanent fortifications. The Germans made no mistake, however, in basing one of their strongest defensive systems, the Hunding-Brunhilde Stellung, on the Argonne plateau from Attigny to Grandpré (Fig. 55), fronted by the natural moat of the Aisne valley and the transverse trench of the Aire.

THE AIRE LOWLAND

The narrow lowland which is drained northward by the Aire as far as Grandpré, then southward by the Agron and northward by the Bar, need not detain our attention long. None of the longitudinal streams mentioned is of any great size, nor are their tributaries of great importance. But the open lowland, with the exposed floor of the river valleys, sometimes marshy and frequently flooded, is a natural moat which adds to the defensive value of the rugged Argonne behind it.

No continuous railway traverses the length of the depression, but parts of two standard-gauge and several narrow-gauge lines, and for a short distance a canal, serve its transportation needs, in so far as these are not met by the national highway following the Aire for nearly fifty miles and by the other roads which cross it in various directions. Since the headwater portion of the Aire River lies in the imperfect lowland eroded in front of the Barrois plateau some miles east of Bar-le-Duc and crosses obliquely northwest into the Aire lowland in front of the Argonne plateau, its valley forms an easy although far from direct connection between the two depressions, of which advantage is taken by a highway and narrow-gauge railway, included among those already mentioned. It is evident that the Aire lowland is no great route of travel and traffic and that it could have no such commanding importance as a line of communication in time of war as might the Aisne lowland. Nevertheless, as a route of invasion from the north and as a minor supply line for an army during occupation, the cleared lowland offers distinct advantages over the almost impassable terrain of the Argonne on the west and the partially wooded Barrois upland on the east.

Among the strategic points within the Aire lowland are the towns commanding the entrances to the transverse gateways or defiles through the Argonne barrier: Le Chesne, Châtillon, Noirval, Grandpré (long the guardian strong point defending the principal gateway), and Clermont. Boult-aux-Bois is the eastern terminus of the ridge road via La Croix-aux-Bois, and Varennes occupies a similar position with respect to a road crossing into the Biesme valley at Le Four de Paris. Small places all, scarcely more than villages; yet destined to play their part in the drama of the World War.

THE BARROIS PLATEAU

From under the clays and sands of the Aire lowland rises the layer of resistant limestone forming the subsidiary cuesta lying on the western flank of the great Meuse plateau (Fig. 78). A suitable name for this upland is difficult to find. Popularly the name Argonne is extended over all the district, although it is better restricted to the Argonne Forest upland already described. The asymmetrical plateau formed by the limestone layer is a very striking feature farther south, in the district around Bar-le-Duc known as the Barrois; and, since there is no break in the plateau from that region to the section east of the Argonne Forest, we may call the whole of this single, continuous cuesta "the Barrois plateau." Like the other cuestas of the Paris Basin, its asymmetry is strongly marked, the eastern scarp dominating Cuisy, Malancourt, Esnes, and other villages nestling at its base with a frowning face which at Sivry-la-Perche becomes almost a cliff. It is not difficult to guess that this long line of villages nestles at the base of the scarp to get the water issuing as springs from just above the clay beds underlying the limestone; and, as they lie in small ravines cut back into the face of the upland, protected on either side by projecting spurs which offer facilities for extended observation, it was inevitable that the sites of the villages should become of tactical importance in any fighting for control of the plateau crest.

Although much dissected, the Barrois plateau shows no such rugged terrain as makes the Argonne redoubtable. Between Verdun and Varennes the gentle backslope is well preserved, and the extensive woods of Montfaucon, Malancourt, Avocourt, Cheppy, and Hesse clothe a surface in which deep ravines are comparatively rare. As the soil is less sterile and the terrain less rugged than in the Argonne, the amount of land abandoned to forest growth occupies much less than half the surface, the remainder showing a succession of cultivated fields and meadows covering rolling hills and giving a landscape of pleasing beauty. Over such a country roads may pass in every direction, and only at the steep eastern scarp will difficulty be encountered. Even here Nature has prepared some easy passages. The Andon River, tributary to the Meuse at Dun-sur-Meuse, has cut its way obliquely southwest back into the upland, so far that its head is found well down the gentle western slope. It is by this favorable defile that the main highway runs from Varennes to Dun, connecting the Aire lowland with the trench of the Meuse. Farther to the south the Aire and several of its tributaries slash obliquely through the scarp from southeast to northwest, giving a succession of similar gaps. The Verdun-Rheims railway and national highway ascend a side ravine of the Meuse to the scarp, crossing it near Nixeville to drop into one of the transverse trenches at Blercourt; while a narrow-gauge railway and another highway skirt the base of the scarp to reach the oblique gateway of the Aire itself. The cols of Souilly, Heippes, Mondrecourt, Rignancourt, and Chaumont-sur-Aire are all minor gateways of this class. Should the main railway connecting Verdun with Châlons and Rheims be severed by an enemy, it is evident that these more southerly openings through the Barrois cuesta might acquire considerable military importance.

While it is true that deep gorges and steep-walled ridges are rare on the western slope of the Barrois plateau, this does not mean that the region is devoid of military obstacles. The streams are small and easily forded except when in flood; but lines of bushes and trees usually border them and during the war afforded concealment to enemy forces which swept the open approaches with machine-gun fire. Some ravines, deep and narrow, concealed narrow-gauge railways. Still smaller gullies, apparently recent in origin and probably due to excessive erosion following deforestation of the slopes, provided the Germans with excellent machine-gun nests and artillery positions screened by a natural camouflage of bushes.[6]

Both at the extreme north and the extreme south of the Verdun battlefield the Barrois cuesta is more profoundly dissected, giving a rugged topography. In the north this appears to be the result of the nearer approach of the Aire lowland (there drained by the Bar) and the valley of the Meuse, the branch streams tributary to each soon interlocking their headwaters and reducing the upland to a complex of ridges and valleys. Here an important route from Stenay on the Meuse ascends a tributary of the latter river, crosses a ridge divide, and descends another ravine to Buzancy in the Aire lowland. Far away to the south it is the Marne River which trenches the Barrois upland and badly frays it with numerous short branches but which at the same time cuts a pathway through the barrier carrying railroad, highway, and canal (p. 334).

Natural Observatories

Two types of elevations on the Barrois plateau have supreme military importance. First are the spurs projecting eastward from the crest of the escarpment, giving commanding observation over the whole western slope of the Meuse upland, to and beyond the valley of the Meuse River. The spur north of Sivry-la-Perche, due west of Verdun, rises far higher than any point on the Argonne plateau to the west and looks down upon the Meuse valley and its chief stronghold. All the western forts included in the fortress ring of Verdun are commanded by this dominating height, from which an enfilading fire could

[6] Kirk Bryan: Memorandum on the Water Courses and Valleys of the Zone of Advance of the 5th Army Corps, prepared for use of the Corps during the Argonne campaign. Published as part of the same author's paper: The Rôle of Physiography in Military Operations, *Scientific Monthly*, Vol. 11, 1920, pp. 385–403.

be directed upon the parallel ridges and ravines forming the natural defenses of the city (p. 365). Should an enemy succeed in seizing this vital crest, the fate of Verdun and of the whole line of the Meuse would be in jeopardy. South and north of the little village of Esnes the spurs known as Hill 310 and Hill 304 command the northern defenses of the fortress city. Hill 304 also dominates the highly important natural trench formed by the Forges brook, which rises on the face of the escarpment near Malancourt and flows eastward past Béthincourt to join the Meuse near the village of Forges. The Forges trench, one of the barriers against an advance on Verdun from the north, could never be securely held by an enemy so long as the defenders dominated it from the projecting plateau spur of Hill 304. Hence we shall find the slopes of this small spur drenched with blood in the great battle for Verdun in 1916.

Quite different in origin and appearance is the second class of elevations possessing peculiar military value. These are erosion remnants of the Argonne Forest sandstone and underlying clays, left as isolated buttes here and there upon the backslope of the Barrois cuesta. They are of interest to the geologist and geographer as proving the former eastward continuation of the formations now worn back to make the steep face of the Argonne cuesta; but to an army they offer exceptional observatories commanding a wide sweep of country in every direction. Vauquois, southeast of Varennes, is perched on the summit of one of these buttes. To the casual observer its smiling slopes blooming with fields and orchards, and its crest crowned with a picturesque cluster of low houses above which towered the spire of the village church, bespoke no unusual significance. But from the summit a long stretch of the Aire lowland is open to view, as well as a broad expanse of open plateau country to the north and the valleys bounding the Forests of Cheppy, Montfaucon, and Malancourt on the east. So long as the crest was held by the Germans the route northward through the Aire lowland was blocked, Allied observation of enemy movements north of the butte was difficult or impossible, accurate Allied artillery fire on important enemy supply lines was prevented, and long-range German fire on Allied positions and communications was assured. Hence it was most bitterly contested, and concentrated artillery fire soon battered it into a shapeless mass of upturned earth (Fig. 85). It is the butte of Montfaucon, however, which best exemplifies the possible value of such land forms. Situated near the high eastern crest of the Barrois plateau, in the midst of a vast upland clearing from which valleys radiate outward and downward to the east, north,

FIG. 85—The butte of Vauquois, transformed by the Germans into a strong point buttressing their line. Its military importance may be measured by the scars of battle which disfigure its once smiling slopes. (French official photograph.)

west, and southwest (Fig. 84), this outlier of the Argonne "dominates the whole region; its views extend for several leagues in every direction. It is not only a remarkable observatory, but forms a strong point of support of advantage to an army wishing to defend the left bank of the Meuse opposite Vilosnes and Sivry-sur-Meuse."[7] It was here that the Crown Prince was to take his station during the great Battle of Verdun.

THE MEUSE PLATEAU

The thin bed of clays at the base of the Barrois plateau do not give a conspicuous or continuous lowland in front of the escarpment. Only at the south does the upper Aire occupy a

[7] Niox, quoted by Paul Joanne: Dictionnaire géographique et administratif de la France, 7 vols., Paris, 1905, article on Argonne, Vol. 1, p. 149.

FIG. 86—The Meuse plateau west of Verdun, looking northwest from near Fort Marre. The butte of Montfaucon is visible as a low mound on the sky line near the center of the view. (This photograph is continued on the right, with a slight overlap, by Fig. 87.)

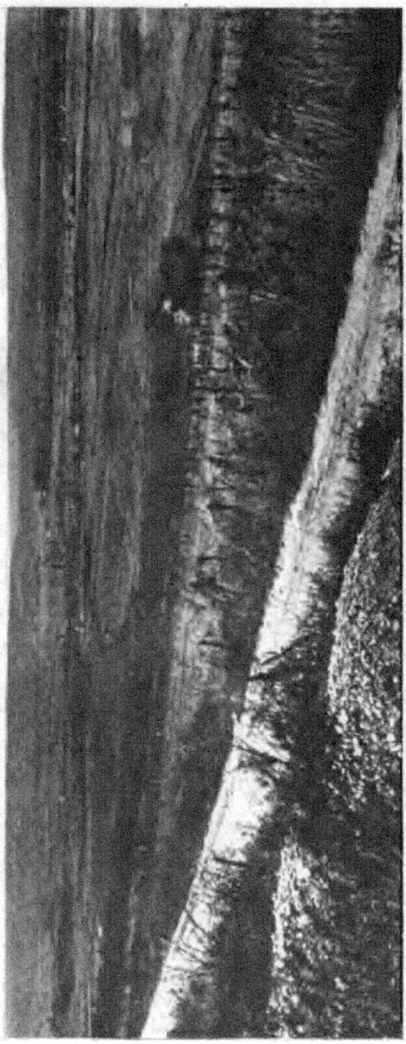

FIG. 87—Looking north from the meander spur bearing Fort Marre to where the Côte de l'Oie meander spur (on the left) overlaps the tip end of the Côte de Talou spur, showing the even sky line of the dissected plateau. The Meuse valley in the middle-ground, barbed wire and trenches of the Fort Marre ridge in the foreground.

FIG. 88.—The irregularly eroded scarp of the Meuse plateau (to the left) facing the Woëvre lowland (to the right). In the right distance are seen the projecting plateau spur forming the Hattonchâtel bastion and, near its base, the village of Vigneulles, where in the last Battle of St. Mihiel Salient the American troops descending from the plateau upland to the west met their comrades pushing north across the level plain of the Woëvre, thus cutting off from escape German forces in the salient.

veritable lowland, and this for no very great distance. We may therefore treat the Barrois plateau as a subsidiary cuesta lapping up over the western flank of the main Meuse cuesta, or asymmetrical plateau, and proceed at once to examine the latter belt of the Verdun battlefield.

A brief glance at any good map portraying the topographic features of northern France is sufficient to show that the massive limestone formation (Fig. 78), of which the Barrois plateau-maker was but a forerunner, has produced one of the most remarkable asymmetrical plateaus not merely in France but in any country. With notable continuity this upland belt extends from southwest of Mézières close to the Belgian border, in a great sweeping arc southeast, south, and southwest, to the borders of the central highlands of France, a distance of some three hundred miles. In the Verdun battlefield its crest rises more than 1,300 feet above sea level and in places dominates the eastern lowland of the Woëvre by nearly 600 feet. The upland, although trenched by numerous short streams, preserves considerable areas of its former flat surface, and in the distant view conveys to the eye all the evenness of sky line that the term plateau may imply (Figs. 86 and 87). The east-facing scarp, notched by countless ravines, is clearly marked, and the transition from plateau to lowland is sometimes sudden and striking in the extreme (Fig. 88 and Pl. IX, *A*). Seldom does Nature mark her topographic boundaries with such precision.

In general the limestone upland is dry, heavily forested, and sparsely inhabited. The most frequent exceptions to this are found north of Verdun, where a layer of marl at the top of the main limestone series has not been wholly eroded. Here enough water is retained close to the surface to supply the needs of a few villages which stand on the upland, in the midst of cultivated fields made possible by the more amenable soil. The unusual height of the villages above the adjacent lowland and valley bottoms is emphasized by the termination "mont" in names made familiar to the world by one of the greatest battles in his-

Fig. 89—The east-facing escarpment of the Meuse plateau (at the right) and the Woëvre lowland (at the left), looking southward from near Toul. Observe the forested upper slope and the more gentle, cleared lower slope of the scarp.

tory: Haumont, Beaumont, Louvemont, and Douaumont. Elsewhere we find vast stretches of forest on each side of the Meuse, two parallel bands of shadow and silence crowning the lonely upland, separated by the valley clearing in which road, railway, and canal bear the life of the country between sloping vineyards and smiling meadows.

The Côtes de Meuse

The east-facing escarpment is called by the French the Côtes de Meuse, an expression difficult to render into English. "Meuse Cliffs" is not satisfactory, for the Côtes do not necessarily show vertical or even very steep slopes. "Meuse Escarpment" is probably the best equivalent. On the Battlefield of Verdun the escarpment usually exhibits a steep upper slope where the limestone formation is exposed and more gentle lower declivities on the clays (Fig. 89). Although occupying an extremely narrow strip of territory, the Côtes form a zone of more habitable country between the dark forests of the dry plateau above and the muds and marshes of the wet plain below. Located along the spring line at the contact of the fissured limestone with the underlying impervious clays, villages on the well-drained slopes near the base of the Côtes find sufficient water for their needs, but not enough to prove a plague. "The slopes have a more equable climate, are less exposed to frost than the valley floor, and are protected from the northwest winds as well. They are sunny, especially when facing southeast; an admirable topsoil consisting of decomposed bits of limestone mixed with clay covers the slopes and is well suited to any crops. Orchards and vineyards thrive here. All conditions are favorable to man. . . The inhabitants can exploit the forests above. . . Quarries in the limestone cliffs furnish building materials for the villages. . . These escarpments have played a great part in the history of the country from a military point of view, for since time immemorial they have been the rampart of France against Germanic invasions."[8]

[8] Blanchard and Todd, Geography of France, p. 133.

That a great plateau, dominating all the low country to the east from a commanding height (Pl. VI), seamed with ravines providing innumerable concealed artillery positions, cloaked with a forest assuring ample secrecy to the maneuvers of the defending forces, and served by a well-protected lateral communication trench carrying men, munitions, and supplies by water, rail, and road to all parts of the front—that such a plateau should prove a well-nigh impregnable position when attacked from the lowland to the west would seem too obvious to require demonstration. It is a remarkable fact that, while the Argonne and other cuestas of the Paris Basin are repeatedly trenched by the cross valleys of numerous streams, there is not a single valley which cuts clear through the Meuse cuesta from one side to the other between the River Bar on the north and the headwaters of the River Marne on the south, a distance of one hundred and thirty miles. In the Middle Ages the forested mass of the plateau was for a time the strong defense of France's eastern frontier, the boundary between her own and the territory of the German Empire running along the Meuse River. Small wonder that the French of a later day have made this plateau and its river trench the northern foundation of their strongest frontier defense system.

Yet not all portions of the Meuse plateau offer equal resistance to an enemy attack. At the north the Meuse trench opens out upon the Woëvre lowland, offering a gateway by which the plateau stronghold may be entered (Figs. 79 and 84). But the Meuse pathway to Verdun is long and crooked, and we shall later see that it does not lack the protection of strong natural defenses. In any case it merely leads into the barrier and not through it. For the next fifty miles southeastward there is not a single break in the wall more serious than the short ravines, like that of Vaux, which cut but a mile or two back into the scarp, leaving the upland barrier of the Hauts de Meuse (Meuse Heights), as the French call the plateau strip between the Woëvre lowland and the Meuse trench, unbroken. The valley of Les Éparges is longer, but, as it soon bends to trend parallel with, and just back of, the scarp, it merely adds another trench to the series of barriers to be overcome in an advance westward (Fig. 90).

Natural Bastions and Curtains

At Hattonchâtel the plateau juts far forward into the Woëvre lowland, forming a great bastion (Fig. 88 and Pl. VI) measuring

FIG. 90—The crest of the Meuse plateau transformed into a formidable stronghold in the Les Éparges sector. In the foreground are the ruins of Combres. (French official photograph.)

ten miles from its apex to the Meuse trench in its rear, which the French call the "Salient of Vigneulles," and which commands a magnificent view of the Woëvre plain in a great sweep from the northwest around to the south. Perched on this strategic point, Hattonchâtel was long a fortified stronghold to which the adjacent country looked for protection. At the base of the vineyard-clad apex of the bastion lies the village of Vigneulles, of which we shall hear more.

South of the Hattonchâtel bastion is a deep re-entrant, or "curtain" as it would be called in an artificial fortress, where the headwaters of the Rupt de Mad, a small stream tributary to the Moselle, have worn their way back almost to the Meuse trench near the town of Commercy (Fig. 91 and Pl. VI). It is in this "Commercy curtain" that we find one of the weak points of the barrier. It may be that in an earlier period, when the Meuse plateau still stretched far eastward, longer and larger streams

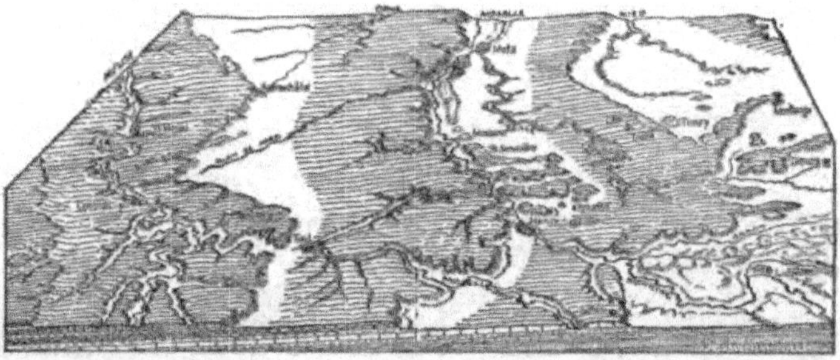

FIG. 91—Portions of the Barrois, Meuse, Moselle, and Saffais plateaus, showing the former course of the Moselle River westward from Toul to the Meuse, and the salients and re-entrants along the several plateau scarps forming natural bastions and curtains. (Modified after Davis.)

than now exist here flowed down its west-dipping slope into the Meuse trench, cutting lateral gorges deep below the upland. As the east-facing scarp was worn back the heads of these rivers would be cut off; and by the time the scarp reached its present position nothing would remain of the beheaded rivers save small rivulets occupying the lower ends of the large gorges. Whether this or some other history is responsible,[9] the striking fact is that the Meuse plateau in the vicinity of the Commercy curtain is repeatedly cut across by deep, stream-carved notches by which one may easily pass from the Woëvre lowland into the Meuse trench.

[9] J. Vidal de la Blache: Étude sur la vallée lorraine de la Meuse, Paris, 1908, p. 63 *et seq.*

Strategic Defiles

The longest of the notches (Pl. VI) begins a few miles southwest of Vigneulles near the village of Creue and continues through the upland past Spada to reach the Meuse at Maizey. It is well known in French military circles as the "defile of Creue or Spada," and Fort Paroches west of the Meuse is designed to block it by pouring an enfilading fire upon any enemy column advancing along the road which runs through the depression. South of St. Mihiel is a shorter notch carrying a road from Apremont in the Woëvre lowland to the Meuse valley, known as the "defile of Marbotte or St. Agnant," from villages in the depression. Just south of it a third and still shorter notch, the "defile of St. Julien or Boncourt," is traversed by a road and by the narrow-gauge railway which skirts the base of the scarp all the way northward to beyond the Verdun district. To stop these two gaps Fort Liouville was placed on the crest of the scarp between them, its guns commanding both the depressions and the approaches to them. A few miles to the southeast the shortest notch of all, the "defile of Cormeville," carries another road through the scarp. Fort Gironville on the crest of the upland (Fig. 81) between the defiles of St. Julien and Cormeville blocks both of them, as well as the national highway leading directly across the upland to Commercy. The longer "defile of Trondes," next to the south, is defended by forts on either side, while the remarkable "gap of Toul" (Fig. 92), through which the Moselle River quite certainly flowed[10] to join the Meuse before it was captured and turned northeast to its present course (Fig. 91), is ringed about with a formidable series of defensive works, including a line of forts on the crest of the main scarp. A whole series of important railways, roads, and canals converge upon this broadest and most important of all the breaches in the outer rim of the Meuse plateau. Farther south there are several other gaps, but they lie beyond the limit of our present field of study.

[10] W. M. Davis: The Seine, the Meuse, and the Moselle, in his "Geographical Essays," Boston, 1909, pp. 587–616.

FIG. 92—The abandoned valley of the Moselle River in the Meuse plateau west of Toul. After carving this valley the Moselle was diverted northeastward to its present course. Roads, railways, and canals converge to pass through this former stream channel, making Toul a point of great strategic importance.

It is evident that the Commercy curtain is a naturally weak sector of the Meuse plateau defenses. But, while an enemy might be tempted to assail this part of the scarp, it is strongly protected by the projecting Hattonchâtel bastion on the north and by the Toul bastion (Figs. 88 and 91, and Pl. VI) on the south. Moreover, as we have just seen, the several defiles and their approaches "are commanded by forts and batteries which stand forth in their brutal majesty or sullenly conceal themselves in the midst of the forest."[11] In front of the scarp a series of lakes and the marshy Forêt de la Reine (Queen's Forest) offer good defensive positions on the plain.

A further element of strength is found in the outlying ridge and butte of Le Mont and Mont Sec (Fig. 93 and Pl. VI) near Apremont, erosion remnants of the plateau which manifest the same northeast-southwest alignment characteristic of most of the ravines, valleys, and ridges of the vicinity, including the three northern defiles and the Rupt de Mad valley in the Woëvre plain. The possessor of Mont Sec and Le Mont enjoys exceptional command of the adjacent plain (Pl. IX, *A*) and is sure of effective control of artillery fire on the approaches to the Commercy curtain. North of Verdun, in the region between Ornes and Damvillers, there are several additional outlying buttes, of much value as advanced strongholds and observation posts before the main escarpment. The double-crested butte near Ornes called Les Jumelles is significant not merely as to form, justifying the translation "The Twins," but to the military geographer its value for observation suggests another meaning of this French word, "The Field Glasses." It is in the Toul sector, however, that the military value of the erosion buttes is most fully utilized. Mont St. Michel stands two miles in front of the main escarpment and is crowned by a powerful fortress which commands the marshy plain of the Woëvre in every direction. Its guns dominate a long sweep of the Moselle River and the roads, railways, and canals converging on the Toul gap. West of it is a smaller butte, fully prepared for the installation of

[11] Bertrand Auerbach: Le plateau lorrain: Essai de géographie régionale, Paris, 1893, p. 129.

FIG. 93—Mont Sec, an outlying erosion remnant of the Meuse plateau (visible in the background) converted by the Germans into a formidable stronghold commanding a broad expanse of the Woëvre lowland. The great natural strength of this position was one of the factors which rendered the reduction of the St. Mihiel salient so difficult. (French official photograph.)

artillery in case of attack. Thus are the outlying plateau remnants incorporated as a vital part of the formidable defensive system constituting the main southern buttress of the fortified line of the Meuse. Even where not crowned in advance by permanent defensive works, it would be difficult to overestimate the possible value of these islands of high, dry land standing forth in the low, wet plain.

The greater ease of communication through the narrowed and trenched plateau in the vicinity of the Commercy curtain gives to that region an unusual importance. Not only do roads converge on the defiles from distant localities, but life in the valley is more active, villages are more numerous, and industries more prosperous. Among the latter the quarrying of building stone from the limestone formation is important, and quarries and caverns, old and new, honeycomb the hills, offering excellent artillery positions and underground shelters from enemy bombardments. In a meander spur near St. Mihiel are grottoes in which primitive man sought refuge, while the ruins of an ancient Roman camp further attest the early military importance of the spot. A modern fort on the same site commands the river trench for some distance to the north and south.

The Valley of the Meuse

For much of its course across the Verdun battlefield the forested Meuse plateau is split lengthwise by the Meuse valley (Fig. 79). This natural trench is a remarkable topographic feature, maintaining a straight southeast-to-northwest general course parallel to the strongly developed belts of the terrain, yet showing in detail a series of serpentine curves. That the present Meuse River did not carve the valley would seem to be indicated by the fact that its curves are of a much smaller pattern, utterly out of harmony with the grand meanders of the gorge (Fig. 94). Evidently the gorge was carved while the Moselle River still flowed into the Meuse through the Toul gap, and the combined volume of the two made a single large river swinging in majestic curves and incising its meanders into the

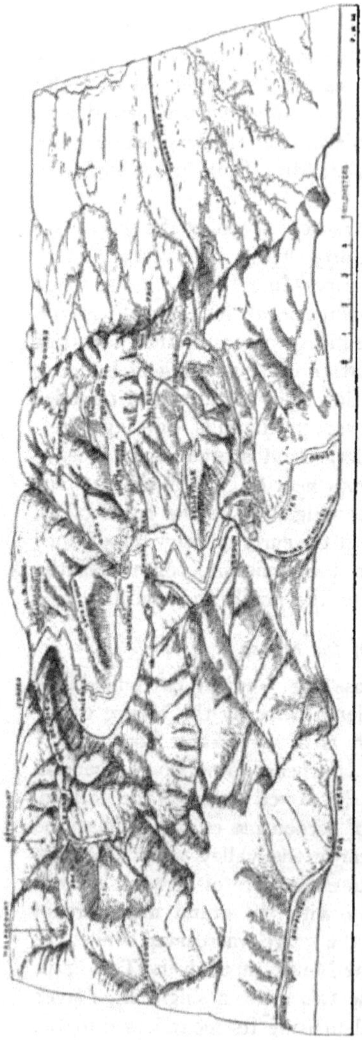

FIG. 94—Block diagram of the Verdun region, showing the salient features of the terrain. Note the small meanders of the Meuse River, the large meander pattern of the valley, and the inter-meander spurs with steep upstream sides facing the observer.

upland surface. Later capture of the Moselle left the shrunken Meuse to wander distractedly along the valley floor, its small meanders and branching channels threading the meadows and marshes without definite intention and betraying a stream ill adapted to the trench in which it finds itself.

We noted that the meanders of the Somme occasionally had some tactical value (p. 148). The greater meanders of the Meuse gorge form an important element in the natural defenses of the Battlefield of Verdun. Since the general direction of the valley is southeast-northwest, the axes of the meanders, and hence of the ridge spurs around which they curve, trend southwest-northeast, or across the grain of the country. Conformable to a law of river erosion, the upstream side of each ridge is the steeper and the downstream side a smoother, more gentle slope. Hence an enemy entering the Meuse gateway at Dun-sur-Meuse and endeavoring to advance up the valley toward Verdun would find himself confronted by a series of ridge spurs projecting into the valley, first from one side, then from the other (Figs. 94 and 84), each overlapping and interlocking with its neighbors and each opposing to him a natural glacis swept by the fire of defenders concealed on the steeper backslope. The spurs, like the valley walls, are deforested and planted with vineyards or orchards which afford little cover, or have open, cultivated fields across which the attacking columns must advance without protection. Thus is the apparent value of the Meuse trench as a pathway along the plateau denied by the excellence of its natural defenses. Only when already held by an army can its value for transport be realized; then the interlocking spurs, preventing distant views along the trench and hence reducing interference from enemy artillery to a minimum, make it safer than the ordinary valley subject to long range domination from some commanding viewpoint. The Côte de l'Oie (Fig. 94), at the western end of which is the famous Mort Homme, or Dead Man's Hill, and the Côte de Talou are examples of the asymmetrical meander spurs which defied the massed strength of the German army and helped transform one of the greatest German offensives into a bloody disaster.

The ridges of the type just mentioned are supplemented and strengthened by ridges of a wholly different origin. Tributary streams, gnawing back into the upland from either side of the main river, have excavated fairly straight lateral ravines separated by ridges trending, like the meander spurs, at right angles to the general course of the valley. The parallel lateral ridges are so many defensive walls, each fronted by its ravine moat, barring the path of an enemy advancing along the length of the valley. To overcome one of them would be a difficult but feasible task; to overcome a whole series of them in succession is a task of stupendous magnitude, when the army of defense is properly directed. It was on the asymmetrical meander spurs and their associated parallel ridges that the forts of Verdun, on both sides of the river north and south of the city, were aligned. The Côte du Poivre (Pepper Ridge), Côte de Froide Terre, and Douaumont ridge are among the ridges of the second type which were mainly concerned in halting the colossal German drive of 1916.

A peculiar feature of the parallel ridges, resulting from the failure of the side ravines to cut clear through the plateau, deserves special notice. Where ravines from the Woëvre lowland head against but do not quite meet ravines from the Meuse valley, the remaining upland ridges have the form of a letter H lying horizontally, thus: ⌶. Additional cases of the same form in succession give the appearance indicated in Figure 95. This is the condition found east of Verdun, where it imposed special tactical conditions upon the military operations. The Germans in seeking to gain the plateau crest at successive points endeavored to push up the ravine between the legs of the H, then to advance over the crossbar to the next ridge. As the crossbars were the bridges which alone made a southward advance possible without descent into the ravine-moat, and as the possessor of them could enfilade the ravines on either side and so take in the rear important enemy positions, they were most bitterly defended and attacked. The little village of Fleury, standing on the crossbar connecting the Côte St. Michel-Fort Vaux ridge with the Froide Terre-Fort Douaumont ridge, was one of the few towns which in the course of the World War was literally pulverized and blown off the face of the earth by long-continued, concentrated artillery fire (Fig. 96).

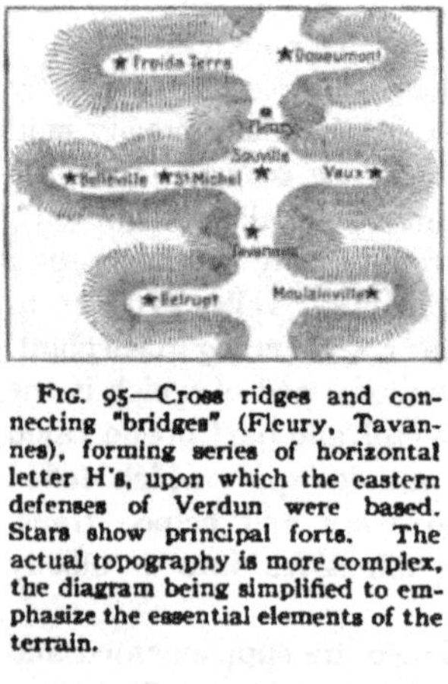

FIG. 95—Cross ridges and connecting "bridges" (Fleury, Tavannes), forming series of horizontal letter H's, upon which the eastern defenses of Verdun were based. Stars show principal forts. The actual topography is more complex, the diagram being simplified to emphasize the essential elements of the terrain.

The Meuse River

The Meuse River itself is an obstacle of no mean proportions under certain circumstances. Its meandering course and branching channels repeatedly interpose an unfordable current before an enemy fighting his way along the valley and unable to follow a single highway or to keep to one side of the valley only. The river's upper branches drain the impervious clays and marls coming up from beneath the limestone formation and hence discharge great volumes of flood waters down the valley in time of heavy rains or melting snows (Fig. 97). So faint is the slope of the valley floor that the floods pass off but slowly, sometimes lasting for several weeks. As the valley is not wide, even a small flood may cover all its floor; when the ground is saturated in winter every light rain is sufficient to swell the river's volume. The valley is therefore often inundated, and the floods are a matter of practical military importance. The German assaults on Verdun were greatly hampered by the fact that the Meuse, swollen by winter floods, separated the two wings of the attacking armies and prevented their efficient co-operation. Even when free of floods, the valley floor, between one thousand and two thousand yards wide, sometimes marshy but usually covered with open meadows, is an obstacle difficult to cross under enemy fire. Taken all in all, the character of the river itself, the form of its valley, and the disposition of its lateral ravines and ridges make of the Meuse an imposing system of defenses rather than an easy route of invasion.

It was the defensive value of the Meuse trench which led the French to place a number of forts linking Verdun with Toul along the river, instead of along the crest of the plateau farther east. There the thick woods would make it impossible for the different forts to support each other effectively in case of attack, whereas the cleared walls and floor of the river trench would expose all the terrain between each pair of forts to the fire of one or both of them. Thus the three forts of Génicourt, Troyon, and Camp des Romains dominate all of the Meuse trench from Verdun to south of St. Mihiel, as well as the principal

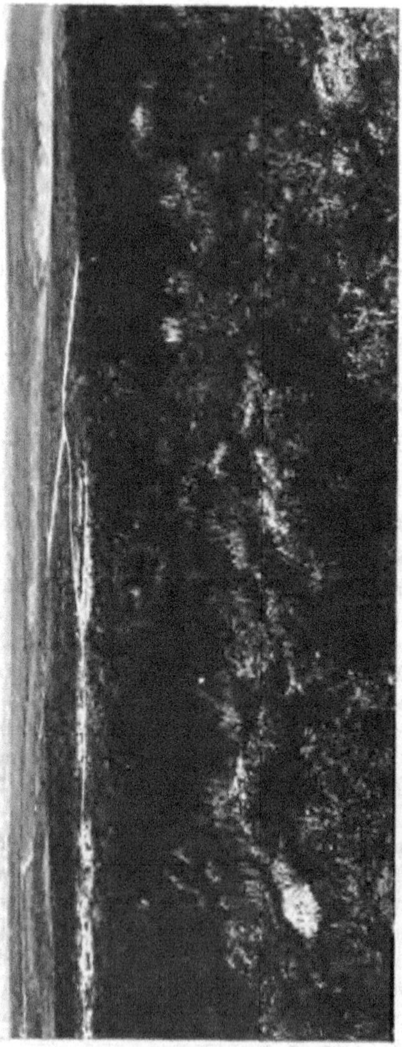

FIG. 96—The Meuse plateau east of Verdun. Douaumont ridge in the middle distance, with Fort Douaumont at right. Crest of Louvemont ridge just showing beyond. The white, freshly shell-torn slopes this side of Fort Douaumont mark the head of Vaux ravine, while just beyond the road intersection near the center was the village of Fleury, on the plateau "bridge" connecting Douaumont and Vaux ridges.

exits from the forest on the east. The danger due to the nearness of the forest cover to the forts is in a measure counteracted by special methods of constructing the defensive works and of placing the guns. It was against this barrier, first in the vicinity of Forts Troyon and Génicourt, then at Fort Camp des Romains, that the Germans in 1914 twice tried in vain to breach the Meuse barrier.

In the immediate vicinity of Verdun the barren ridges, interlocking monotonously to give a sky line devoid of any irregularity, form a dreary and repellant landscape. To one standing on the meander spur bearing Fort Marre and looking northward, the Côte de l'Oie on the left and the Côte de Talou on the right seem almost to merge in one, and the valley below disappears into the plateau wall as if by magic (Fig. 87). Or if the eyes be turned northwest, the successive ridges blend like gray waves into an even horizon, above which Montfaucon rises miles away, at this distance looking like a low mound which modestly conceals its dominating importance (Fig. 86). Seen from Fort Souville the famous Douaumont ridge just in front can none too easily be distinguished from the Louvemont ridge beyond, although the mound of the fort swells noticeably against the horizon (Fig. 96). Barren of trees, devoid of topographic variety, shell-torn and silent, the sinister landscape of Verdun (Pl. IX, *B*) today deserves more than ever the oft-quoted description of Ardouin Dumazet: "A rather sad landscape, no variation in the form of the hills, no woods, a sort of immense amphitheater formed by arid slopes covered with vineyards and fields of grain. . . In the center of the immense semicircle, little Verdun sleeps about its cathedral, at the foot of its citadel. No military site, not even the mountain fortresses like Grenoble, Briançon, or Besançon, is so impressive as Verdun. It is formidable and appalling."[12]

THE WOËVRE LOWLAND

No small part of the military value of the Meuse barrier derives from the topographic character of the Woëvre lowland

[12] Ardouin Dumazet: Les grandes manoeuvres de l'est en 1891, Paris, 1891.

which it dominates. The plain of the Woëvre is very broad, because the clay formation on which it is eroded (Fig. 78) is very thick, in places exceeding six hundred feet. It is difficult to assign a definite breadth to the Woëvre, not only because it varies from a minimum width near Toul and to the south to a maximum several times as great farther north, but also because its eastern limit grades insensibly into the rising limestone upland of the Moselle plateau (Fig. 79). Throughout much of its length, however, the lowland varies from ten to fifteen miles in breadth. It stretches far away to the north and south as a vast, monotonous expanse of level clay plain, the even horizon usually unbroken by anything more striking than occasional forest patches, or villages commonly of dimensions meriting rather the term of hamlet (Figs. 80 and 81). Only on the west does the view encounter the low wall of the Meuse plateau (Fig. 93); and there the level sky line of the plain is merely exchanged for the equally level plateau horizon just above. Eastward the land rises as the next underlying limestone formation emerges to play its rôle in the topographic scheme; but the rise is almost imperceptible and attracts the eye only when the attention has first been directed to seek for it.

It is the clay which places its ineffaceable stamp upon this lowland country. It holds the rainfall on the surface, necessitating extensive drainage operations, and at the same time furnishes the material out of which the drain tiles are made. In its depressions stagnate the countless shallow ponds and lakes, some natural, many artificial, which diversify the green surface with shining patches of silver when seen from the plateau crest. When the winter rains are heavy, it is the clay which turns the whole plain into a vast slough in which neither man nor beast can find a solid footing. And when the summer sun shines hot, it is the same clay which bakes to a natural brick, requiring five horses to the plow which the peasant farmer would drive through the hardened mass.[13]

Man finds on the lowland surface an indifferent place of

[13] Auerbach, Le plateau lorrain, p. 145.

abode. The stagnant waters breed fever in the land, the labor of drainage is a heavy burden, the soil despite broad areas of loam usually requires enrichment, stone for purposes of construction is lacking, roads are bad, and material for their improvement is far to seek. By dint of unremitting toil the better lands are made to yield harvests, especially of cereals, while

FIG. 97—Flooded valley of the Meuse River in the Meuse plateau. The frequent inundations to which the river is subject seriously hampered the German military operations during the Battle of Verdun. (French official photograph.)

the outcrops of purer clays remain covered with forests. But just as in early days the plain of the Woëvre remained long an uninhabited wilderness of marsh and woodland, while man colonized Mont Sec (Dry Mountain) and other islands of dry land or sought the warmer slopes of the Côtes de Meuse, so today the higher lands beckon alluringly to the toiler in the clay lowland, and villages crowd each other along the slopes of the plateau scarp, while for the most part mere hamlets dot the surface of the more sparsely inhabited plain.

As on other low clay plains, the rivers are obstacles, especially in time of heavy rains. Under ordinary conditions they wander sluggishly among marshes and bogs, bordered by reeds and willows. But heavy rains, denied escape underground because of the imperviousness of the clay, flow over the surface to swell the streams with a rapid increase of volume which their channels cannot contain. Then the flood waters spread far and wide, and large areas become impassable. The Orne receives such abundant contributions from the Woëvre that its level often rises ten feet or more above low water, and other streams have a similar unstable régime.

For a hostile army to maneuver on the clay plain is of itself a difficult task, especially when the surface is soft with winter rains. To force a path across it in the face of a determined enemy skillfully utilizing rivers, lakes, and forests as defensive barriers is a task of more serious proportions. But to use it as a base of operations against the formidable plateau stronghold to the west, to break the preliminary defenses of fortified lake and forest masses, like that of the Forêt de la Reine in front of the Commercy curtain, to maneuver with every daylight movement open to enemy observation, to advance along roads accurately registered for enemy fire, and to assault the heights under a hurricane of steel accurately directed from above—this is a task demanding almost superhuman power.

THE MOSELLE PLATEAU AND SEILLE LOWLAND

In a later chapter the features of the Moselle plateau and Seille lowland will be studied in connection with the military operations primarily based upon them. The fighting in the Battlefield of Verdun touched only the western slope of the Moselle cuesta, and it will suffice for the moment to state quite briefly the salient features of its topography.

From under the thick clays and marls worn down to form the Woëvre lowland a more resistant limestone formation rises gradually to give the gentle western slope of the asymmetrical Moselle plateau (Fig. 78). Much of the Woëvre lowland drains northeastward to the Moselle River by means of tributary streams like the Rupt de Mad and Orne cutting obliquely through the rising upland to reach the Moselle trench (Fig. 79). Since these tributary rivers grow constantly deeper at the same time that the plateau is continuously rising, the relief becomes bolder and the scenery more imposing, until on the crest of the Moselle gorge one sees the river flowing more than 600 feet below him.

The Haye Region

The higher, more dissected eastern portion of the Moselle plateau is quite generally forested; although only at the south, where we find the great Forest of Haye on the broad upland east of Toul, are the wooded areas nearly so extensive as on the plateau of the Meuse. All of the southern portion of the Moselle plateau west of the Moselle River, including not only the Forest of Haye but extending northward to the vicinity of the Rupt de Mad, is known to the French as the Haye region. It is a dry land in which the streams sink deep into the fissured limestone, occasionally disappearing into underground channels; a rugged land in part, where different layers of the limestone formation terrace the surface or where stream erosion and solution have gashed chasms and ravines in the sides of hills; yet to some extent a cultivated land, for a surface coating of whitish loam yields fair crops of oats and wheat. Quarries are numerous, especially along the Moselle valley, because certain layers of the limestone furnish excellent material for building, paving, and other purposes. Yet neither agriculture nor industry have wholly tamed the Haye, and it remains a region difficult for an army to traverse.

With the Moselle trench and the bridgehead of the Grand Couronné terminating eastward in the steep scarp of the Côtes de Moselle (Fig. 78), the Moselle plateau presents an almost insuperable obstacle to an enemy advance westward from the Seille lowland on the east. Only the fact that the German frontier lay close in front of the barrier south of Pont-à-Mousson and actually breached it to the north of this point, added to the fear that Germany might regard the fortification of the obstacle as an unfriendly act, prevented the French from establishing an elaborate system of defensive works on the exceptionally favorable terrain. Steps were, in fact, being taken for the fortification of the Nancy portion of the plateau, where only one or two forts existed previously, when the World War broke out. Against a hostile advance from the west the plateau presented a long natural glacis, backed by a deep trench and behind that a steep-faced wall, all of which the Germans strongly fortified in the vicinity of Metz.

When American troops first took over a sector of the European battle front, it was across the sloping western side of the Moselle plateau in the Haye region and the wet plain of the Woëvre, between Pont-à-Mousson in the Moselle trench and Apremont at the base of the Meuse escarpment, that they confronted the enemy. That they were in the Haye they could not well forget, for the names of villages close to the front constantly reminded them of the fact by the frequent recurrence of the termination "-en-Haye:" Domèvre-en-Haye, Villers-en-Haye, Fey-en-Haye, Regniéville-en-Haye, Viéville-en-Haye. Through the western, or Woëvre, end of their front passed the Rupt de Mad, which flows northeastward across the open, marshy plain with sluggish mien until the western edge of the limestone plateau is reached. Here, a few miles above Thiaucourt, the stream quickens its gait and goes swirling down through a winding limestone gorge which, ever deepening as the plateau rises, only ends when it opens into the Moselle trench as a narrow valley cut 600 feet deep into the plateau. The Rupt de Mad gorge was to the Allies a barrier against a northwestward advance and for the Germans a valuable corridor carrying railway and highway from the entrenched camp of Metz through the plateau to the margin of the Woëvre at Thiaucourt.

The Briey Plateau

Beyond the gorge, to the north, the asymmetrical Moselle upland takes on a new aspect. At first the country is more open, forest patches are restricted to smaller limits, and the soil seems to yield more readily to the labors of man. Careful examination will show that a layer of marl mantles the more resistant limestone over much of the region. But it could hardly be called an agricultural country, and its productivity is not sufficient to make it a rich military prize, like the fertile plain of the Somme. Topographic forms are much the same as in the Haye: the same sloping plateau trenched by streams which begin in shallow ravines on the west to end in deep canyons at the east; the same steep valley walls or hill slopes gashed here and there by the narrow clefts or larger chasms typical of limestone countries; the same abundance of quarries, especially along the Moselle. But certain features make it a region apart, among them a series of southwest-northeast faults, or breaks in the earth's crust, which reveal themselves by a somewhat larger number of ravines and ridges taking that course than in the Haye; by a series of spring lines where waters from underground sources escape to the surface; and occasionally by bringing to the surface and rendering more accessible certain valuable mineral deposits. It is these deposits, consisting of beds of iron ore in the plateau rocks, increasing in importance around Briey and farther to the north, which give the region its strongest stamp of individuality. Here the landscape takes on an industrial aspect; mines, furnaces, and smelters dot the countryside, factories crowd into the towns, foreign laborers replace the French peasants, and the sleepy plateau awakens with the pulse of a new life. This is the great mineral region of Briey, containing one of the richest iron ore deposits in the world. From it France has for some time derived nine-tenths of her iron. The region gives its name to a large portion of the Moselle plateau north of the Haye, and makes the "Briey plateau" a military prize of the very highest importance. That Germany would instantly strike to seize it in case of war could not reasonably be doubted.

We have seen that the Champagne, Argonne, Barrois, and Meuse cuesta scarps all trend from southeast to northwest across the Battlefield of Verdun. The Moselle cuesta at first trends almost due north, until it is a score of miles beyond the town of Briey; then, as if just discovering it was getting lost from the procession, it turns sharply west to resume its place in line parallel with its neighbors. Just north of it is the Sedan lowland or corridor, already referred to (p. 327), which is the equivalent of the Seille lowland. As in the south rivers like the Ache, Rupt de Mad, and Orne rise in the Woëvre lowland and flow obliquely northeast through the rising plateau to the Moselle trench or to the continuation of the Seille lowland, so in the northern section the Loison and Othain rise in the Woëvre and flow obliquely northwest through the rising plateau to the Sedan lowland, joining on the way the Chiers River, which gathers their waters and those of the Crusnes (in the plateau) into a single current uniting with the Meuse near Sedan. The fact that each of these rivers is parallel with the Meuse plateau scarp for much of its course gives them special value as advanced defensive positions in front of the main Meuse line. The Chiers curiously enough rises in the Sedan lowland north of the plateau scarp, flows into the plateau through a gateway carrying a railroad and guarded by the old fortress of Longwy, only to turn westward through the plateau past Montmédy and flow out again into the Sedan lowland below the junction of the Loison. It was at Longwy, the "Iron Gate of France," perched picturesquely on a cliff-rimmed spur of the plateau and dominating a vast stretch of country to the north and east, that the Germans first knocked with peremptory demands for the surrender of the fort when they attacked the outlying defenses of Verdun early in August, 1914, just as more than a hundred years before the first blows of the invader had fallen upon this same strategic point.

Still farther to the north lies the massive wooded barrier of the Ardennes, trenched by the wonderful serpentine gorge of the Semois River. The latter is also roughly parallel to the plateau and lowland belts and therefore to be considered in connection with them as a part of the outlying natural defenses of the Battlefield of Verdun.

CHAPTER VIII

MILITARY OPERATIONS ON THE BATTLEFIELD OF VERDUN

THE BATTLE OF THE FRONTIER

Early in August, 1914, after a series of encounters between patrols of enemy and Allied troops, considerable forces of the German Crown Prince's Fifth Army, directed against Verdun, advanced to the north-facing scarp of the Moselle plateau and vainly summoned the defenders of the Longwy gateway to surrender. Part of these forces pushed south over the plateau into the Woëvre lowland, where they encountered French advance guards on the line of the Loison at Mangiennes. A brisk engagement ensued, and the Germans were thrown back over the Othain. French mobilization being now completed and the necessity of meeting a strong German advance through Belgium becoming more and more apparent, the French Fourth Army under De Langle de Cary (Army of Sedan) was by the middle of August thrown forward upon the Ardennes upland north of the serpentine gorge of the Semois, the Third Army under Ruffey (Army of the Northern Woëvre) on its right extending from the vicinity of the Semois southeastward across the Moselle plateau to skirt its western base as far south as Conflans, where its right rested on the Orne River. The forts of Metz on the crest of the plateau east of Conflans denied the French a good topographic position in that sector; the valuable mineral fields of the Briey region were held by strong enemy forces; the crest about Longwy was already in the hands of enemy troops, and the fortress itself was invested; while De Langle's Fourth Army stood on a fairly open upland with a river gorge at its back. For one reason or another neither the Third nor Fourth French Armies occupied a strong defensive position when the Battle of the Frontier wa engaged.

To the west the French Fifth Army under Lanrezac (Army of the Sambre and Meuse) took up a defensive position in the Ardennes behind the deep gorge of the Meuse from the Givet* fortress northward to Namur,* thence westward behind the Sambre gorge to the vicinity of Thuin, beyond Charleroi;* while farther westward the British army continued the front behind the Haine River and the Mons-Condé Canal which runs through its broadly open valley. Despite a weak link where the front crossed from the Sambre near Thuin to the canal east of Mons,* this advanced position was in many respects exceedingly strong; and since the result of the preliminary battle for possession of the Meuse-Sambre front directly affected events on the Verdun battlefield to the southeast, a further word concerning the terrain may not be amiss. "In this portion of its course the Sambre is a strongly meandering river which follows a winding trench cut three hundred feet or more below the upland surface. The flat floor of the trench is five or six hundred yards in breadth, and covered with open meadows. As a rule the southern wall of the trench is steep and forested. An enemy advancing from the north would find it difficult to cross the exposed meadows, bridge the river, and dislodge the defenders from the wooded heights beyond." The main gorge of the Meuse between Givet and Namur is even more formidable. "Less strongly winding than the Sambre, it is cut deeper below the Ardennes upland and has steeper walls. The river is deeper and broader, and practically fills the bottom of the trench. Precipitous cliffs of bare rock rise in places several hundred feet from the water's edge. Where the slope is less steep, the walls are (usually) heavily forested." [1]

This position not only blocked the highly important Meuse-Sambre route through the Ardennes into France, but it also flanked the route of the main German advance over the Belgian plain farther north, compelling the invaders to turn and fight on a line selected by the defenders. But it possessed one marked

* Givet is 25, Namur 50 miles N of Mézières. Charleroi is 19, Mons 41 miles W of Namur. Cf. Fig. 55.

[1] D. W. Johnson: Topography and Strategy in the War, New York, 1917, p. 27.

element of weakness. The junction of the Meuse gorge with the Sambre gorge makes a little less than a right angle, with the result that Lanrezac's Fifth Army occupied a dangerous salient. The apex of the salient being the junction of two principal natural pathways through the mountains, it was the locus not only of an important town but also of the most valuable crossings over the two rivers. Thus the strategic value of Namur was very great, and for its protection there had been constructed a ring of concrete forts. It was believed that much of the weakness of the salient was offset by the strength of its apex, which apparently furnished a solid support for the right wing of the main Franco-British front facing north along the Sambre gorge-Mons canal line, as well as for the subsidiary French front along the Meuse line connecting (but imperfectly) with the Fourth Army in front of the Semois gorge.

It is evident that while the Allied center, consisting of the French Third and Fourth Armies, was placed on the margin of the Verdun battlefield in a position poorly adapted to withstand a heavy enemy attack, as has already been shown, the Allied left wing, composed of the French Fifth and the British Armies, held a strong defensive position. This was not inconsistent with the plan of General Joffre that his center and left should now (August 21) pass to the offensive, the center to attack at once, the left to remain on the defensive temporarily until the success of the center's offensive was assured. The French Third Army began its sweep up the backslope of the Moselle plateau with great *élan*, only to be checked and later thrown back by well-directed enemy fire from artillery concealed in forest and ravine, and which the French could not locate in the thick mist prevailing during the attack. The offensive of the Fourth Army was even more effectively halted. Instead of awaiting an enemy attack in a carefully selected and strongly organized position, these forces were advanced far into the difficult terrain of the Ardennes Mountains, a rolling upland partly cleared in some places, densely wooded in others, and trenched by the deep gorges of many streams. Here the Germans had quickly and skillfully established strong lines in front of the principal strategic points, defended by barbed wire entanglements and deep trenches. Reconnaissance of the wooded and dissected country was difficult, and not even the air scouts were able properly to locate the enemy defenses and forces. French columns in full march were overwhelmed and routed, others attacking hidden barriers were ruthlessly slaughtered, others fled precipitately. The Fourth Army, definitely defeated, fell back behind the Semois, and the engagement, sometimes called the Battle of the Semois, registered a victory for the enemy which had easily crushed an imprudently conducted advance into a formidable terrain.

Meanwhile the forces farther west were not only unable to initiate their offensive, given the failure of the center to advance, but were being desperately put to it to hold their defensive positions. As early as August 15 German forces were assailing the Meuse barrier near Dinant, 15 miles south of Namur. A first attempt to cross at Bouvignes north of Dinant was repulsed, likewise a second attempt still farther north. Late in the afternoon a stronger effort near Dinant itself was temporarily successful, but the enemy was soon thrown back in disorder, losing large numbers in the swift current of the treacherous stream.[1] A week later, on the day following the initiation of the French center's unsuccessful offensive, the enemy succeeded in forcing the Meuse barrier south of Namur and in widening the gap by pushing southward along the west bank toward Dinant. Another crossing south of Dinant placed the enemy well in the rear of the Allied line along the Sambre gorge and the Mons-Condé Canal. The Namur forts were already melting under a hurricane of high explosives, and in an incredibly short time the supposedly strong apex of the dangerous salient was in ruins and the Germans were pouring over the crossings to aid in outflanking the rest of the defensive position.

This position had, however, already become untenable. In the desperate battles raging for the passages of the Sambre gorge farther west, during the course of which Charleroi changed

[1] Pierre Dauzet: De Liège à la Marne, Paris, 1917, p. 29.

hands no less than five times, the enormous numerical superiority of the enemy was beginning to tell, and the French Fifth Army was losing its grip on the natural trench. Along the Mons-Condé Canal the British had found it impossible to maintain their bridge-heads north of the barrier against the increasing enemy pressure and so withdrew to the south side, destroying the bridges behind them. At Mons the canal makes a loop convex to the north, creating a dangerous *tactical salient* like the famous Tervaete bend on the Yser barrier (p. 57). When the danger here was increased by an outflanking move of the enemy farther east, where he was forcing the weak sector connecting the Sambre and the canal barriers, preparations were made for the possible abandonment of the salient. But, before the situation at this point became critical, the British Commander-in-Chief received from General Joffre the astounding news that the French Fifth Army on his right was in retreat, having lost the Sambre position; that overwhelming German forces were massing in his front; and that his left was threatened with a dangerous enveloping movement by the enemy. It was only too true that, the Namur fortress ring having fallen, the Sambre crossings having been forced, and the enemy having crossed the Meuse gorge far in the rear, the French Fifth Army was rapidly being flanked out of its entire defensive system. The battle of the northeastern frontier had been lost all along the line, and the heartbreaking retreat was already under way. The British and French Fifth Armies fell back by stages to the line of the Oise, as detailed in Chapter IV. De Langle de Cary's Fourth Army, its left wing in the air as a result of the Fifth Army's failure to hold its Sambre-Meuse defensive position, continued its retreat from south of the Semois barrier to the Chiers-Meuse line in the Sedan lowland and later entirely behind the line of the Meuse. The Third Army fell back behind the line of the Othain River, from Montmédy to Spincourt, where the enemy pursuit was halted and a decided reverse inflicted upon him. But the retirement of the Fourth Army behind the Meuse barrier compelled the Third Army to continue its retreat until it also stood behind the same obstacle.

On August 28 the enemy assaulted the Meuse barrier on a front of 40 miles from west of Mézières past Sedan and Stenay to the vicinity of Dun. For two days a violent struggle raged all along the line, German forces crossing at certain points only to be hurled back, failing completely in attempts to pass the obstacle at others, and maintaining with great difficulty precarious footholds on the southwestern bank at still others. At Cesse, close to Stenay, the invaders were thrown back over the river at the point of the bayonet, while near Dun an enemy regiment seeking to force the well-defended natural trench was almost completely annihilated.[3] Fierce combats at Donchery, Marfée Woods, Angecourt, Yoncq, Beaumont, and Jaulnay Wood were well sustained by the French. When General de Langle de Cary received Joffre's order for the retreat southward which prepared the stage for the Battle of the Marne, he was reluctant to forsake the strong defensive line which his troops had held so effectively, and asked leave to stand his ground. The reply was a permission to hold the barrier for another day, to complete the success gained and to demonstrate that the retreat was strategic and not forced; but also an insistence that retirement must then take place, and the line of the Meuse north of the entrenched camp of Verdun thus be sacrificed in conformity with the new plan of campaign.

The Battle of the Marne

The retreat to the Marne had its eastern pivot on the massive Verdun buttress, which we have found was so fashioned by Nature as to be almost impregnable from whatever direction assailed. The French Third and Fourth Armies in swinging back from the Meuse trench toward the southwest, then the south, and finally the southeast, like a spoke in a slowly turning wheel the axle of which was at Verdun, first found their combined front parallel to the successive topographic belts of the Verdun battlefield. The generalissimo's "Instruction Générale" of August 25 having prescribed that "the movement will be covered

[3] Dauzet, De Liège à la Marne, p. 70.

by rearguards left behind favorable depressions of the terrain, in such manner as to utilize every obstacle to arrest . . . the enemy's march or at least to retard it," good use was made of these natural barriers in holding the enemy in check and assuring an orderly retirement of the Allied forces to the line selected by Joffre as the basis of a new operation. Along the northwestern extension of the Meuse and Argonne plateaus, where they appear to overlap in a single upland having but one strong northeast-facing scarp, well clothed with forest, very important fighting occurred in the Signy l'Abbaye-Launois region 13 miles southwest of Mézières, in the last days of September, "zouaves, colonial troops, and native sharpshooters making marvelous use of the terrain." The Fourth Army held the Aisne-Argonne defenses from Attigny to Grandpré for a brief period. As the retreat continued, the front crossed the grain of the terrain at a marked angle, and only the minor east-west valleys and ridges, or sections of larger valleys, could be employed for defensive purposes. Foch's Ninth Army, put into line to fill a gap formed between the Fourth Army retreating southward and the Fifth Army withdrawing more toward the southwest, fought a brilliant action between Faux and Bertoncourt, along a low east-west ridge just north of the Attigny-Rethel sector of the River Aisne, then fell back south behind that part of the Aisne barrier. Farther east the Third Army, standing on the strong terrain dominated by the butte of Montfaucon, successfully repelled violent enemy assaults before resuming its retirement. Both the Third and Fourth Armies continued the retreat southward along the cleared lowlands between the forested plateaus, the Third Army utilizing the Aire lowland, the Fourth Army that of the Aisne.

When the order was received to cease retiring and pass to the offensive, the Third Army stretched obliquely across the Barrois plateau from the western exit of the Ornain gateway at Revigny to the eastern crest of the sloping upland near the minor gateway at Souilly, whence the mobile forces of the Verdun camp continued the front northward for some distance along the crest of the Barrois plateau scarp. It thus faced roughly west, toward the Triaucourt gateway around the southern end of the Argonne plateau, the lake-forest-hill country of the Little Argonnes just to the south, and the Revigny gateway between these hills and the forested heights south of Sermaize. If it should debouch westward through the gateways into the southward continuation of the Aisne lowland, it would take in the flank the German forces pushing southward along that lowland to attack the Fourth Army. This latter body of troops was facing nearly north, in a strong position behind the transverse barrier formed by the interlacing channels of the Saulx and Ornain Rivers, the Marne-Rhine Canal, and their adjacent meadows and marshes, backed by the forested heights of Sermaize.

General Joffre's orders to the Third and Fourth Armies for the Marne offensive directed that the Third Army, now under General Sarrail, should "debouch toward the west to attack the left flank of enemy forces marching west of the Argonne." Thus the favorable situation offered south of the Argonne barrier for a flank attack upon the southward-marching Germans was to be exploited. At the same time the Fourth Army was to engage the enemy in front. As we have previously noted in the more general account of the Battle of the Marne (p. 269), both the Third and Fourth Armies were compelled to support repeated enemy assaults of extreme violence and could develop their own offensive but slowly. Ground was lost at Revigny on the first day of the battle, although at the same time the French forces holding the crest of the Barrois plateau north of Souilly were able from their favorable position to push southwest down the backslope some distance and threaten the flank of the Crown Prince's forces, which were trying to drive southeastward and isolate Verdun.

On the second day of the battle the German threat to complete the isolation of Verdun by an attack from the east began to develop. To the north and east of Verdun there was no French army standing in front of the entrenched camp to defend it. The Third Army was on the west, but the Second Army, as we shall see later, was far to the southeast defending the Moselle plateau, with its left near Pont-à-Mousson (Fig. 79). A gap of more than 60 miles was practically undefended, save for the great barriers erected by Nature—the exposed plain of the humid Woëvre lowland, the imposing wall of the Côtes de Meuse, the wooded Meuse plateau, and the winding trench of the Meuse River—strengthened by the chain of fortresses guarding weak points and defended by the fortress garrisons and small covering forces. On August 7 General Sarrail learned that a serious enemy attempt to pierce the formidable Meuse barrier was under way. German forces from the entrenched camp of Metz crossed the Woëvre lowland with the object of scaling the plateau, capturing one or more of the forts, and breaching the line of the river south of Verdun. Success in this would not only isolate Verdun but would take in the rear General Sarrail's Third Army and possibly cause the collapse of the whole Marne counteroffensive.

That day and the following, while the battle raged furiously all along the Third and Fourth Army fronts across the Wet Champagne, the Little Argonnes, and the Barrois plateau, and ground was being lost at one place and gained at another without any very significant change, Sarrail prepared to meet the eastern attack on the Meuse barrier. Despite his need of troops on the main front, he sent one or two divisions to help hold the Meuse line south of Verdun and ordered the bridges over the Meuse River destroyed. Additional troops were despatched from the Toul forts to assist in meeting the new menace. Already the Germans, finding the plateau practically unguarded, had reached the upland near the ravine of Les Éparges and begun the bombardment of Fort Troyon, north of the western exit of the Creue defile. Their siege guns were so effectively concealed in the deep, wooded ravines dissecting the upland that the guns of the fort had great difficulty in locating them to put them out of action.[4] On September 9 heavy Austrian siege artillery opened on another of the forts, Génicourt north of Troyon; and

[4] F. E. Whitton: The Marne Campaign, London, 1917, p. 193.

Fort Troyon itself, practically silenced, was being assaulted by enemy infantry. But the German forces engaged in this raid were fatigued by their long march across the Woëvre and by their uphill advance into the difficult wooded plateau country, and failed to press the advantages which the absence of a large defending army offered them. French reserves from Toul arriving next day helped to drive off the attackers, and, while the heroic garrison suffered further enemy attempts during the next three days, the effort to breach the Meuse barrier was for the moment defeated. German troops on the two sides of the Meuse plateau had come within less than 15 miles of meeting and making the isolation of Verdun complete. Had the reinforcements which tardily arrived but been on the ground when the plateau was first approached, no such progress across the difficult terrain would have been effected.

Direct attacks on the entrenched camp of Verdun, defended by the fortress-crowned plateau scarp and transverse spurs and ridges, produced little effect. Two forts (Bois Bourru and Marre) on the spur projecting northeastward into the first eastward meander of the valley north of Verdun, were heavily bombarded; later Fort Douaumont on the crest of the scarp overlooking the Woëvre was attacked. But the formidable nature of the terrain and its military defenses discouraged any major operation directly against Verdun until very large forces and a crushing weight of artillery could be assembled for the almost superhuman task.

Meanwhile the Third and Fourth Armies had been beating off a series of terrific attacks against the Saulx-Ornain-Sermaize heights barrier which threatened for a time to force the French front from the protection of this important obstacle, and had been worrying the Crown Prince's army by launching attacks down the backslope of the Barrois cuesta against his communications in the Aire lowland. Two offensives undertaken by the two opposing armies in this sector largely neutralized each other. By September 10 the German offensive was weakening, and the great retreat induced through defeat of the German right flank on the Marne plateau was imminent. Slowly and methodically the

Germans withdrew northward along the Aisne and Aire lowlands, yielding to the French the Triaucourt, Les Islettes, and (after some delay) La Chalade defiles through the Argonne barrier. The enemy front finally came to rest transversely across the topographic belts of the Verdun battlefield, firmly entrenched in the difficult terrain of the Argonne plateau south of the Grandpré gateway, buttressed on the Montfaucon butte on the Barrois plateau which the French assailed at heavy cost without avail, and facing the natural and artificial defenses of Verdun in the Meuse plateau. The Battle of the Marne was ended.

That this battle terminated in an Allied victory is due in no small measure to the topographic features of the Verdun battlefield. Had the Germans succeeded in any one of their efforts to break and roll up the front of the French Fourth Army, or to pierce the Third Army front and surround Verdun, or to breach the fortified line of the Meuse from the east, the victory gained on the Marne plateau might have been largely dissipated. The first enemy objective was defeated on the Saulx-Ornain-Sermaize heights barrier. In aiming at the second the Crown Prince was seriously hampered by French control of the Barrois plateau crest trending obliquely across his flank and exposing his communications to enemy raids launched down the western slope, and by the fear of an enemy attack on his rear from the Verdun bastion. The third objective failed of achievement as soon as adequate forces were arrayed in defense of the naturally strong Meuse position. Joffre's strategy could not achieve its purpose if the strong buttress on which he pivoted his eastern front should give way. The immensely difficult terrain of the Battlefield of Verdun, almost impregnable against assault from any quarter, was equal to the demand which the generalissimo made upon it. Yielded in part in accordance with the larger strategic plan, it still remained a strong point of support when the command to pass to the offensive came. Projecting like a rocky peninsula into the gray German sea, Verdun stood unshaken while the waves of war lashed themselves into red foam against its flanks. The storm died down, and the sullen waves receded.

The First Battle of St. Mihiel Salient

Scarcely had the enemy's retreat been fully accomplished, and the battle along the Aisne trench was still raging, when the Germans launched another attack on the Meuse defenses. This time much larger forces from the Metz stronghold pushed across the Woëvre plain, penetrated the poorly defended plateau along the defile of Creue, or Spada, and succeeded in reaching the western bank of the river at St. Mihiel. Fort Camp des Romains, with the subterranean caverns of the meander spur on which it was built, fell to the enemy and formed a strong protection for the apex of the wedge which was thus driven into the natural defenses of the Meuse front. Vigorous efforts to enlarge the gains west of the Meuse were defeated by an effective defense of the river trench, while troops hurrying southward from Verdun and northward from Toul garrisoned the other topographic barriers of the region and checked the enemy's attempts to enlarge the wedge.

Although the danger was thus held within bounds, the Germans could henceforth profit from the strength of those obstacles they had seized while inadequately garrisoned. The southern side of the "St. Mihiel salient" was defended by the outlying ridge and butte of Le Mont and Mont Sec (Fig. 93), commanding a vast stretch of the marshy Woëvre lowland, and by the dissected and wooded upland of eastern Haye; while the northwestern side was solidly placed on the partially detached spur at Combres, defended on the west by the moat of Les Éparges valley (Fig. 90). Thus the projecting Hattonchâtel bastion was securely in German hands; the road, railway, and canal connecting Verdun with Toul were severed; the highly important Paris-Toul-Nancy railway was brought under fire; the investment of the Verdun stronghold on the north, east, and south completed; and a menacing wedge so firmly lodged in the Allied front that nearly four years would pass ere the danger was removed.

Local Combats

In the Argonne the less dissected portion of the plateau's gentle western slope covered by Grurie Wood and the dissected walls of the Biesme trench to the south were hotly contested for many months. Here the Germans pushed southward again, but the bloody struggles which made all the world acquainted with the names of small subdivisions of the forest—the Woods of Bagatelle, Fontaine-Madame, Four de Paris, and others—showed at how dear a price any advance in this savage country must be purchased. The defile of La Chalade was long the scene of a bitter struggle, each side seeking to possess and use this important pathway over the plateau. Violent German attacks in June and July, 1915, preceded by heavy bombardments and the copious use of poison gas, may have aimed at a rupture of the French front and the development of operations on a larger scale; but despite the fact that "certain negligences of the local command on the French side" gave the Crown Prince undue advantages, he was unable to exploit his initial gains in a country where Nature was the strong ally of the defense.

The Germans, having succeeded in holding the butte of Vauquois, two miles southeast of Varennes, had converted it into an apparently impregnable buttress of their line across the Barrois plateau. Together with the even more commanding Montfaucon butte, where the Crown Prince had his headquarters and which the French attacked both in 1914 and 1915 without success, it seemed to guarantee stability to this sector of the enemy front. The French Third Army assaulted the Vauquois buttress about the middle of February, 1915, in a vain effort to carry it by storm. But it was too formidable an obstacle to be taken without serious preparation. Ten days after this preliminary attack a violent bombardment paved the way for an assault in force pushed home with the greatest energy. After several days of murderous combat the French secured the coveted height (Fig. 85).

The Second Battle of St. Mihiel Salient

Early in 1915 the French undertook the task of crushing in the St. Mihiel salient, in order to remove this menace to the Meuse barrier defenses and to free the corridor of the Meuse from German interference. Both sides of the salient were vigorously attacked in February, March, and April. The strong natural trench and ridge positions at Les Éparges and Combres, buttressing the northwestern side of the salient, were partially conquered after repeated costly assaults. Little progress was made on the rest of the forested upland, but in the lowland to the northeast the attacking troops plowed through the mud for larger gains before brought to a halt east of Fresnes. On the south an attempt to push north and seize the railway which from Metz followed up the Rupt de Mad to Thiaucourt, the center of supply for most of the salient, was brought to a standstill before either the mud of the Woëvre or the dissected plateau of the Haye had been traversed for more than a mile or two. The long and bitter struggle to gain the partially detached plateau upland northwest of Pont-à-Mousson, bounded on the east by the Moselle trench, on the north and south by two of its lateral tributary gorges, and covered by a forest (Bois le Prêtre) made famous by many bloody combats in its maze of quarries and caverns, did not terminate favorably to the French until the second week in May; and a further advance over the similar terrain beyond was not deemed feasible. However hazardous to the German troops within it the St. Mihiel salient might appear, and however tempting to attack, it was in reality so strongly protected by natural obstacles that nothing less than an overwhelming superiority in artillery fire and man power could crush it. After suffering very heavy losses and registering but slight gains, the French abandoned the attempt.

The Battle of Verdun

On the morning of February 21, 1916, there suddenly opened against the northern and eastern defenses of Verdun a bombardment more terrible than anything hitherto experienced in war. For a complex of reasons which it is not here necessary to examine but which included among them the desire to crush the formidable bastion which might at any moment serve as the base for a dangerous Allied operation against the Briey plateau and its mineral deposits, or against the entrenched camp of Metz, the German high command had deliberately decided to strike at one of the strongest points on the whole Allied front. It was hoped that the suddenness and overwhelming power of the assault would offset the natural difficulties of the terrain, which latter the Germans fully understood. The permanent fortresses were less to be feared than the barriers of Nature, against which the strongest explosives were impotent. Experience at Liège, Namur, Maubeuge, and elsewhere had demonstrated that no masonry fort could withstand for long the crushing weight of fire developed by heavy siege artillery; but against plateau scarps, massive limestone ridges, and deep natural moats no impression could be made. Only in case the defenders of the terrain could be destroyed or demoralized by the fury of the bombardment, and the plateau then overrun by resistless waves of men, could victory be achieved. The German army had long been assembling the vast stores of munitions and the masses of men necessary for such an undertaking, building the dozen or more new railways necessary to feed them rapidly to the front of attack, training the forces of assault, and rehearsing the operations in which they were to participate.

The French, equally realizing the danger of trusting to obsolete forts for the defense of a given point and appreciating the superior value of natural obstacles, had organized the Verdun terrain with much skill. They could still learn from the Germans valuable lessons in the art of utilizing to the utmost the defensive advantages offered by a favorable topography, and in the Battle of the Somme would soon purchase such instruction at a heavy price. But what was lacking in art on the Battlefield of Verdun was more than compensated by Nature. The plateau scarp, the asymmetrical meander spurs interlocking along the Meuse trench, the cross ridges between lateral ravines, the narrow bridges of the upland surface connecting one parallel ridge with another (p. 366)—all were seized upon and strengthened by an elaborate system of field works, including trenches, tunnels, dugouts, and wire entanglements, linking up forts, fortified villages, quarries, and subterranean caverns. Pushed beyond the outer limits of the fortress ring were field works designed to break the force of the enemy onslaught on one line of ridge and spur after another, before the forts themselves could be assaulted. Big guns were taken from the forts and moved to better positions on the far stronger natural barriers. To the limestone hills of Verdun, rather than to walls of steel and concrete, was the safety of France to be confided.

The terrific bombardment of February 21 lasted only nine hours, but was delivered by 2,000 great guns throwing shells of medium and largest size charged with high explosives. The whole surface was transformed into a labyrinth of deep craters. Then the infantry waves leaped forward to the assault. Enough Frenchmen had lived through the inferno to defend the plateau scarp and cross ridges with tenacity. A scant footing only was gained on the former north of Ornes, opposite the outlying butte of Les Jumelles already held by the Germans. The attack from the north was quickly checked on the northern part of the Caures Wood cross ridge (which is the eastward continuation of the Côte de Talou meander spur), after the Germans had overrun all of the minor Haumont ridge just to the north except the village of Haumont itself. A second and a third day of terrific bombardment, effectually preventing the supply of food, munitions, and other aid to the defenders, and repeated infantry assaults were necessary to complete the conquest of the Haumont ridge and to secure full use of the connecting spur or "bridge" joining it to the Caures Wood ridge. Two French divisions, thinned by heavy losses and worn almost to the point of exhaustion, were sufficient on this remarkable terrain to hold five divisions of picked German shock troops, supported by one of the heaviest bombardments in history, to very moderate gains. Snow and mud made the lot of the defenders more pitiable, but increased the difficulties to be overcome by the advance. Two more days of the hurricane of explosives, steel, poison gas, liquid fire, and all the other murderous implements of modern scientific warfare, supplemented by prodigal sacrifices of human lives, carried the enemy over the Caures Wood ridge and along the broad Beaumont "bridge" connecting it with the Louvemont-Côte du Poivre ridge next to the south.

In five days of the most terrible warfare ever known, two of the northern defensive barriers of Verdun had been captured. The unexpectedness of the attack (the French knew preparations were in progress, but various facts indicated it was not imminent) and the unheard-of fury with which it had developed made the progress of the enemy disquieting, although the German high command was deeply chagrined that their "colossal" efforts had failed to smash through to immediate victory. On the night of the 24th Joffre approved General de Langle de Cary's plan to have his forces in the Woëvre lowland, menaced by the German advance southward along the plateau, fall back upon the solid barrier of the Meuse escarpment east of Verdun, but decisively rejected the suggested possibility of withdrawing the forces north of Verdun to the west bank of the Meuse lest, caught between overwhelming numbers of the enemy in front and the river at their backs, they should meet disaster. De Castelnau, whose keen appreciation of the defensive value of topographic barriers had made him the savior of Nancy (p. 481), was sent to examine the situation on the ground, and assured himself that the defense could be successfully maintained on the cross ridges. Pétain was then placed in charge of the further conduct of the battle.

The difficulties of defending the Verdun terrain were enormously augmented by the lack of proper transport facilities. With the only railroad to the south cut at St. Mihiel and the only railroad to the west, via the Les Islettes defile through the Argonnes, rendered useless by enemy artillery fire, there remained but the narrow-gauge line and the highway southward along the poorly developed lowland at the base of the Barrois plateau scarp. The highway, greatly broadened and improved, was the main supply line, and along this "Sacred Way" there passed 6,000 auto vehicles daily, transporting 90,000 men and 50,000 tons of material between Bar and Verdun every week.[5]

[5] Victor Giraud: Histoire de la Grande Guerre, Paris, 1919, p. 334.

The fierce struggle continued without pause, and new assaults carried the meander spur of Côte de Talou and most of the Louvemont-Côte du Poivre ridge (Fig. 94). At last the enemy was knocking at the doors of Douaumont, the northernmost of the Verdun forts. But it was the "bridge" leading to the Douaumont-Côte de Froide Terre ridge and the low knoll in the center of this ridge upon which the fort was located (Fig. 96) rather than the fort itself which were the objects of the bitter fighting. Those who may have imagined that it was the fortress ring rather than the natural obstacles on which the defense was based may find confirmation of the limited rôle of the old forts in the fact that Fort Douaumont was entered by a German patrol which found the drawbridge down and the tiny garrison of 23 men, left to serve the one gun not dismantled for use elsewhere, sound asleep. The enemy installed himself in the fort without firing a shot.[6] Other portions of the permanent defenses were stripped of their guns and garrisons for use in the field, and surrendered to the enemy without special resistance. There is reason to fear that this dismantling of the fixed forts was carried too far, for with their protected guns and their deep subterranean shelters they can prove most valuable supplements to field works.

For more than a week the battle raged around the high crest of Douaumont, about the dismantled fort. The enemy's first assaults had been favored by the covered woodland patches, but these were now shot away. Still he sought to penetrate the plateau by the eastern ravines and work westward and southward by utilizing the advantageous positions of the plateau "bridges." In the end the eastern portion of the Douaumont-Côte de Froide Terre ridge was held by the enemy; but his rate of advance across the ridges and along the connecting "bridges" was becoming dishearteningly slow. Bloody assaults and the repulse of repeated French counterattacks were required for every foot of ground gained. Another difficulty of the terrain was beginning

[6] Joseph Mangin: Comment finit la guerre, *Rev. des Deux Mondes*, Vol. 56, 1920, pp. 481–520, 721–762; Vol. 57, 1920, pp. 241–285, 481–537, 774–815; Vol. 58, 1920, pp. 74–101; reference in Vol. 56, p. 729.

to make itself felt more and more: the narrowness of the topographic belts on the Verdun battlefield. On the east the zone of attack was limited by the plateau scarp of the Côtes de Meuse. Under other conditions the French troops facing north would have been exposed to a galling fire in their rear and to the constant danger of a flank attack. But with the enemy on the exposed plain of the Woëvre, the French positions on the plateau were secure. German artillery in the plain was unable to control by direct observation its fire on objectives concealed from view on the upland beyond the crest; and, if it placed guns for this purpose, they could soon be located by the dominating observation posts on the heights, and put out of action. To the west the German front of attack was limited by the deep winding trench of the Meuse, which made it difficult for troops on the two sides to co-operate effectively, the more so as at this season of the year the valley floor was more or less constantly a mire of mud or a vast sheet of flood water. From the scarp crest to the valley trench was a front of only five to eight miles, and, with such a limited sector to defend, the French, now adequately warned and taking advantage of the admirable defensive features of the terrain, were equal to the worst which the Germans could attempt. Furthermore, the advanced German line on the east side of the river was now being fired upon from the rear by the French on the west side, who retained their old positions farther north. The Crown Prince was at last compelled to widen his zone of attack despite the disadvantages caused by operating astride the river barrier; but this decision was too tardy to permit his reaping the advantages which might have come from a broad attack pushed home with vigor while surprise was possible.

On March 6 the Germans attacked the natural trench of the Forges brook, which was protected on the south by the dominating meander spur of Mort Homme-Côte de l'Oie, continued westward by the point of the Barrois plateau scarp known as Hill 304 which dominated the whole (Fig. 94). The stream trench and this line of heights formed a single strong defensive system which it was necessary to conquer by a single blow if the initial operation was to be successful. The eastern end of the meander spur, forming the Côte de l'Oie, subjected on all sides to German fire, was captured by the following day; but wave after wave of the assaulting troops swept up the northern glacis of the Mort Homme only to be massacred wholesale. Day after day the frightful slaughter continued, and still the grim summit of Dead Man's Hill rose unconquered above its mantle of corpses. In the meantime simultaneous attacks were in progress east of the river. Efforts to capture the Côte du Poivre and so to complete the conquest of the Louvemont-Côte du Poivre ridge were defeated; while farther east enemy columns entered the face of the escarpment by the Vaux ravine, capturing part of the village of the same name as they worked up toward the ravine head, where they hoped to debouch on the plateau at the Fleury "bridge" (Fig. 95) and around Fort Vaux. In this last attempt they were checked, a vigorous French counterattack throwing back enemy units which reached the plateau bridge at Fleury and were pushing south to the Côte St. Michel-Fort Vaux ridge.

Baffled by the resistance encountered, the Germans returned to artillery bombardment, which raged with extreme fury all along the northern front for two days. Then, on March 14, the assaulting waves once more hurled themselves against the solid rock defenses. Again the slopes of Mort Homme flowed blood, and the scarp at the Vaux ravine became an inferno in which charge after charge broke impotently upon the steep slopes. Soon the attack extended to the western end of the Forges brook trench and its guardian Hill 304, the latter being assaulted from the shades of the forest about Avocourt on the western slope of the Barrois plateau. Slight advances were made here and there during a week of murderous fighting, but at a cost out of all proportion to the gains scored. The main defenses stood unshaken.

Having failed to carry the natural fortress by the first grand assaults on the two sides of the river, the Crown Prince now set about to batter down the walls by a continuous series of sledge-hammer blows. By the end of March he had secured the western end of the Forges trench at Malancourt (Fig. 94); early in April the central portion at Béthincourt. The entire trench was now his, but it could be held only at a terrible price so long as the French on Mort Homme ridge and the plateau spur of Hill 304 poured high explosives into it with deadly precision. At whatever cost, these dominating heights at the south must be conquered. Hence on April 10 and 11 a furious assault was directed against the positions from the north and west. Mort Homme was smothered by the gray waves and passed to the enemy, buried to its summit in the mounting shroud of German dead. The vital height of Hill 304, which looked down upon Mort Homme and so prevented the Germans from utilizing their gain, withstood every shock. Several weeks more of intense fighting were required to give the Germans a foothold on this important spur, and even then they failed to gain complete control of it. East of the Meuse the Côte du Poivre defied German efforts to complete its conquest, and the Vaux ravine, of vital importance as a key giving access to the plateau "bridge" of Fleury and the upland around Fort Vaux, was still but partially in German hands, notwithstanding innumerable costly attacks. At one moment (April 2) German units had succeeded in debouching from the ravine and in pushing southward from Douaumont ridge along the Fleury "bridge" (Fig. 95), thus threatening the Côte St. Michel-Fort Vaux ridge just to the south; but a vigorous French counterattack restored the situation. Pétain having been named commander of the central group of armies, Nivelle, ably seconded by Mangin, now directed the heroic defense.

During April and May the bitter struggle continued without appreciable success on the part of the attacking forces. Defeated by the strength of topographic barriers which were defended with unflinching determination and unfaltering courage by the heroic soldiers of France, the Crown Prince dared not confess defeat. So he continued the unequal contest. Douaumont ridge and the ruins of Fort Douaumont, the latter better defended by the Germans than formerly by the French, were partially recaptured by a French counterattack but quickly lost again. Early in June, after a violent bombardment of several days' duration, the escarpment near Damloup was assailed and the eastern end of the Côte St. Michel-Fort Vaux ridge, including Fort Vaux itself, captured by the Germans after days of bloody combat during which the Germans held the upper part of the fort while the French resisted heroically in its subterranean galleries.[7] At the same time the enemy pushed doggedly forward from Douaumont ridge over the Fleury "bridge" to attack the Côte St. Michel-Fort Vaux ridge in its center, about Fort Souville (Fig. 95). Again the danger of disaster in case the defenders should be overwhelmed in front while blocked by a deep river trench in their rear was discussed by the French command, and Pétain for a second time called the attention of Joffre to the gravity of the situation. The latter, imperturbable, replied that "it was necessary at any cost to continue the struggle on the right bank."[8]

The Crown Prince now prepared a supreme blow against the Côte St. Michel-Fort Vaux ridge, the last barrier protecting Verdun on the northeast. In the scheme of the fortified camp four forts were aligned along its crest from west to east: Fort Belleville, Fort St. Michel, Fort Souville, and Fort Vaux (Fig. 95). The latter work, together with the eastern end of the ridge, was already in German hands. Could the center of the ridge about Fort Souville be conquered, the western end with the remaining forts could be taken in the flank and soon disposed of. Then the defenders would be forced to escape to the western bank of the Meuse River across bridges under direct enemy fire. The stage was carefully set for a French disaster, and about the 20th of June the "colossal" bombardment began. In one day more than 100,000 shells of asphyxiating gases fell on the narrow front, in addition to the deluge of high explosives. On the 23rd the field-gray waves swept over the Fleury "bridge," deluged the village and its defenses, and beat about Fort Souville. The western, or Froide Terre, end of the Douaumont-Côte de Froide Terre ridge, which still held out against the enemy, was assaulted

[7] Mangin, *Rev. des Deux Mondes*, Vol. 56, p. 735.
[8] *Ibid.*, p. 736.

in vain; and even the western extremity of the Côte du Poivre ridge farther north resisted complete conquest. But "the situation was grave, for the German wave was now ready to beat against the Belleville ridge (Côte St. Michel), the last dike separating it from Verdun. It had reached the heads of the ravines descending from Froide Terre toward the Meuse, and the Côte du Poivre was in danger of being submerged with its defenders taken in the rear." For the third time Pétain represented to the generalissimo the danger of the situation, and once again the imperturbable Joffre commanded that Verdun should continue to be defended on the right bank of the Meuse.[9]

A few days later the great Franco-British offensive on the plain of the Somme (p. 144) burst in all its fury, with the relief of pressure on Verdun as one of its objectives. In July, and again in August, the German assault waves were to break against the Côte St. Michel-Fort Vaux ridge about Fort Souville at the southern end of the Fleury "bridge," but in vain. The eyes of the German army were turning anxiously toward the danger on the Somme, and the storm at Verdun died away. Germany's supreme effort against the formidable terrain of the Meuse plateau had failed of success. Barred from any major assault from the east by the great wall of the plateau's east-facing escarpment, the Army of the Crown Prince was forced to attack southward across a series of parallel ridges and ravines, each of which exacted a price in German blood which left the enemy measurably weakened. New levies could be thrown into the struggle and further advances purchased at the price demanded; but in the end the natural defenses of Verdun were as a whole too strong for the assembled might of Germany to overcome. And when, a few months later, the French took advantage of Germany's preoccupation with the menace of the Somme to launch at intervals several brilliant attacks, the Côte de Froide Terre-Douaumont ridge and part of the Côte du Poivre-Louvemont ridge were successively recovered.

[9] Mangin, *Rev. des Deux Mondes*, Vol. 56, p. 737.

But the cost to the French of assaulting the ridge positions was too great to permit unlimited progress, despite the fact that the Germans at first had but a single line of defense with inadequate outposts, while the mechanism of the French artillery preparation and of the advance itself were superbly planned, with an ingenuity never matched by their Teuton antagonists. The difficult terrain, now more formidable than ever by reason of the labyrinth of shell holes filled with mud and water, could only be reconquered piecemeal, albeit in larger slices than the enemy was able to seize. The common statement that in two or three short engagements the German gains of ten months of bloody struggle were wiped out, and the enemy thrown back on his lines of departure, is a gross exaggeration. A large part of the territory gained in the enemy's offensive remained in his hands. But the French recovered all that was necessary to make Verdun reasonably secure and to demonstrate to the world, and particularly to the German people, that the bloody campaign against the impregnable stronghold was a complete failure as well as a ghastly blunder.

The last flame of the Battle of Verdun flashed up brightly in August and September of the following year, when as one element in a series of combined offensives the French by a succession of brilliant strokes practically completed the recapture of the Côte du Poivre-Louvemont ridge and gained the Côte de Talou meander spur forming part of the Caures Wood ridge next to the north. On the western side of the Meuse the famous Mort Homme-Côte de l'Oie ridge was seized (Fig. 94), as well as the part of Hill 304 gained by the enemy the previous year; and the Army of the Crown Prince once more found itself down in the Forges brook trench under a deadly fire from the French. The operations of these two months put one more complete ridge barrier between the enemy and Verdun, and furious German counterattacks were unable to recover the lost terrain.

The Third Battle of St. Mihiel Salient

Ever since the abortive attempt of the French to extinguish the St. Mihiel salient early in 1915 the Allies had been menaced with the possibility of an extension of the bridgehead which the Germans held on the eastern side of the Meuse barrier. The defense of Verdun had been rendered immensely difficult by the presence of this thorn in the side of the French, which not only cut an important artery carrying life to the entrenched camp but threatened at any moment to open a wound which might prove fatal. Even after the collapse of the Crown Prince's disastrous attack on the entrenched camp, the possibility of an expansion of the salient had always to be faced. To prevent this the French had organized a series of strong defensive positions along the Meuse trench and the plateau upland and scarp on the northwest; while first the French and later the Americans prepared four main defense lines facing the southern side of the salient across the Woëvre plain and Haye plateau from the base of the scarp near Commercy to the Côtes de Moselle east of the Moselle trench. Of these lines, most of which had variants, the first was based in part on the marshes and lakes of the upper Rupt de Mad, and the butte of Mousson in the Côtes de Moselle; the second in part on the lake-forest massif of the Forêt de la Reine in the Woëvre, the Forest of Puvenelle on a plateau tableland in the Haye nearly surrounded by deep gorges, and the Ste. Geneviève mesa in the Côtes de Moselle; the third on lake barriers in the Forêt de la Reine, the deep gorge of the Ache River through the Haye plateau, and the valley trench and plateau tableland next south of Ste. Geneviève; the fourth on the Meuse plateau scarp near Fort Gironville (Fig. 81), the Terrouin River trench to its junction with the Moselle at Aingeray, thence the trench of the Moselle and the Forest of Haye eastward to where the river bends sharply northward, north of Nancy. Thus was the danger of the enemy's exploiting the St. Mihiel gain minimized by a skillful use of such topographic barriers as ran transverse to the general grain of the terrain. But the danger remained.

When the progress of the Victory Offensive in the summer of 1918 made desirable a northward advance of the French and American armies holding the front from west to east across the Verdun battlefield, it became essential to remove the menace of the St. Mihiel salient; or, to borrow the figure and words of another, to "cut the thread, or rather the cable, which for too long a time had held us tied by the foot in the region of the Meuse."[10]

The task was confided to the American forces under General Pershing, assisted by important French units. Originally planned earlier in the summer as an operation of considerable magnitude, the undertaking had later been limited to the simple suppression of the salient in order to disengage the Verdun-Toul and Paris-Nancy railways and to prepare a good base of departure for subsequent more important operations. This end was to be accomplished by having the First and Fourth American Corps forming the southern side of the salient swing northward and northeastward, the left wing to reach the point of the Hattonchâtel bastion at Vigneulles and the right to pivot on the solid buttress of the dissected eastern edge of the Haye region where trenched by the Moselle gorge, crowned by the famous stronghold of Bois le Prêtre, and defended by the outlying barrier of the Côtes de Moselle (Fig. 78). At the same time the Fifth Corps on the margin of the Meuse plateau to the northwest should attack the strong ridge position at Combres buttressing the salient near Les Éparges (not completely conquered by the French in the Second Battle of St. Mihiel Salient) and endeavor to crush in this side of the salient by advancing southeast to Vigneulles. As these two jaws of the trap closed to meet at the point of the Hattonchâtel bastion, the French Second Colonial Corps, holding the apex and much of the west side of the salient, would close in on St. Mihiel.

General Pershing was under no illusions as to the formidable character of a position which had defied two previous attempts at its reduction. "Its primary strength," he writes, "lay in the natural defensive features of the terrain itself."[11] The wooded

[10] Louis Madelin: La Bataille de France; 21 mars-11 novembre, 1918, *Rev. des Deux Mondes*, Vol. 52, 1919, pp. 798–853; Vol. 53, 1919, pp. 59–99, 270–310, 533–569, 785–828; Vol. 54, 1919, pp. 64–108, 314–363; reference in Vol. 53, p. 794.

[11] Final Report of General John J. Pershing, Commander-in-Chief, American Expeditionary Forces, *Annual Rept. War Dept. for 1919*, Vol. 1, Part I, pp. 547–642, Washington, D. C., 1920; reference on p. 587.

heights of the Meuse plateau as well as the ravine of Les Éparges and the Combres ridge, buttressing the western side of the salient, have already been described. Along the southern side the attacking troops would have to advance over the difficult terrain of the wet Woëvre plain. Here "the front lines lay in a low, marshy valley, so that the trenches were always awash with mud, and each night the demand for working parties required almost the entire garrison to repair and drain these trenches." "On account of the swampy nature of the country" the American Commander-in-Chief considered it "especially important that the movement be undertaken and finished before the fall rains should begin, which was usually about the middle of September." In describing the original American front along the south side of the St. Mihiel salient, Captain Shipley Thomas observes: "But the one great feature of the landscape was the German position. The German front lines were also in the marsh, and their support lines on the rising ground beyond, somewhere in the edge of the woods, but above all was Mont Sec. Immediately in front of the Toul sector stood this hill—solitary, conical like a sugar loaf, 457 feet above the valley bottom—and from it the Germans saw every move that was made in the sector and were able to adjust their artillery fire with the greatest nicety. Mont Sec was just 2,000 yards behind the front line, and, with its concrete dugouts and tunnels and observation posts, it commanded the entire country for miles around—a fortress unassailable by any force at our command."[11] Now the American forces were enormously augmented, and outflanking operations were to reduce that fortress.

On September 12, after a bombardment of four hours, the infantry launched the assault. With remarkable precision the newest army of the Allies executed its orders as if on maneuver. The American forces on the southern side of the salient, "plodding almost knee-deep in the awful quagmire, . . . followed the barrage which splashed in the mud in front of them" and swung round northeastward on Thiaucourt, extending their left toward Vigneulles; while American and French troops on the north-

[11] Shipley Thomas: The History of the A. E. F., New York, 1920, p. 58.

western side rushed the defenses of the Les Éparges ravine and Combres ridge, defended by an Austro-Hungarian division which, as Ludendorff mildly expresses it, "might have fought better," and pushed into Vigneulles as directed. The imposing positions of Le Mont and Mont Sec, outflanked by the American advance and assaulted by the French, fell at a blow.

At last the dangerous wedge which had harassed the Allies for four years had been blotted out by the employment of a weight of men and material sufficiently great to overcome all obstacles and by the handling of these elements in a manner which earned for the American Commander-in-Chief the warm commendations of Marshal Foch. The troops assembled for this one short operation made an army nearly four times as large as Grant's Army of the Potomac at its maximum strength, three times Napoleon's Grand Army at Leipzig, nearly twice the German army at Sedan in 1870, and much larger than either the Japanese or Russian armies at Mukden, the largest on record before 1914.[13] Unfortunately, the enemy had ample warning of the impending blow and was actually in process of evacuating the salient when it fell. Both American and French officers talked much and indiscreetly of the forthcoming American *première*, Paris papers hinted at it, and Swiss papers discussed the probable date of its occurrence. When the Germans observed the enormous concentration of troops, outnumbering them more than five to one on this front, they began the withdrawal of guns, material, and men. Hence the weakness of the enemy resistance, the low casualty list due to his gunfire, and the disappointingly small number of prisoners captured (14,500) in view of the suddenness with which the salient was pinched out. Yet the victory was clean-cut, and convinced friend and foe that the new American army was a power to reckon with.

THE BATTLE OF THE MEUSE-ARGONNE

Conditions were now ripe for the launching of the Franco-American drive northward toward Mézières and Sedan which, in

[13] Thomas, The History of the A. E. F., p. 210.

connection with the British drive eastward to Maubeuge, was to cut the vital artery of German communications rimming the Ardennes Mountains, split the German armies in twain on the Ardennes wedge, and consummate the Allied march to victory begun July 18 on the Marne. Marshal Foch had directed that the French Fourth Army, now under Gouraud, and the American First Army under General Pershing should attack in unison about September 20 or 25 on a front extending from the Suippe River in the Champagne to the Meuse River near Verdun, the trench of the latter to serve as a protective barrier for the right flank of the advancing Americans.

One of the Germans' greater defensive positions lay back of their present front before the ultimate objective should be reached. This was the Hunding-Brunhilde Stellung (Fig. 55), based on the Aisne trench from west of Rethel in the Champagne to Attigny, thence southeastward along the Aisne trench and the Argonne barrier to the junction of the Aire, thence eastward through the Argonne along the deep, broad defile of the Aire to Grandpré, and on eastward as the "Kriemhild Stellung" across the badly dissected and wooded terrain of the northern Barrois plateau, to join the Antwerp-Meuse line at the Meuse River. The Kriemhild sector, like the rest of this counterpart of the Hindenburg Line, was really a broad zone of successive defensive lines skillfully sited so as to derive every advantage from the topography. The "Antwerpen-Maas Stellung" would be reached (but not passed) when the ultimate objective in the Mézières-Sedan lowland corridor (p. 327), the vital railway artery on which a large part of the German armies depended for their existence, was finally attained.

In the advance Gouraud's army would traverse the Dry Champagne and the Aisne lowland, starting from a front extending from Aubérive-sur-Suippe to Vienne-le-Château at the exit of the Biesme valley from the Argonne. To the American army fell the almost superhuman task of advancing over the Argonne plateau, in which "the ravines, hills, and carefully sited systems of defense concealed by thick underbrush had hitherto been considered as impregnable." East of the Argonne plateau the narrow Aire lowland, the wooded plateau of Barrois dominated by the buttes of Vauquois and Montfaucon and dissected farther north into a maze of hills and valleys, and finally the parallel ridges and ravines of the Meuse plateau west of the River Meuse and of the Barrois plateau east of the River Aire, made a terrain as formidable as one could well imagine outside of alpine mountains. The wooded hills offered a multitude of concealed machine-gun positions from which cross fire, far more demoralizing than fire from in front, could be directed on troops advancing over the open, cleared spaces. Across such barriers lay the American path to victory, starting from the front Vienne-le-Château–Vauquois butte–Forges ravine.

On September 26 the two armies assaulted the German line along the whole of the selected front. Despite the difficulties of the terrain the Americans swept forward several miles over the Barrois plateau to the base of the stronghold of Montfaucon and across the broad ridge north of the Forges ravine. The German defenses had literally been crushed by a hurricane of steel and high explosives, and enemy resistance was for the moment paralyzed along most of the line. In the Argonne plateau, as might be expected, the advance was a little less marked; but in the open Aisne lowland the French made good gains. The crest of Vauquois butte was quickly carried, but a first assault on Montfaucon was hurled back. After brief preparation for a second attack the Americans, with magnificent dash, stormed the dominating height. Let a Frenchman speak:[14] "They captured Montfaucon, the eagle's nest, and its butte, considered impregnable since we, in 1914 and 1915, expended against it so much effort and so much blood. It was a magnificent conquest. Montfaucon (342 meters high) dominates the whole region from the Meuse to beyond the chain of the Argonne. This peak constituted ever since 1914, I can testify, a painful obstacle in the eyes of those defending the northern part of the camp of Verdun. To have carried Montfaucon by storm was a magnificent exploit, a new

[14] Madelin, *Rev. des Deux Mondes*, Vol. 53, p. 808.

proof of the bravery of American fighters." In two days the Americans had conquered as much ground as the French high command had believed possible in the whole fall campaign on this difficult terrain.[15] With the strategic point of Montfaucon in its possession, the American First Army pushed on for further gains during the next few days. West of the Argonne the French advanced steadily in the Aisne lowland.

In the Argonne, however, the story was very different after the first day's surprise attack. Here the Americans confronted the most nearly impregnable natural defenses of the whole terrain. To the ridges and ravines, dense forest, and denser underbrush were added a labyrinth of trenches, invisible wire entanglements, machine-gun nests in the trees, on the surface, and under ground, and a system of tunnels and subterranean galleries which made this section of the front renowned for the military engineering works carefully perfected by the enemy during four years of comparative quiet. Rain began to fall, and the shell-torn soil containing much clay became a mire in which the movement of men and materials was painfully difficult. Both on the east and the west the battle line would bend far northward with large daily gains. But in the Argonne the gains must be small and dearly purchased, for neither bulldog determination nor dashing bravery could beat down the walls which Nature opposed to the advance. It was accordingly decided that the principal attacks should be delivered east and west of the Argonne, with the object of pushing well forward on either side of the barrier and, by threatening from both flanks and the rear the enemy forces in the forest, compelling them to evacuate their strongest positions. Thus would the incalculable slaughter involved in a major operation through the forest itself be avoided. "The Argonne Forest was in fact so strong that the entire scheme of attack of the First American Army and the Fourth French Army west of it was governed by the necessity of forcing its defenders from it by outflanking rather than by direct attack."[16]

It is evident, however, that such a flanking operation must

[15] Pershing, Final Report, p. 586.
[16] *Stars and Stripes*, quoted by Thomas, The History of the A. E. F., p. 237.

prove a double-edged sword capable of cutting both ways. As the French on the west and the Americans on the east pushed farther and farther north of the German line on the Argonne plateau, not only could they assail the enemy's flanks, but the Germans in turn could from their commanding strongholds on the forested upland pour a galling flank fire into the rear of the advanced Allied positions. This soon proved a source of much embarrassment to the Allies, and made necessary some frontal attacks in the Argonne Forest in order to expedite the enemy's withdrawal. Thus General Gouraud, depending on the advance of his center and right to turn the formidable Moronvilliers massif east of Rheims (p. 268) from which an accurately directed artillery fire was holding his left in check, considered that the increasing danger to his right flank and rear from the Germans remaining well to the south in the Argonne made the cessation of his offensive necessary by September 30. That portion of the American army east of the Argonne, caught between the Germans holding fast well in their rear along the forested Argonne plateau on the one side and the Germans left far to the south on the wooded Meuse plateau east of the Meuse River on the other side, suffered heavy casualties until the Argonne and Meuse plateaus were cleared. The detailed history of the Meuse-Argonne offensive [17] is one continuous and striking record of the influence of terrain upon the tactics of a great battle. Along the center of attack we find the advancing forces delayed by the hill fortresses of Montfaucon, checked at Gesnes Creek and the deep Exermont ravine, held in front of the group of hills just north of these natural trenches, "which seemed to have been piled in there together to make this place impassable to any assault," and which "was the key position commanding the whole front of the American First Army." In advancing over a part of these hills the famous 1st Division suffered the heaviest casualties of all divisions in the Meuse-Argonne offensive. On the right wing one division after another executes perforce the delicate maneuver of wheeling to its right under fire in order to take position facing east along

[17] Thomas, The History of the A. E. F., pp. 227–377.

the deep trench of the Meuse, the natural protection of the American right flank. But it is on the left wing that the natural difficulties are greatest. The 77th Division, allotted the formidable task of pushing northward over the Argonne plateau as rapidly as the outflanking movements on either side reduce the enemy's resistance, "could make but little progress" after the surprise attack of the first day. "Long and bitter work it was in those deep wooded ravines bound with wire where the German rearguard could fight off any force which tried to rush that line." On the third day of the attack the 77th "was slowly pushing its way forward;" on the fourth day it "made a little progress through the dense Argonne forest;" on the fifth day "slowly the 77th followed the German rearguards. It was very difficult work. The dense foliage, the deep ravines, and the barbed wire made progress extremely slow." On the sixth day there was no attack, but the day following the 77th returned to the assault. "In this dense forest the troops became much involved, while the artillery was practically powerless. The battalions on the front had orders to press on, regardless of flanking units, but none were able to make the slightest advance, save one." Major C. W. Whittlesey with six companies pushed down a deep ravine nearly a mile, only to find himself surrounded by the enemy and completely isolated from his division. That night the 154th Brigade and two days later the whole of the 77th Division attacked to rescue the "Lost Battalion" "but made no progress." On the ninth day of the struggle "the 77th Division, in the Argonne, made another vain attempt to advance its lines in the forest to rescue Major Whittlesey and his six companies, but the Germans repulsed each advance." On the tenth day the division "attacked again to relieve the Lost Battalion but failed to advance." By the eleventh day of the conflict the Germans in the forest were badly outflanked on either side and were compelled to retreat to avoid capture when the Allies should close in on the pass of Grandpré in their rear. That day the 77th Division broke the resistance of the enemy rearguard, rescued the Lost Battalion, and began to catch up with their far-advanced comrades.

The 28th Division, on the right of the 77th, was condemned by the nature of the terrain to a series of difficult and delicate maneuvers. "The mission assigned to the 28th Division was the most difficult of all. It had to maintain *liaison* on the right with the division which was advancing rapidly down the Aire valley, while on its left the 77th, with which it had also to maintain *liaison*, was in the forest itself, and would advance but slowly as the Germans were flanked out. The real mission of the 28th was therefore to swing around, pivoting on its left, and to outflank the high plateau of the Argonne, and yet always to keep advancing down the valley of the Aire. It practically amounted to facing the forest and then side-stepping along its front. The Argonne plateau which these troops were to face was [characterized by] a series of promontories sticking out towards the east, and it meant constantly pulling the southernmost troops out, marching them along the front, and putting them in on the north of the line." The spur or promontory of Le Chêne Tondu, and Hills 223 and 224, protecting Châtel-Chéhéry, about five miles southeast of Grandpré, were among the strong natural positions which menaced the American left flank and which had to be captured before the Argonne could be cleared of the enemy. Until this was done the American center on the Barrois plateau dared make no further advance, for it already formed a dangerous salient flanked on the east by the Germans on the heights across the Meuse River and on the west by the Germans still in position far south on the Argonne barrier.

It was on October 4 that the American First Army initiated the second phase of the Meuse-Argonne offensive by a new assault against the imposing obstacles which had been holding it in check. With irresistible fury it drove forward for a gain of two miles in places. Then, as the enemy recovered from the first onslaught, began the heartbreaking task of battering down his resistance on one natural obstacle after another. The attack westward across the Aire trench and against the steep eastern face of the Argonne plateau at Le Chêne Tondu and Châtel-Chéhéry was a most difficult operation, but helped to dislodge the enemy from his strong positions on the upland. For a week the bloody contest raged in the somber shades of the rugged plateau and along its frowning escarpment, the khaki-clad figures surging against ravine wall and ridge slope in brown waves which might be broken and hurled back, but which returned to the assault until the defenders, outgeneraled and outfought, yielded one barrier after another. On the 9th the enemy's resistance was badly broken, and during the 10th he fell back northward behind the transverse gorge of the Aire, part of his Hunding-Brunhilde Stellung.

During this difficult and costly advance the forces east of the Argonne pushed northward over the Barrois upland to reach the eastward continuation of the Hunding-Brunhilde position, the Kriemhild Stellung; while east of the Meuse trench a sharp attack had gained the Haumont cross ridge, north of the Caures Wood ridge, and some additional terrain, thus widening the front of attack, relieving the American right flank, and giving greater freedom of movement for the main advance. To the west the French were forging ahead in the Aisne lowland and the Dry Champagne, when the German high command decided to withdraw to the Hunding-Brunhilde position on the Aisne all along the front as far northwest as Rethel and Château-Porcien.* The stage was now set for the battle to break the last main German defense system west of the Antwerp-Meuse Line.

In continuing the offensive the American right wing was directed to press east and north on the eastern side of the Meuse trench until it had secured possession of the whole of the Meuse plateau barrier, forcing the enemy down into the muddy and marshy plain of the Woëvre. This ultimately involved an advance of certain divisions across the great trench of the Meuse in face of heavy fire; but the difficult maneuver was successfully performed, usually at night since ordinarily "it was impossible for a man . . . to move in the river bottom during daylight, so intense was the fire from the hills."[18] On the west the left wing was to keep in close touch with the main body of the French Fourth Army beyond the Argonne barrier through the defiles of

[18] Thomas, The History of the A.E.F., p. 344.

* Five miles W of Rethel.

Grandpré, Quatre Champs, and others to be conquered in the advance. As the western limit of the American sector ran approximately from south to north while the belts of terrain trend southeast-northwest, the Argonne barrier largely passed out of American hands in the latitude of Grandpré, into the sector to be conquered by the French in their northward progress.

The last half of October was spent in assaulting the formidable defenses of the Hunding-Brunhilde-Kriemhild position. To cross the great Aire trench through the Argonne and get it firmly consolidated in American and French possession was a matter of days instead of hours, and the gain was purchased at a heavy price. Northwest of the defile the French on either side of Vouziers forced a passage of the Aisne by a surprise attack and delivered blow after blow against the western wall of the Argonne plateau. By October 20 they had gained a foothold on the western edge of the barrier and the Americans had advanced a short distance north of the defile on the eastern side. But the formidable Aisne-Argonne-Aire barrier was as yet unbroken. The Germans were still resisting with the fury of desperation in their Hunding-Brunhilde-Kriemhild position. The Allies could hold the enemy's counterattacks and seize bits of terrain here and there by costly assaults; but days of heavy fighting brought little real gain. October 31 found the front but slightly changed from its location on October 15.

A new and more violent offensive was necessary to shatter the enemy's opposition and force him to retreat northeastward behind the Meuse barrier. That accomplished, the time would be ripe for the supreme blow, destined to hurl the invader from this, his last organized defensive position in France, and to throw him back, beaten and despairing, upon his own soil. The natural bridgeheads east of the Meuse and Moselle (pp. 330 and 482) were perfectly adapted to this maneuver, and the plan was already matured in the brain directing the Victory Offensive. "The operations in progress," wrote Foch, "are designed to throw the enemy back upon the Meuse. . . To shatter his resistance on this river, it is desirable to prepare attacks on both sides of the Moselle, in the direction of Longwy-Luxemburg on the one hand, in the general direction of the Sarre on the other."

It was on the first day of November that the American and French armies on the Verdun battlefield were hurled afresh against the Brunhilde and Kriemhild positions. The Americans burst through on the less rugged terrain of the Barrois plateau and by evening had advanced northward five miles; but against the Argonne barrier their left flank turned sharply south practically to the original starting point. On the western side of the plateau, the French smashed through the forest toward the eastern side for gains which were unexpectedly large, considering the character of the country, until held up by a bitter defense of the Fournelle defile at Quatre Champs. But with the Americans pushing rapidly north over the Aire lowland and the Barrois plateau, and the French attacking eastward along the defiles of the Fournelle, La Croix-aux-Bois, and Le Chesne, the Germans in that part of the Argonne plateau bounded by the Aire defile at Grandpré on the south and the Le Chesne defile on the north were in imminent danger of being cut off. They began a hasty retreat. On November 3 the left of the American front had advanced ten miles north of Grandpré and the following day was opposite the eastern entrance to the Le Chesne defile. The Germans had fled from the Argonne as from a trap.

Henceforth the Allied advance northward over the Battlefield of Verdun was rapid and consistent. The demoralization of the German armies was proceeding apace, and the despairing appeals of their commanders were insufficient to excite them to the resistance of which they had been capable a few days before. Overwhelmed by the series of unrelieved disasters along the entire front, the steel of certain defeat thrust deep into his soul, the German soldier could no longer be trusted efficiently to execute the orders of his leaders. However, machine-made discipline could not disappear in an instant; and along the gorge traversed by the canal west of Le Chesne, as along other topographic barriers, stiff resistance greeted the pursuers. These were, however, the last imperfect responses to stimuli of a dying organism.

On November 4 Ludendorff ordered a retreat along the whole front to the Antwerp-Meuse line. But lost confidence, indiscipline, open mutiny were rapidly sapping the strength of the German army to a point where neither the strong Meuse barrier nor any other obstacle could stave off a catastrophe. By November 8 the Americans and French had reached the Sedan lowland from Mézières eastward to the junction of the Chiers with the Meuse and had cut the vital railway artery which so long maintained the German armies in France; the Americans were pressing the enemy along the Meuse barrier from Sedan to Stenay and farther south had driven him from his last foothold on the Meuse plateau down into the plain of the Woëvre. Farther southeast in the Woëvre the Second American Army was initiating an advance designed to outflank Metz from the northwest. From the solid base of the Moselle plateau De Castelnau was preparing to launch the combined Franco-American offensive which on November 14 would sweep northeastward beyond Metz, turn that stronghold from the east, outflank the Meuse-Antwerp barrier, and give the *coup de grâce* to the fast decaying German army.

The blow never fell, for the armistice of November 11 intervened. Sedan, symbol of disaster and epitaph of empires, marked the last stand of German militarism. When the gray-clad invaders turned their backs on the fated city, it was to return with a ruined army to a fallen Empire.

CHAPTER IX

THE BATTLEFIELD OF LORRAINE: THE CUESTA-AND-MOUNTAIN BATTLEFIELD

The Moselle plateau, rising on the eastern border of the Battlefield of Verdun, marks the western limit of the Battlefield of Lorraine. It is the sixth asymmetrical upland traversed in the journey eastward from the plateau of the Marne; and still the repetition of this striking topographic form continues. For the last time resume the long march to France's easternmost frontier, and follow the stately procession of uplands and lowlands to its end. Far away to the southeast the crest of the high Vosges is silhouetted dimly against the horizon. Make this your goal.

From the forested heights of the Moselle plateau (Fig. 98) descend into that smiling lowland of the Seille where somber forests of fantastic pattern mottle the green carpet of fields "spreading afar and unfenced o'er the plain." Here in the shades of Champenoux Forest (Fig. 100) many an invader died "for the glory of the Fatherland" as shrapnel rained from the heights to the west in those memorable first weeks of September, 1914. Southeastward the land slopes gently upward as another limestone layer rises from beneath the clays of the lowland (Fig. 99). Looking from the crest of a new escarpment, which stretches away southward in the heights of Saffais (Fig. 101), the floor of yet another lowland is seen at your feet (Fig. 102). Wide expanses of forest, weirdly irregular lakes, and a maze of rolling hills diversify its surface. This is the plain of Lorraine, so broad that a long, hard day of forced marching will barely suffice to cross it.

Toward the Vosges the plain rises gently, occasionally dropping abruptly over some minor limestone ridge (Fig. 99) only to rise as before till the name "plateau" seems more appropriate than

NOTE. For Chapters IX and X the reader should constantly consult the detailed map of the battlefield in the pocket (Pl. V) and the block diagrams (Pl. VI and Fig. 106).

"plain." A dark band appears on the horizon. Another of the forest patches so numerous in this belt? Its unusual continuity denies this explanation. Without a break it stretches northeast and southwest as far as the eye can reach. Climb some hill in the plain, and the majestic sweep of this vast woodland belt only becomes the more impressive. You remember the Argonne, and watching the ground beneath your feet you plunge into the gloomy wilderness.

Quickly the question, half formulated in your mind, is answered. It *is* sandstone upon which you are walking, just as in the Argonne Forest. Unconsciously other points of comparison suggest themselves, and the eye dwells upon the same steep-sided ridges, the same tabular uplands, the same dark ravines which made the other sandstone plateau so formidable. This time, however, the forest is of spruce; and, when the crest is reached, there is nothing to suggest the remarkably straight front of the Argonne escarpment. Rather the margin is frayed into a ragged complex of bays and peninsulas, beyond which outlying masses are numerous and of varied dimensions.

The lowland beyond is quite unlike anything seen before. It is a valley rather than a lowland, and its eastern wall is not only abrupt but reveals curiously irregular and hummocky slopes. Descending the forested sandstone scarp, you read the answer to the riddle in the rock ledges projecting through the soil. You have passed the last plateau and lowland of the sedimentary rocks and stand on the fundamental crystalline basement upon which the sediments were deposited in layers many millions of years before. From the plateau of the Marne (Fig. 99, key section) across the Battlefield of Verdun and finally over the Battlefield of Lorraine you have descended without a break from the higher beds of the Paris Basin to the lowest, and now behold the platform of crystalline rocks which underlies them all. Like the sediments above, the platform too is tilted downward toward the west and rises eastward in a slope so gentle that you may fail to realize for the moment that you have in fact crossed the "sandstone Vosges" mountains and stand in the first valley of the

FIG. 98—Generalized sketch map of the Battlefield of Lorraine. White areas are lowlands, ruled areas uplands, and cross-ruled areas mountains. For the topographic details and place names referred to in the text see Pls. V and VI and Fig. 106.

"crystalline Vosges" which culminate at elevations of nearly 5,000 feet a few miles to the southeast.

The last stage of the journey is the most beautiful of all. Up the long valley of one of the larger mountain streams which open a pathway to the crest, you pass between green fields flanked with sloping walls of forest or pasture. At first the walls are low (Fig. 103) but gradually they rise higher, terminating above in even-crested summits which suggest a trenched plateau much more strongly than a mountain range (Fig. 104). Verdant meadows dotted with pleasant shade trees and picturesque dwellings rise high on the open valley walls or slope down to the very

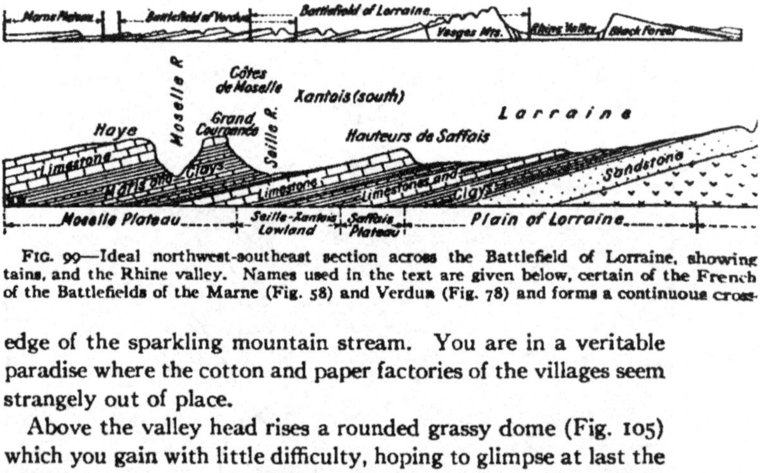

FIG. 99—Ideal northwest-southeast section across the Battlefield of Lorraine, showing tains, and the Rhine valley. Names used in the text are given below, certain of the French of the Battlefields of the Marne (Fig. 58) and Verdun (Fig. 78) and forms a continuous cross-

edge of the sparkling mountain stream. You are in a veritable paradise where the cotton and paper factories of the villages seem strangely out of place.

Above the valley head rises a rounded grassy dome (Fig. 105) which you gain with little difficulty, hoping to glimpse at last the mountain summits that seem forever to elude your view. Instead, the same pleasing but uniform landscape stretches far to the north, west, and south. Eastward, you seem to look off into space. Nowhere are the mountain peaks of the "Hautes Vosges" visible. You stroll across the grassy knob toward the east—and an abyss yawns at your feet! Rocky precipices fall sheer to where a tiny lake sparkles in the depths far below, a gleam of turquoise in the dark forests mantling the base of the crags. Short valleys cut the base of the range into shreds and open out upon the plain of the Rhine. Far beyond, the Belchen and the Feldberg rear their heads above the Black Forest. Without any "mountain climbing," indeed without appreciating the higher elevations attained, you have ascended the gentle western slope of the Vosges to its main crest, and stand on the brink of its steep eastern face.

On a grand scale Nature has here repeated for the last time the pattern she evidently adopted for all of northeastern France: an asymmetrical upland with a gentle slope toward Paris and a precipitous scarp toward Germany. We shall discover, however,

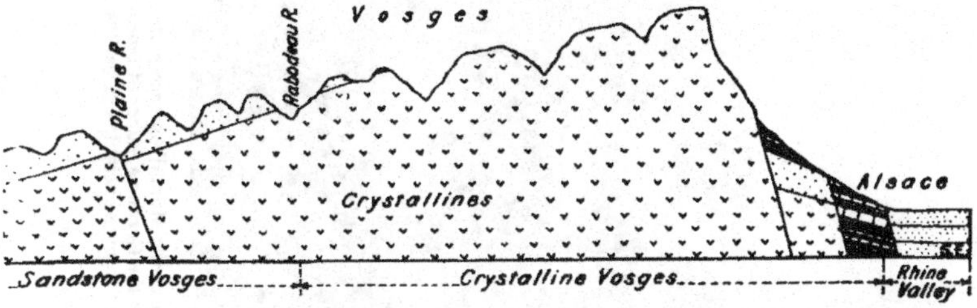

relation of rock structures to asymmetrical plateaus (cuestas), lowlands, the Vosges Moun-
regional names above. The key shows the relation of this cross-section to the cross-sections
section from the Marne plateau to the Black Forest.

that the outermost wall of the natural defenses of Paris owes its asymmetrical form to a geological history strikingly different from the simple story of erosion on inclined beds which sufficed to explain the asymmetrical plateaus farther west.

Strategic Position of the Battlefield of Lorraine

The Battlefield of Lorraine lies at the intersection of several great lines of movement across the continent of Europe. From the higher land within its limits the broad basin of the Rhine leads northward toward the North Sea and the shores of the Baltic, that of the Saône-Rhône depression southward to the Mediterranean. Westward the rivers point the way to Paris and the Atlantic Ocean, while the Danube gateway to the east has its western threshold just beyond the valley of the Rhine. Between the south and the north, between the east and

FIG. 100—The Forest of Champenoux in the lowland at the base of the Mont d'Amance plateau, as seen from the French position on the crest of the escarpment. From the protection of the forest the Germans launched repeated attacks during the Battle of the Grand Couronné, all of which were broken by accurate fire directed from the heights against the exits into the open plain.

the west, from time immemorial important routes of migration, invasion, and commercial traffic have traversed the rectangular area within which lie headwaters of the Rhine, Seine, and Rhône. The natural corridors of this terrain have echoed to the tramp of armed hordes from the earliest dawn of history. The region "was, in a sense, the commercial and political axis of the Gauls."[1] Today canals connecting the Rhine with the Seine, the Rhine with the Rhône, and the Seine with the Rhône, cross the lowland divides which flank the Vosges highland, and several lines of transcontinental railway follow the valleys and plains. The Battlefield of Lorraine thus lies at a nucleus of international routes of more than ordinary importance.

The Vosges Mountains fail to meet the Jura Mountains and the border of Switzerland by some 15 or 20 miles. Thus is formed the famous Belfort gateway, easily traversed by a canal, railroads, and several highways. "This pass, broad and low (305 meters, or 1,148 feet) marks the insignificant summit in the great historic route of travel between the Mediterranean and the North Sea, from the days of ancient Etruscan merchants to the present. This was the route of the invading Teuton hordes which the Roman Marius defeated at Aquae Sextiae and, later, of the Germans under Ariovistus, whom Caesar defeated near the present Mülhausen. Four centuries afterward came the Alamannians, Burgundians, and other Teutonic stocks, who infused a tall blond element into the population of the Rhône Valley. The Pass of Belfort is the strategic key to Central Europe. Here Napoleon repeatedly fixed his military base for the invasion of Austria, and hither was directed one division of the German army in 1870 for the invasion of France."[2]

On the north the Vosges descend to a comparatively low gap of much greater breadth which separates them from the mountains of western Germany and the Ardennes of Belgium. This is the Lorraine gateway, the most dangerous opening on the French frontier, and the route utilized by the Germans for their principal

[1] P. Vidal de la Blache: La France de l'Est (Lorraine-Alsace), Paris, 1917, p. 14.
[2] E. C. Semple: Influences of Geographic Environment, New York, 1911, p. 540.

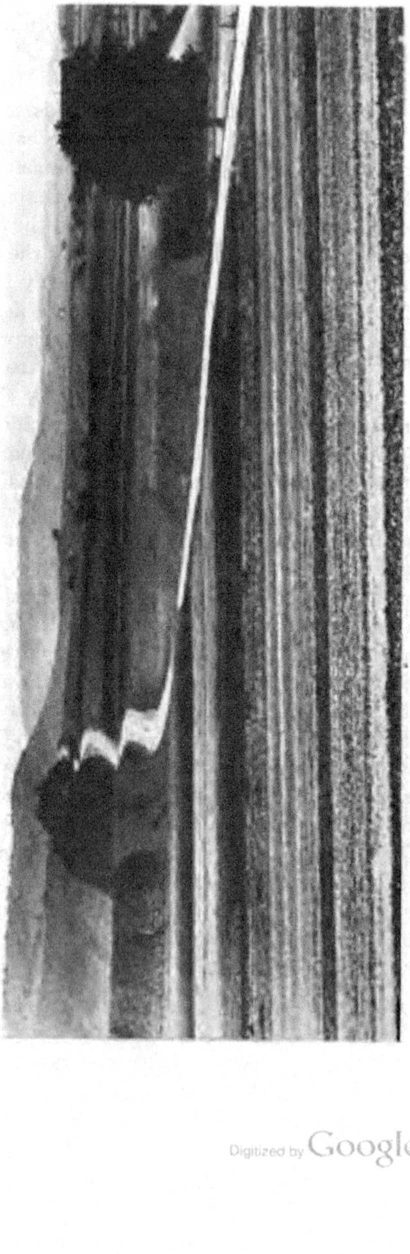

Fig. 101—The Saffais plateau escarpment and the lowland east of it, looking north from near Méhoncourt 16 miles southeast of Nancy. It was the effective defense of this barrier which defeated one phase of the great German offensive against Nancy and the Grand Couronné. The lone tree at the right, appearing again in Fig. 115, will enable the reader to compare the two views.

FIG. 102—Looking southeast over the lowland from the crest of the asymmetrical Saffais plateau near Rosières to the Forest of Vitrimont in the distance, showing the advantages of observation and artillery control enjoyed by the French forces on the upland.

invasion in 1870. Since that unhappy chapter in history it has been a fundamental thesis of French military reasoning that in a new war the Lorraine gateway would certainly be one of the chief openings, if not the chief, through which their enemy would strike. To meet the menace French strategists proposed an offensive in Alsace launched through the Belfort gateway and the Vosges passes, coupled with a simultaneous offensive through the Lorraine gap.

The Battlefield of Lorraine, in the broad sense in which it is used here, includes both the Belfort and Lorraine gateways and the mountain barrier which separates them. It was thus by nature destined to be the scene of important maneuvers in the first days of the war, through which the grand strategy of the opposing staffs gradually revealed itself. On this field France quickly massed the bulk of her forces, and from it she promptly launched her armies through the Belfort and Lorraine gateways. And while events were soon to demonstrate that the latter gap was not the chief danger spot, neither in its southern portion east of Nancy nor yet on its northern border from Metz to Luxemburg, where the principal enemy concentrations were first supposed to be massing, as one of the important routes of attack it claimed the attention of the world until the supreme effort of the eastern German armies to enter France by this door was finally and decisively defeated.

Facing the Lorraine gateway on the west are the frowning scarps of the Meuse and Moselle plateaus, backed by other barriers the great strength of which we have considered in earlier chapters. Thus, while the gateway was wide open, the route directly west was barred. Southwestward, however, between the walls of the plateau barrier and the Vosges Mountains, stretches the broad plain of Lorraine and the comparatively low country of the subordinate plateau and lowland at the base of the Moselle scarp. In this direction lay an easy route to the Saône-Rhône depression, containing some of the vital industrial centers of France, and to the great arsenals and arms factories of the Creusot district. The position of the Franco-

German frontier, obliquely athwart the grain of the terrain (Fig. 98) from the northern end of the Vosges to the backslope of the Moselle plateau, placed the German armies squarely in front of the "Charmes Trough." This natural runway, because of its length better called a trough than a gateway, is one of the most striking strategic features of eastern France; and that there must be a "Bataille de la Trouée de Charmes" was long an accepted principle in French military reasoning.

In an earlier chapter (p. 328) we have seen that the whole scheme of defense of France's eastern frontier was based on the fortification of the Meuse-Moselle plateau massifs on the north (Verdun-Toul line) and the Vosges massif on the south (Épinal-Belfort line), leaving the Charmes Trough as the only practicable route for the enemy invasion (Fig. 83). With the enemy confined to a known route and menaced on both flanks by dangerous strongholds sheltering large forces of mobile troops, the main armies of France could deal with the invasion to the best possible advantage. To meet the shock of enemy masses launched along the Charmes Trough, the French could utilize a succession of strong defensive positions provided by the river trenches, particularly those of the Meurthe, Mortagne, and Moselle, which traverse the lower country from the Vosges mountains to the plateau barriers. Entrenched behind these natural moats, which in the cases of the Mortagne and Moselle are bordered on the southwest by higher ground commanding the approaches to the barriers, the Armies of the Republic could hope to contain the enemy offensive, and then from the strong natural positions on the plateaus and in the Vosges to launch counteroffensives against the enemy's flanks which would complete the victory. Thus in a variety of ways was the Battlefield of Lorraine designed to be the key to the defense of France.

When the invasion of Belgium opened the northern frontier of France to the enemy, the French staff was compelled to alter the whole plan of the national defense. The Battlefield of Lorraine occupied a different but important place in the new scheme. Joffre's conception of a strategic retreat to the Seine while

FIG. 103—Valley of the upper Moselle four miles north of Remiremont, showing fairly open floor of the trench and its forested walls of crystalline rocks. The river flows in an inner trench cut 25 to 50 feet below the main valley floor seen in the view. This is the natural trench forming the basis of the famous Épinal-Belfort line of defense, and this part of it is swept by the guns of Fort Parmont, on the south wall of the trench at Remiremont.

Fig. 104—The upper valley of the Moselle River as seen from the base of the Ballon d'Alsace, 12 miles north of Belfort, showing cleared valley bottom and walls and forested uplands. The forts of the Épinal-Belfort defensive barrier crown the heights on the southwest (left) side of this great natural trench. The remarkably even sky line is an old erosion surface (peneplane) on complex crystalline rocks, forming the gentle western slope of the range.

FIG. 105—The Vosges Mountains looking north from the Ballon d'Alsace. In the foreground the rounded, grassy summit of the Ballon. Center and left distance show the even sky line of the gentle western slope of the range, while to the right begins the labyrinth of ravines and ridges formed on the steep eastern slope.

pivoting on Verdun could pave the way to ultimate victory only in case the right wing of the Allied line was not turned by an enemy break-through south of Verdun or on the Lorraine plain. On the other hand, the German plan to roll up the Allied front by a turning movement at both ends depended for its success as much upon the outcome of the eastern attack as upon Von Kluck's great sweep around to the west. For both of the belligerents, therefore, one of the keys to the strategy of the Marne lay along the Moselle. That France won the key, and hence the victory, was due in no small measure to an intelligent use of topographic barriers on the important battlefield located on the Lorraine frontier. We may now turn our consideration to the several belts of terrain which together constitute the exceptionally interesting Battlefield of Lorraine.

Surface Features of the Battlefield of Lorraine

THE MOSELLE PLATEAU

Let us recall for a moment the salient features of the asymmetrical plateau known to geographers as the "Moselle cuesta." As we noted in our survey of the Battlefield of Verdun, the plateau owes its existence to a resistant limestone formation (Fig. 99) rising gradually from beneath the clays of the Woëvre lowland to the west and terminating eastward in a precipitous escarpment. Deeply trenched by the gorges of the Ache, the Terrouin, the Moselle, and the headwaters of the Meuse, and by innumerable smaller ravines into whose narrow but deep recesses the rays of the sun never penetrate, the limestone upland is cut into more or less isolated tablelands of varying size. The forms common to calcareous rocks in other regions reassert themselves in the topography of the Haye, as the plateau is called from south of Nancy to the northern limit of the region now considered, and in the plateau of Langres, as we may call its more southern portion. Sink holes occur, though not so abundantly as in some limestone regions. Subterranean caverns and grottoes abound, some of them of great extent, like the famous cave of Sainte Reine and the cave of the Celts southeast of Toul, probably used as places of refuge by prehistoric man. Profound chasms or narrow clefts 50 to 100 feet deep with vertical walls seam the surface, many of them renowned as natural wonders, like the "Depths of Toul," the "Pocket," and the "Valley of St. Amond" trenched in the plateau west and southwest of Nancy.[3] Like the upland surface itself, some of the chasms and ravines and occasionally larger valleys may be dry and arid, because the rain which falls on the plateau quickly escapes downward through fissures and caverns until checked by the marls and clays below. A detailed map will reveal a number of these deserted valleys, especially in the sector between the Moselle gorge near Nancy and the headwater branches of the Meuse, where the dry valleys of the Arr and the Aroffe are noted examples. The "lost rivers" which pursue their courses in subterranean passages may reappear as streams in the lower parts of the valleys or simply as abundant springs gushing forth in some neighboring gorge which cuts down to the contact of the impervious marls and clays. It is evident that a plateau dissected into tablelands traversed by such a series of gorges, ravines, dry valleys, vertical chasms and fissures is a topographic obstacle of serious proportions.

The course taken by the Moselle River, while cutting a double gateway through the plateau, does not materially decrease the military value of the barrier. South of Nancy the river enters the face of the escarpment and traverses the whole breadth of the barrier in a wild, narrow gorge on its way to the Toul gap where it formerly entered the Meuse River (Fig. 91). Just before reaching Toul it turns sharply back to the northeast, retraversing the plateau in another gorge determined by the stream which cut through the barrier and diverted the Moselle to its present course. As a result of the river capture we find this portion of the Moselle plateau protected by a great natural fortress of triangular shape, each side 12 miles long, the southwestern and northwestern sides defended by deep natural moats through which runs an unfordable river, the eastern side by the steep escarpment facing

[3] Bertrand Auerbach: Le plateau lorrain: Essai de géographie régionale, Paris, 1893, p. 166.

an exposed lowland crossed by a canal south of Nancy and by the gorge of the Meurthe farther north. Completely surrounded by water, its arid upland difficult of access, seamed with chasms and cloaked with the vast Forest of Haye, the triangular massif which in earlier days was the secure retreat of wild beasts and lawless men, is in modern times a solid buttress of an important line of defense against hostile aggression. Its strategic importance is much increased by the fact that it lies at the intersection of highly important routes of travel, each of the three sides being skirted by railroad, highway, and canal. At each corner of the triangle we find one or more modern forts. Those at the west form part of the defenses of Toul. Fort St. Barbe or Pont St. Vincent, near the southeast corner but across the river on a cleared plateau spur, sweeps the railways, roads, and canals approaching the entrance to the gorge. Fort Frouard at the northeast corner controls another plexus of railway, road, and canal routes converging on this entrance to the gorge and dominates the course of the lower Moselle for many miles.

Grand Couronné de Nancy

After contouring the triangular Haye Forest massif, the Moselle in its northward course does not lie in the great lowland east of the plateau scarp, but, like the Meuse, trenches the backslope of the upland just west of the escarpment in such manner as to leave a narrow strip of the highland forming a natural bridgehead east of the river (Figs. 79 and 98). As the Moselle lies closer to the scarp than does the Meuse, the detached plateau strip is much narrower than in the Meuse case; at the north it is a mere ridge, often broken by erosion into detached buttes and mesas; but nearer Nancy it suddenly expands into an eastward projecting bastion measuring half a dozen miles from the river to its apex (Fig. 106). This outlying strip of the Moselle plateau is known among French military geographers as the "Grand Couronné de Nancy." The incalculable strategic value of this natural bridgehead fortress in the defense of France's eastern frontier will appear in the sequel.

FIG. 106—Block diagram of the Nancy reg

wing the salient features of the terrain.

FIG. 107--The valley of the Natagne traversing the Mont d'Amance plateau bastion, looking south from the Ste. Geneviève-Mont Toulon ridge to the next east-west plateau strip beyond. Both the parallel valleys and the intervening plateau strips or ridges were principal elements in the French defensive system. The wire entanglements in the foreground and zigzagging away to the right are part of the French second line position based on the Ste. Geneviève-Mont Toulon ridge.

FIG. 108—Ste. Geneviève plateau, part of the Grand Couronné, forming northern face of the Mont d'Amance natural bastion. Note the steeper wooded upper slope and the smooth, open lower slope, well exposed to grazing machine-gun fire from positions at the base of the steeper slope. It was the effective defense of this barrier which broke the great German offensive against Nancy.

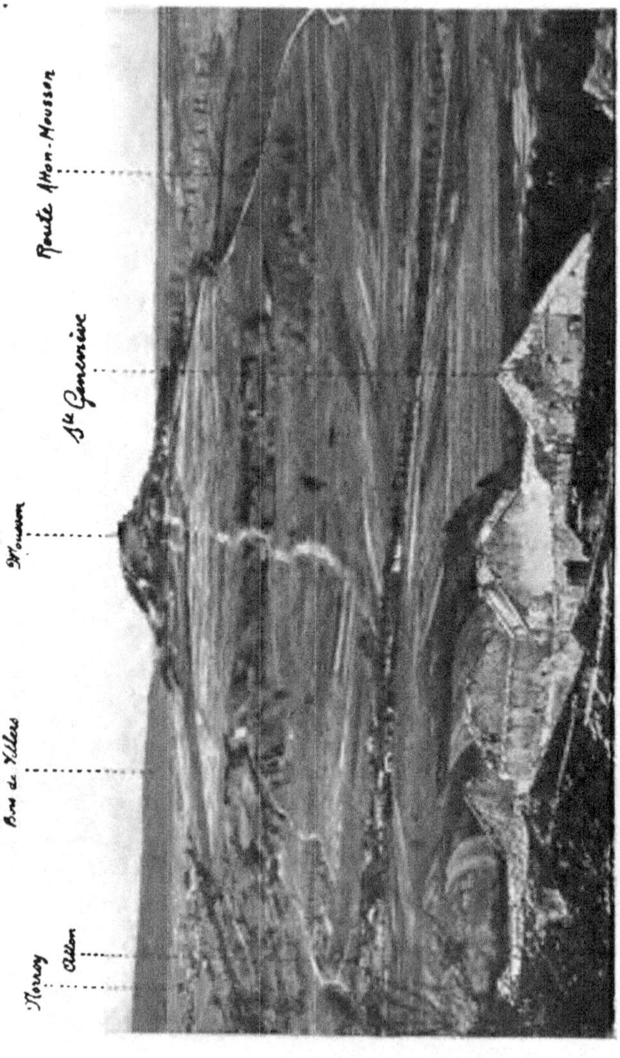

Fig. 100—The butte of Mousson, with the forested Moselle plateau in the background. In the foreground is the low gap forming the Pont-à-Mousson gateway from the Seille lowland into the trench of the Moselle River. (French official photograph.)

Like the Meuse plateau bridgehead in the vicinity of St. Mihiel and Commercy (Pl. VI), the Grand Couronné is repeatedly breached by transverse stream valleys. Two of these just north of Nancy carry railways and highways from the lowland on the east through the projecting upland bastion to the Moselle gorge (Fig. 106). Next to the north the valley of the Natagne (Fig. 107) almost but not completely breaches the bastion, and a road crosses from east to west through the depression. North of the Natagne the remaining slice of the bastion forms Mont Toulon ridge and Ste. Geneviève plateau (Fig. 106). Then comes a broad opening through which an important highway leads from the frontier to the river at Pont-à-Mousson. We may call this gap the Pont-à-Mousson gateway, and the Ste. Geneviève plateau (Fig. 108) and Mousson butte (Fig. 109) the two pillars guarding it on either side. From Mousson northward the Grand Couronné is continued as a narrow strip of plateau divided by erosion gaps into the butte of Xon and two north-south ridges (Fig. 91), the northernmost of which carries two of the forts defending Metz.

The analogy with the Meuse plateau barrier can be traced further and reveals itself in the military operations as well as in the form of the terrain. Just as Verdun was defended against an attack coming from the north by the transverse ridges or upland strips and the ravines paralleling them which resulted from the erosive work of small streams tributary to the Meuse, so Nancy has as its system of northern defenses the succession of east-west plateau strips and valley trenches carved by tributaries of the Moselle. And just as the Germans first threatened Verdun from the east by moving against the face of the Meuse escarpment, and then tried an advance from the north upstream across the transverse lateral ridges, so the original attack on the scarps defending Nancy on the east was followed by an upriver advance directed from the north against the parallel strips of the Moselle plateau. In both cases the second operation, like the first, was a failure. The most striking difference in form between the two terrains is the absence of a continuous plateau barrier east of Nancy. As a consequence of this greater exposure to an attack from the east

FIG. 110—Eastern portion of the Mont d'Amance natural bastion, part of the Grand Couronné 7 miles northeast of Nancy. In the distance are seen the partially cleared plateau surface, and, at the right, the steep, forested upper portion of the marginal escarpment and the more gentle, cultivated lower slopes. Machine guns at the base of the wooded scarp swept the open slopes with a grazing fire, while indirect artillery fire from rows of batteries concealed in one of the plateau valleys, under the camouflage coverings visible in the center and left foreground, was accurately directed, by observers hidden along the wooded crest, upon the enemy on the plain beyond.

the campaign against Nancy involved more important operations from that direction, with the result that the bastions and curtains of the Moselle plateau played a much more important rôle in the fighting than did those of the Meuse plateau.

The Moselle trench, like the Meuse, provides a valuable lateral line of communications behind the defensive positions on the Grand Couronné and, in case of the loss of the latter, a natural moat behind which an army might long defend itself. Along the trench run a railroad, canal, and two important highways bordered by furnaces, factories, and quarries and dominated by vineyard-clad slopes terminating above in wooded cliffs. Toward Metz, as also toward Nancy, the trench is a narrow gorge; but between these sectors is a more open valley where the river wanders over a flood plain a mile or two broad covered with natural prairies. Whether narrow or broad, the Moselle trench is a difficult obstacle to cross under enemy fire from the western crest; and the fact that this crest is so largely forested deprives the holder of the eastern bank of part of the advantage he would enjoy from the fact that that bank is often the higher of the two. As the Moselle trench is not nearly so serpentine as that of the Meuse, projecting points give observation and artillery control over long stretches of the river and its crossings.

Isolated mesas and buttes, detached from the Moselle plateau by erosion, stand in front of the main escarpment at many points. The butte of Vaudémont, about 20 miles southwest of Nancy, the V-shape of which cannot fail to impress the observer with its bastion form, rises more than 700 feet above the lowland. It was early selected as the site of a castle which dominated a vast expanse of the plain below and became the political center of the region.[4] Mont d'Amance and the Pain de Sucre (Fig. 106), outlying mesa and butte guarding the approaches to the Grand Couronné, played a leading rôle in the defense of Nancy in 1914.

Form of Plateau Profile

The form of the plateau profile renders it admirably adapted to defense. The limestone cap weathers into steep slopes, vertical

[4] Auerbach, Le plateau lorrain, p. 235.

walls, or even overhanging cliffs, which make the level upland extremely difficult of access. This development of a cornice just above the gentler slopes eroded on the weak marls and clays is much more striking in the Moselle plateau scarp than in any of the other examples we have studied. From Nancy southward the scarp is double, an intervening bed of weaker material separating it into an upper and a lower terrace. On the other hand the lower declivities may be more gently inclined and of more uniform slope than usual, while the soil conditions are not unlike those encountered along the Marne plateau scarp. Limestone fragments from the cap rock mix with the clay below to give a surface deposit admirably adapted to grape culture, fruit growing, and vegetable gardening. Even when not steep enough to deserve the adjective "precipitous," the upper slope is usually much too steep to be cultivated and is therefore covered with trees and bushes (Figs. 108 and 110). This assures to the defense several important advantages. Observation posts excavated in the steep rock face and reinforced in front for greater safety are hidden by a natural camouflage of trees which prevents the enemy from concentrating fire for their destruction. The enemy knows that he is being watched from that crown of woodland; but where along its length the eyes are hidden he cannot tell. Hence, the observers can survey enemy defenses, batteries, and troop operations, and direct their own artillery fire in comparative quiet (Fig. 111). The one chance shot out of thousands which might happen to strike a vulnerable spot in one of these posts does not cause much anxiety. At the base of the steep slope the woods offer concealment for machine-gun positions, from which the cleared expanse below (Fig. 108), sloping gently downward to the plain, may be swept with a hail of bullets driving parallel with the surface. Such a grazing fire being far more deadly than a plunging fire from the heights above, it is essential that the base of the steep slope should be made easily accessible from the upland. This is readily accomplished by tunnels through which the machine-gun detachments can quickly man their positions in case of attack or retire safely at the last moment in case they fail to check the enemy (Fig. 111). The forest cover, which owes its continued existence to the steepness of the upper slope, affords valuable concealment to the men carrying on the tunneling operations and to the tunnel exits and gun emplacements when completed. If the surface of the plateau is cleared, the trees on the steep slope often rise high enough to form a screen concealing from direct enemy observation all movements on the upland. Such a screen is also well developed along parts

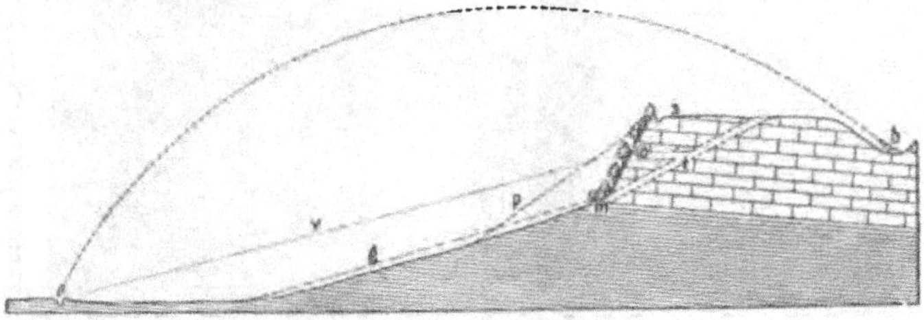

FIG. 111—Diagram to illustrate how the forest cover of the steep upper slope of the Moselle plateau scarp conceals from the enemy's direct view: all operations on the plateau surface (s); observation posts (o) from which there is an unobstructed line of vision (v) to the enemy's positions, permitting perfect control of indirect artillery fire from batteries (b) hidden in ravines in the plateau; and machine-gun nests (m) at the base of the cliff which are connected by tunnel (t) with the upland, and from which the gentle lower slopes may be swept by a grazing fire (g), much more effective than a plunging fire (p) from the crest.

of the Saffais and Meuse plateau margins (Fig. 112) and in the case of the Marne plateau led Sir John French to complain of the advantages it afforded the Germans occupying the upland north of the Aisne.

Natural Bastions and Curtains

In plan the Moselle plateau, like that of the Meuse, shows a succession of projecting points, or "bastions," alternating with reentrant angles, or "curtains" (Fig. 91 and Pl. V). Starting from the north we have first the Pont-à-Mousson curtain, corresponding in more than one respect with the Commercy curtain of the Meuse plateau. Then comes the bastion which we may name the Amance bastion from the vitally important mesa, Mont

FIG. 112—A cleared portion of the plateau surface, showing the thick screen of trees along the brink of the escarpment, especially characteristic of the Moselle and Saffais plateaus. This view shows the Meuse plateau surface, looking southeast from Fort Gironville, 4 miles northeast of Commercy.

d'Amance, near its apex. The Nancy curtain is bordered on the south by a slightly projecting spur of the Haye Forest massif referred to by French military writers as the Heights of Ludres, and which for convenience we may call the Ludres bastion. Next south is a broad curtain dominated by the centrally located Mont d'Anon butte, beyond which a great bastion projects far forward to terminate in the Vaudémont and other buttes. We may call this the Vaudémont bastion, including by this designation not merely the bastion form of the triangular butte itself but the whole forward-projecting mass of the upland. Beyond the curtain of Châtenois to the south the border of the plateau is of more even alignment. Against a frontal attack these natural bastions and curtains would oppose many of the difficulties which on a smaller scale the similar features of an artificial fortress were intended to occasion.

If we add to all that has been said above these further facts: that the Moselle plateau throughout the region now under discussion is almost everywhere heavily forested, the cleared areas occupying but a very small fraction of its upland surface; that the hill slopes and valley walls are frequently honeycombed with quarries and caverns from which stone for building, paving, and other purposes has been extracted for centuries; that the plateau scarps are "crowned with a precious diadem" of iron ore just below the limestone cap which not only turns the lowland and valleys into industrial regions dotted with mines and furnaces but provides along the margins of the upland underground galleries and chambers of very great value as elements in military defense works; and that the rugged, dry, infertile upland repels man to the point of causing him to seek even the margin of the inhospitable Woëvre for a smoother surface, accessible water supplies, and less intractable soil; if we add these elements to our picture of the Moselle plateau, it must be apparent that it presents to a hostile army an obstacle of impressive magnitude. From the earliest days the defensive value of this remarkable terrain has always appealed strongly to the imagination of man. There is evidence that different points along the Côtes de Moselle were occupied and forti-

FIG. 113—The Côte de Delme seen from the crest of the Moselle plateau scarp. In the foreground are the wire entanglements on the lower slopes of the Grand Couronné; in the middleground, the Seille lowland. (French official photograph.)

fied in the Stone Age and in the Iron Age. The Celts and other early tribes built defense works, traces of which are still visible on the projecting bastion of the plateau near Ste. Geneviève.[5] Doubtless the Romans tested the defensive value of the terrain during the conquest of Gaul; and certainly the castles of the Middle Ages took advantage of the strong points which Nature offered, ready-made, all along the scarp. When the modern invaders assaulted this barrier, they were but repeating on a grander scale a drama frequently rehearsed upon the same stage.

THE SEILLE-XANTOIS LOWLAND

Eastward from the base of the Moselle

[5] Auerbach, Le plateau lorrain, p. 283.

plateau scarp extends a lowland some five to ten miles broad in the northern part, less than half that breadth from Nancy southward. This lowland is a unit as to origin, for throughout its whole extent it is developed on the weaker marls and the clays or shales coming to the surface from under the Moselle plateau limestone (Fig. 99). It is a unit also in its military significance, for it is the exposed belt which an enemy attacking from the east must everywhere traverse under the dominating observation and artillery fire from the western heights before he can even begin the attack on the Moselle plateau barrier. But the northern and southern parts of the lowland show different breadths, are drained in different directions, and exhibit certain differences of surface form.

The Seille lowland, extending from Metz southward to the vicinity of Nancy, resembles the Woëvre plain in its general appearance, especially when looked down upon from the plateau scarp. There are the same broad expanses of plain mottled with large and small areas of forest (Fig. 113); the same white ribbons of roads, often bordered by rows of evenly spaced trees offering excellent ranging marks for artillery (Figs. 82 and 114); the same sluggish streams wandering through natural prairies or marshy flats. But the ground is more undulating than the typical Woëvre plain, and a closer inspection reveals the fact that the Seille River occupies a meandering valley like that of the Meuse, except that it is entrenched very slightly below the general level of the plain.

The numberless lakes which diversify the humid region of the Woëvre are lacking in the lowland of the Seille. Not, however, that the clay fails to perform its normal rôle of retaining moisture near the surface. In wet seasons the roads of the Seille depression are boggy, and the forest patches frequently mark areas where the cold, resistant clay projects entirely through the thin surface covering of loam, giving lands wholly unfit for agriculture.

The Seille River drains a part of the lowland northward to the Moselle at Metz. From Nancy southward the lowland is for the most part drained by streams flowing transversely across it, and is known as "the Vermois" region for the narrow section between

FIG. 114—Looking southward from the "Amance bastion" of the Grand Couronné, across the lowland re-entrant forming the "Nancy curtain," to the higher land of the "Lunéville bastion" dimly seen on the sky line. At the left is seen a westward extension of Champenoux Forest, while the line of trees regularly spaced from left to right across the center of the view accurately locates the main highway from the German border to Nancy. It was the wonderful observation from the dominating height of Mont d'Amance, insuring perfect control of artillery fire upon the enemy maneuvering on the plain below, which defeated every effort of the Germans to advance over the lowland into the Nancy curtain.

the Meurthe and Moselle, and as "the Xantois" south of the Moselle. Here the lowland is not only much narrower than in the Seille region but is less marshy because the rivers and their branches have cut more deeply into the surface. Instead of being a flat plain, the lowland is here a maze of low rounded hills rising to a common level, dominated on the west by the much higher plateau scarp. The rivers Madon and Brenon, which drain parts of the Xantois northward, have cut their channels between 100 and 200 feet into the floor of the depression.

Better drainage and other factors combine to render the Xantois-Seille lowland a more productive country than the Woëvre lowland to the west. Layers of limestone included with the clays enrich the soil, and phosphate nodules are sufficiently abundant in places to be mined for fertilizer. Sandy alluvium, including débris washed down from the Vosges, extends as a covering over parts of the surface, and above all is a thin coating of loam. Thus we have a richer subsoil than usual, and a fertile porous topsoil which permits water to sink below the surface, yet is kept sufficiently moist by the presence of clays a few feet down. The favorable soil conditions and the fairly good drainage assured through the moderate trenching of the floor of the lowland by stream valleys, together make this belt "the granary of Lorraine." Most of the lowland is cleared and cultivated, and an army occupying it for a period of years, as did the Germans from 1914 to 1918, could count upon large quantities of food from the fields and of fodder from the natural prairies on the river flood plains.

THE SAFFAIS PLATEAU

Rising gradually toward the east from under the weaker formations of the Seille-Xantois lowland is a bed of resistant impure limestone (Fig. 99) which, following the usual rule, forms an asymmetrical plateau having a gentle westward slope and terminating eastward in a steeper escarpment. This plateau is lower in elevation, more irregular in form, and much less striking in appearance than its western neighbors and has no single name by which it is commonly known. One section of it which played a prominent rôle in the Battle of the Grand Couronné is called the Heights of Saffais (Fig. 101); and for the sake of convenience in referring to the topographic feature as a whole we will call it the Saffais cuesta or plateau, fully recognizing, however, that the French use the term for a much more restricted area.

The crest of the Saffais plateau is usually between 100 and 200 feet lower than the Moselle plateau and rises only 200 to 300 feet above the lowland east of it. Yet even this modest elevation is sufficient to assure commanding observation over the lower country, as any one looking eastward from the crest will quickly perceive (Figs. 102 and 115). Unlike the higher plateaus on purer and more massive limestone formations, the Saffais upland is neither extremely dry nor extremely infertile and is not made difficult of access by wild, deep gorges. Consequently much of its surface is cleared and cultivated, although a fringe of trees on the upper slopes of the east-facing scarp often rises high enough to conceal forces maneuvering on the upland from the view of an enemy advancing across the lowland (Fig. 112). Finally, the scarp is ordinarily only of moderate steepness (Fig. 101) and dissected into a succession of spurs which bear only a mild resemblance to the great walls of the Côtes de Meuse and Côtes de Moselle. Assuredly the Saffais plateau, while a serious military obstacle, is far less formidable than the examples previously studied.

Natural Bastions and Curtains

The extreme irregularity of the eastern margin of this plateau often makes it difficult to tell where to draw the boundary between it and the lowland farther east. Long, irregular spurs and innumerable buttes, isolated and in groups, might be regarded as carrying the upland far eastward toward the margin of the Vosges in places; just as broad and deep valley re-entrants force it back westward close to the Moselle plateau in others. If we restrict consideration to the more perfectly developed portions of the plateau, we may recognize that its margin contours a succession of projecting bastions and re-entrant curtains of very pronounced type (Fig. 98). Thus southeast of Metz the upper Nied and its tributaries (Fig. 106) have eroded a re-entrant which we may conveniently call the Nied valley curtain. Then from the vicinity of Château-Salins there project a dozen miles or more northeastward into the plain two remarkable plateau spurs, long and narrow in form and separated by a branch of the Seille River called the Little Seille. As the town of Morhange (Mörchingen) lies at the southern base of the northern spur, and Dieuze just south of the southern spur, we may call the whole projecting mass formed by the two walls and their included valley the Morhange-Dieuze bastion. To the south the scarp swings back to the west of Arracourt, giving a deep curtain which we may designate by the name of that town.

The plateau crest continues southwestward as the "Rembétant" to the gateway of the Meurthe River at St. Nicolas-du-Port, beyond which it is equally distinct in the Heights of Saffais. But if we restricted our consideration to the most perfectly developed part of the scarp, we should ignore an extremely important outlying mass of hills, detached remnants of the plateau however much erosion may have altered their form, which from near the edge of the scarp at Dombasle project eastward to the Forest of Parroy with an advance guard in the ridge north of Manonviller. As the apex of this bastion is defended by the outlying Fort Manonviller, located near the eastern end of the ridge referred to and dominating the main route from Strassburg and Saarburg into France, we may call the projecting mass the Manonviller bastion. On the north this bastion is protected by the natural moat of the Sanon valley followed by the Marne-Rhine Canal, on the south by the valley of the Vezouse River, a stream which pursues its devious, serpentine course through open meadows on a marshy flood plain.

In the deep re-entrant which follows to the south are located the towns of Lunéville, Blainville, and Rosières. As Lunéville is by far the most important, we will call this re-entrant the

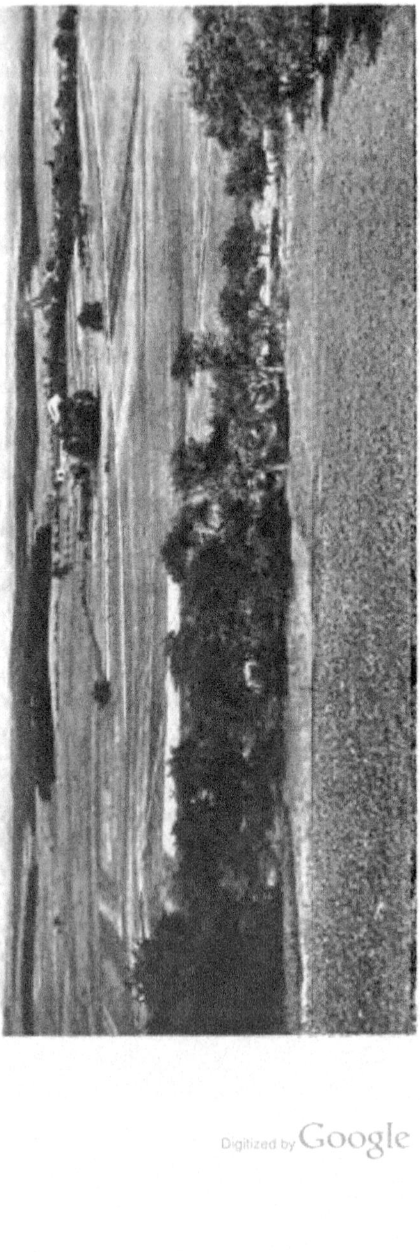

Fig. 115—Lowland east of the Saffais plateau as seen when looking eastward from the face of the plateau scarp west of Méhoncourt. The German advance reached the village at the right but broke down under French fire accurately directed from the crest of the scarp. The lone tree by the road, to the right of the center of the view, is the one appearing in Fig. 101.

Lunéville curtain. It is followed on the south by a badly dissected portion of the plateau projecting southeast from the Heights of Saffais to terminate in the high Côte d'Essey protected from erosion by intrusions of igneous rock, and in the hills southwest of Gerbéviller. This Essey bastion is protected on the northeast by the natural moat of the Meurthe-Mortagne valley and is in its turn succeeded by the very deep curtain of Charmes (Fig. 118), almost filled by the great Forest of Charmes, just as the Lunéville curtain is largely covered by the Forests of Vitrimont (Fig. 102), Mondon, and others. Next comes the massive, though much dissected, Bouxières bastion, terminating in a high hill just south of the town of Bouxières and protected on the north by the formidable trench of the Moselle River, on the south by a branch of the Madon. Beyond this bastion the escarpment pursues a well-defined and approximately straight course toward the southwest as far as Lamarche, where another series of salients and re-entrants begins. These, however, are of less immediate interest to us.

The strongly developed bastion-and-curtain form of the Saffais cuesta materially increases its defensive value. Instead of attempting to hold the entire length of the escarpment with uniformly distributed forces, thereby weakening the defense against a concentrated enemy attack, the bastions may be strongly held at the expense of the curtains. An enemy attempting to continue over the low plain into the less strongly defended curtain would subject himself to a flanking fire on both sides from the bastion heights commanding all the roads across the lowland. As he could not continue until the menace to his flanks and rear was removed, he must perforce attack the defenders on the exceptionally favorable ground selected and organized by them, and must launch his attack from a disadvantageous position on the lower plain. It remains to be noted, however, that the presence of large forests in the curtains of Lunéville and Charmes, where the usual condition is reversed and a *cleared plateau* overlooks a *forested lowland*, detracts from the ease of defense by providing the attackers with cover for their maneuvers. Here the forests are indeed bandages over the eyes of the defenders.

Côte de Delme

One feature of the Saffais plateau northeast of Nancy deserves particular attention. There is an erosion outlier of the higher plateau to the west, left stranded on the very crest of the Saffais cuesta and known as the Côte de Delme (Figs. 98 and 113). Its exceptionally steep western slope seems to be the result of erosion along displacements in the rocks, while its unusual elevation is due to the preservation of the higher formations. With its bold western front and its commanding observation over the Seille lowland, it possessed considerable strategic value for Germany; and it is said that in this region the frontier of 1871 was purposely drawn so as to assure to the Empire the military advantages which possession of the Côte de Delme would confer. The Tincry mesa just to the east also retains remnants of the higher formations upon its summit. Once in possession of the French, the Côte with its steep eastern scarp, together with the mesa, would effectively supplement the defense of the Saffais plateau scarp against an attack directed along the Nied valley curtain. Its chief potential value, however, has been supposed to lie in the possibility of utilizing it for defending the flank of a French army invading German Lorraine against a sortie from Metz. So long as the Delme ridge was in French possession, an army debouching southward from Metz to attack the flank of French forces would itself be threatened in the flank.

THE PLAIN OF LORRAINE

The lowland east of the Saffais cuesta is as a whole less perfectly developed than any of the major lowlands heretofore studied. In the first place the overlying limestones belonging to the Saffais upland are not completely removed. Numerous mesas, buttes, and masses of rolling hills formed of the higher rock are scattered over the plain far east of the principal scarp. As noted on an earlier page, it is often difficult to tell where to draw the line between the projecting bastions of the Saffais plateau and hills which would better be considered merely as erosion remnants on the plain. Secondly, there are limestone layers of medium hardness, and at least one of fairly strong resistance, interstratified with the marls and clays of the lowland. These have had the effect of preventing stream erosion from reducing this series to a uniform plain. Instead, numerous minor terraces or cuestas, and one of fairly pronounced type, diversify the surface. In the third place, the near approach to the Vosges uplift is betrayed by folds and fractures in the rocks which here and there have disturbed the gently inclined beds and caused erosion to etch them into relief forms not common to typical lowlands. And, finally, the uplifting of the Vosges has permitted streams to entrench themselves so deeply that "plateau" seems a more fitting term than "lowland" for the areas on either side of the valleys.

In view of all these circumstances the reader will not be surprised to find that what is here called the "plain of Lorraine" is by others called the "plateau of Lorraine," despite the fact that it begins on the west at the base of the Saffais scarp (Fig. 115). Some who recognize the northern and southern parts of the belt as parts of the same lowland nevertheless designate the central portion as the "plateau of Meurthe and Moselle." Nor will the reader fail to understand why different authors in describing this region fail to agree on the limits of the different belts, particularly if he studies a relief map and notes that west of Baccarat a subsidiary escarpment connects the eastern points of the Essey and Manonviller bastions, almost blocking the entrance to the Lunéville curtain and carrying the higher land practically unbroken from the Saffais upland to the Vosges Mountains; that only a short distance south the lowland is equally continuous to the margin of the Vosges; and that still farther south the same surface drops by successive terraces into an amphitheatral depression, the dissected margins of which are called the Monts Faucilles. Fully recognizing, then, the difficulties of delimiting and describing a "lowland" which is often rather high land, let us fix in our mental image of this region the salient features of its varied landscape.

Effect of Clay Deposits

Some features are common to the entire belt from its northern to its southern limits. The underlying rocks usually give an infertile soil containing much clay. In rainy weather roads are boggy, and the fields wet and cold. After a series of sunny days the earth bakes hard, and as in the Woëvre a number of draft animals must be hitched to a single plow to drag it through the resistant material. Many patches too difficult to cultivate are abandoned to forest growth. Where sands and gravels have washed down from the Vosges in large quantities, the deposit proves too infertile for agriculture, and again we have large areas of forest land. The great Forests of Charmes, Vitrimont (Fig. 102), and Mondon, as well as the long belt of forest west of the Sarre valley from Lörchingen to Saarunion, appear to be of this origin. Altogether the plain of Lorraine has a much larger proportion of forest land than the Saffais plateau to the west. Such land as is cleared is best adapted to cereals, while the natural prairies along the valley bottoms produce excellent hay. It is only natural, therefore, that stock raising should form one of the principal occupations of the inhabitants, and hence that an invader overrunning the country could count on replenishing his supply of draft animals and meats, rather than his stores of other farm products.

Where the clay is close to the surface and not deeply dissected by stream valleys, stagnant water forms numberless ponds and lakes, as in the Woëvre. This is particularly true in the northern part of the plain, where large, irregular expanses of water make the region very difficult to cross. These lakes are utilized as reservoirs to feed canals traversing the plain, and, when a succession of lakes in the midst of a forest belt are linked together by canals, as is the case west of the Sarre valley, the whole forms a military barrier of great strength. The fact that basket-making is an important industry in the plain, as for example in the country north and west of Baccarat, tells the story of marshy and peaty valleys, overgrown with rushes and bordered by willows; while the porcelain factories at Lunéville, Sarreguemines (Saargemünd), and other points, as well as the lakes, remind one of the vast clay deposits underlying the plain.

Saline Deposits

Some of the most characteristic features of the Lorraine plain depend on the fact that the rocks contain important deposits of various salts. These reach the surface in the form of mineral springs famous for their medicinal values, of saline waters which may be evaporated to recover their salt contents, or through artificial workings for the exploitation of beds of rock salt and gypsum, often made more accessible by the faulting or displacement of the formations. In the south, where that portion of the lowland bordering on the east the southern extension of the Saffais plateau is called La Vosge (not to be confused with the mountains called Les Vosges), it is the mineral waters which make famous such resorts as Vittel, Martigny, and Contrexéville. Farther north quarries and mines of gypsum and salt dot the country, and even the map reveals the situation to the observant. Rosières-aux-Salines, Château-Salins, the Canal des Salines repeat in various combinations the telltale termination; while the word "saline" alone at many points indicates the presence of deposits which have given name to no particular town. We encounter the same root in other forms in the name Saulnois given to a large portion of the northern part of the plain and in Seille, the name of the river which drains it. On the ground itself one quickly notes the important industries to which the deposits give rise—the plaster and soda factories, salt mines and refineries, and chemical establishments of various kinds, upon which Rosières, St. Nicolas du Port, Varangéville, Dombasle, Einville, Vic, Dieuze, Saaralben, and other villages thrive.

River Barriers

Across the plain of Lorraine, from the Vosges on the southeast to the plateaus on the northwest, a series of rivers flow in roughly parallel courses. Of these the Moselle is the most important, but the Mortagne, Meurthe, and Vezouse carve trenches but slightly less impressive. South of the Moselle, the Madon, and, north of the Vezouse, the Sanon have smaller valleys capable, if need be, of serving as lines of defense, particularly that of the Sanon, which carries the Marne-Rhine Canal. But it is the valleys of the three major M's, the Moselle, Mortagne, and Meurthe, which provide the transverse barriers of chief military importance, an importance which was greatly enhanced by the fact that they lay parallel to the Franco-German frontier across the plain and traversed one of the main topographic depressions (the Charmes Trough) inviting hostile invasion.

Strategic Gateways

Toward the northeast the plain of Lorraine, rising gently, laps around the northern end of the Vosges Mountains and into the low saddle separating these mountains from the highlands of western Germany and the Ardennes. This is the "Lorraine gateway" by which the Germans invaded France in 1870. Completely across the gateway, from one mountain pillar to the other, runs the trench of the Sarre River, cutting its way directly through the Sarre coal basin and its bordering hills, which are due to deep erosion of a much disturbed portion of the rock series of the plain, just beyond the northern limits of the battlefield. Toward the southeast the plain skirts the southern end of the Vosges, entering the narrow "Belfort gateway" between the Vosges and Jura mountains. This easily defended defile is strengthened by the ringed fortress of Belfort, and neither in 1870 nor in 1914 could it be breached by the Germans.

Between these two gateways the plain of Lorraine slopes gradually upward to merge into the western slope of the great Vosges barrier. From beneath the marls and clays with their interstratified limestone beds there rise massive beds of sandstone which lap up over the crystalline rocks of the range (Fig. 99). The sandstone occasionally terminates in an eastern scarp, sometimes in two scarps when the formation is broken and dislocated. But it forms no continuous, easily recognizable plateau separated from the Vosges by a distinct lowland. On the contrary, it is often badly dissected into a maze of sharp-crested ridges which merge almost imperceptibly with the crystalline hills beyond; or it forms the summits of hills having crystalline bases below. The sandstone gives a soil so infertile that little attempt is made to cultivate it. Hence, crossing on to the sandstone formation one crosses into the Vosges Forest. We will draw the eastern limit of the Lorraine plain, therefore, where the sandstone ridges and unbroken forest begin.

THE VOSGES MOUNTAINS

Beneath the great series of sandstones, limestones, marls, chalk, and clays responsible for the topographic features we have studied on previous pages lies the basement of massive crystalline rocks. In the region of the Vosges and eastward the earth's crust was raised in a great arch. The crest of this arch was broken by two parallel lines of fractures trending northeast-southwest and the long, narrow block included between the two fractures dropped down several thousand feet. There resulted two mountain ranges; the remaining western limb of the arch formed the Vosges Mountains, the eastern limb the Schwarzwald, or Black Forest Mountains (cf. small index diagram, Fig. 99). The down-dropped block between the two constitutes the valley of the Rhine from Basel to Mainz.

Each of the two mountain ranges has strongly contrasted slopes. The declivities facing in toward the Rhine trough are steep and forbidding, because formed by erosion of the precipitous fracture faces. On the contrary, the slopes representing what remains of the former western and eastern sides of the gentle arch naturally show gentle declivities. This explains why the Vosges has a gently inclined backslope on the west and a steeper eastern face, whereas the Schwarzwald has its gentle backslope on the east and its precipitous scarp toward the west (Fig. 61). With the Franco-German boundary thrown back on the Rhine as a result of the World War, an equilibrium of strategic advan-

FIG. 116—Looking eastward from the crest of the Vosges down one of the short valleys cut in the steep eastern face of the range. The flat plain of the Rhine lies just beyond the farther hills. For four years the battle front, trending obliquely across the range, crossed this valley in the middle distance. Note the ease with which artillery can enfilade the only road leading up this valley to the crest.

tages is re-established in this region. Each country presents to the enemy a mountain wall of formidable steepness, while retaining for itself the ability to maneuver on the more gentle backslope.

Extensive erosion has stripped much of the sedimentary cover from the backslope of the Vosges, exposing to view the underlying crystallines. It is probable that a goodly part of this erosion was accomplished before the arch was lifted to its maximum height and before the central block was dropped to form the Rhine valley, the original dome first being worn down to a surface of faint relief beveling across sediments and crystallines alike. Then, when renewed uplift and the fracturing gave the present form, the old beveled surface formed a sloping plateau, or "peneplane" (almost a plane surface), as the geographer would call it, into which renewed erosion has cut deep canyons. It is because of this history that the backslope of the Vosges fails to show a jagged sky line of craggy peaks of varying height but reveals instead, even on the folded and broken crystallines—granites, gneisses, and a variety of eruptive rocks—a simple sky line beveling indifferently across the most complex structures (Fig. 104). The Vosges Mountains are therefore, at least on their gentle western slope, of a form less forbidding than many other ranges. Had it not been for the stream erosion which has so deeply trenched the level upland, one could traverse the western slope in any direction with comparative ease.

Not so the eastern scarp. Here the original fracture surface, had it been preserved, would doubtless have proved an absolutely inaccessible rock wall. It is only thanks to the erosive action of streams, which have cut back into it and reduced its smooth face to a labyrinth of ravines and ridges (Fig. 116), that one may, at cost of considerable labor, scale its acclivities. Notwithstanding the modifying and mollifying action of the streams, the eastern scarp still remains in strong contrast with the backslope and justifies the description which one traveler gave of the crest near the Hohneck more than half a century ago:

Yesterday in fact, in order to reach the summit of this massive and imposing mountain, we were forced to scale it from the Alsatian (eastern) side over enormous barren crags, passing along the base of granite walls almost 200 meters high; today, on the contrary, we ascend its (western) slope by walking over a smooth field covered with rich pastures and reach the crest almost without fatigue. The Lorraine side, where we are now, presents almost everywhere this gentle slope—no escarpments and few rock ledges; . . . . on the Alsatian side, on the other hand, scarcely is one at the foot of the rounded half-dome which forms the summit of the mountain, when its declivity drops abruptly to a depth of between 100 and 300 meters; the wooded slopes which succeed the escarpments are themselves very steep. As a result the valleys come to crouch, so to speak, under the very feet of the spectator.[6]

Effects of Glaciation

Prior to the glacial period the crest of the eastern slope had acquired a somewhat less precipitous character than it shows today, and the highest summits had been softened by the wearing influence of the weather to the rounded forms which the natives today call *ballons* (German: *Belchen*). This term rather appropriately suggests the form to which it is applied, for one of these domelike summits (Fig. 105) resembles not a little the top of an inflated balloon: yet, contrary to the popular idea, there is no connection between the term *ballon* applied to the summits of the Vosges, and the French word *ballon* meaning "balloon." In the former connection the word should perhaps be spelled *bâlon*, and, while its derivation is disputed, it may have originated in the ancient Celtic cult of Bel or Belus, celebrated upon certain of Vosges summits.[7]

Be this as it may, when the glacial period gripped the Vosges Mountains in its icy grasp, the heads of the eastern valleys, close up under the *ballons*, became a favorite gathering ground for glaciers. As these masses of ice grew larger and began gnawing into the mountain flank, they excavated amphitheaters, or cirques, whose steep head walls undermined the rounded slopes above. But before glaciers from several sides of the domes could carry their work to the point of removing all of the smooth upland surface and leaving only a jagged peak or horn, like the Matter-

[6] H. de Peyerimhof, quoted by G. Bleicher: Les Vosges, Paris, 1890, p. 23.

[7] O. Barré: L'architecture du sol de la France: Essai de géographie tectonique, Paris, 1903, p. 113.

horn of the Alps, a milder climate intervened and the ice melted away. Today we read the record of the incompleted task in the smoothly rounded remnants of the preglacial upland surface, sharply broken into, especially on the east, by the steep head walls of glacial cirques. It is in the bottoms of the cirques that we find the little glacial lakes, or tarns, such as Lac Noir and Lac Blanc, which with their wild rocky basins and towering precipices recall the glacial scenery of the Alps. In some of these rock-rimmed depressions safe emplacements were found for guns which during four years waked the mountain echoes; in others elaborate stone-walled encampments for troops obtained secure shelter under the high cliffs.

We have now in mind a general picture of the great earth block forming the Vosges mountain mass, with its gentle western slope beveling across crystalline rocks at the higher levels, across the overlapping sandstone on the lower flank, and dissected into a mountainous topography by west-flowing streams and their branches; and with its steep eastern scarp frayed into ravines and ridges and often terminating aloft in the rock walls of glacial cirques which undermined the lofty domes of the *ballons*. We shall find it profitable to look more closely at the forms of this mountain mass, to discover in what manner it might serve as a barrier against enemy invasions, and where and by what means a hostile army might traverse it. For it can hardly be doubted that a range rising to an elevation of nearly 4,700 feet, or about 4,000 feet above the plain of the Rhine, and spreading over a breadth of nearly 40 miles, must constitute a military barrier worthy of our attention, whatever its detailed form may be. Let us, therefore, first examine the lower western flank, then traverse the higher valleys to the crest, observe the character of the passes, and descend the eastern scarp far enough to satisfy ourselves regarding the salient features of the military geography of the Vosges.

The Sandstone Vosges

In ascending the valleys of the Meurthe, the Mortagne, or the Moselle, one enters the sandstone Vosges just beyond Baccarat,

gives the hills a tabular form, the steep lower slopes often being crowned with nearly or quite vertical walls, occasionally weathered into fantastic shapes resembling the ruins of medieval castles. Nor could castle walls prove more difficult to scale than some of these architectural forms designed by Nature. They would be more impressive were they not repeated with such monotonous regularity throughout the whole belt of the sandstone Vosges, which may accurately be described as difficult to traverse and little inhabited. Sands suitable for glass manufacture and wood for fuel and the wood industries have not proved enough to redeem the most of the terrain from its primitive condition as a wilderness.

The Forest Cover

Over all is spread the dark mantle of a dense spruce forest, which owes its existence primarily to the infertility of the sandy soil resulting from the disintegration of the sandstone beds, and secondarily to the relative inaccessibility of its steep slopes and isolated tabular uplands. With but the rarest exceptions it is only along the narrow valley bottoms that open fields and meadows relieve the gloom of the continuous band of close-set trees guarding the western approaches to the Vosges passes. Toward the north the sandstone laps farther and farther eastward, until it reaches the crest of the range in the vicinity of the Donon and from there northward forms the summit. In the Lower Vosges, as the sandstone plateau north of the Pass of Saverne (Zabern) is called, it was always the dense forests rather than the form of the surface which made the region a barrier so difficult to traverse that it formerly served well as a natural frontier.

It must readily appear that an escarpment heavily cloaked with forest from summit to base cannot so well serve the purposes of defense as those we have studied farther west. Extensive clearings would have to be made before satisfactory observation was possible, and even then one part of the escarpment might not be able effectively to support another subjected to attack, because of intervening woods. Add the fact that the lowland which fronts the escarpment is neither broad, flat, nor fully exposed, and an appreciable proportion of its importance is gone. But it must not be imagined that such a maze of steep-sided hills and narrow valleys, fortified with an almost impenetrable forest cover, can under any circumstances lose its military value. In the somber shadows of the forest near the crest of the escarpment south of Raon l'Étape, at the Col de Trace or de la Chipote, the traveler passes a recently erected monument hewn from red sandstone. This monument marks the spot where the German invasion of 1914 was halted after having reached the col itself. It is a memorial to those heroes who perished in denying the enemy access to the outermost rim of the Paris Basin. It might also be regarded as a tribute to the sandstone ridge from which it was carved: for it was no mere accident that the repulse occurred on this escarpment. At Épinal it is the dissected sandstone scarp and upland which give the excellent series of positions utilized in fortifying the solid buttress upon which rests securely the left of the defensive "Line of the upper Moselle."

The Crystalline Vosges

East of the sandstone belt the crystalline rocks reach the surface, first appearing in the imperfect lowland eroded along the base of the irregular sandstone scarp. Quickly the country rises to mountain heights again, but this time in more knobby masses which lack the architectural forms of the sandstone and which slope more gently as they approach the valley bottoms (Fig. 104). Here and there tabular masses of the sandstone, saved from the ravages of millenniums of erosion, cap the granite mountains, most frequently just east of the sandstone Vosges escarpment. Farther up in the mountains to the southeast the crystallines alone are exposed, and the only suggestions of geometrical lines appear in the straight courses of certain valleys and in the level sky line testifying to the old erosion surface which once beveled across the whole mass. Glaciers formerly occupied a number of the valleys, possibly widening and deepening certain of them, and formed lakes some of which are dammed by moraines, as is the case with Lake Gérardmer. It is the higher, glaciated portion of the crystalline Vosges which is scenically most beautiful and most difficult for an hostile army to negotiate.

Nevertheless, the feature which most impresses the traveler on entering the crystalline Vosges is the extent to which the country is cleared and populated. The contrast with the heavily forested and sparsely inhabited sandstone Vosges is striking in the extreme. Indeed, the crystalline mountains are more densely peopled than the plain of Lorraine. Both valley floors and valley walls are cleared, and fields and meadows mount high on the slopes (Fig. 104). In a few places even the uplands are cleared and occupied; but elsewhere extensive peat bogs and forests cover the level highlands. The forests supply material for the wood industries and for paper manufacturing, as well as fuel. Although the mountain streams flow between grassy banks in valleys which are fairly open, they descend with enough rapidity to develop abundant water power; and into these valleys the cotton industry of Alsace-Lorraine has penetrated so extensively as to transform them into industrial centers of much importance.

Valley Barriers

Notwithstanding the open and smiling landscapes with which the valleys of the crystalline Vosges delight the observer, these natural trenches through the mountain mass have a rôle to play in the grim business of war. They are not merely routes for supply lines leading to the crest of the mountain barrier, but are also lines of defense against an enemy advancing parallel with the axis of the range. The most notable example of the latter use of the valleys is found in the case of the Moselle River, upon which is based the famous fortified "Line of the upper Moselle," the Épinal-Belfort string of fortresses (Fig. 83). From the base of the mountains at the Épinal gateway to the crest at the Ballon de Servance (near the Ballon d'Alsace), the southern wall of the Moselle trench is crowned with forts so spaced as to sweep the valley floor throughout its length and every road crossing the barrier southward toward the Saône basin with accurately directed artillery fire. Except in times of flood the river is easily fordable and near its headwaters is little more than a trout brook; but with the northern approaches so mountainous as to reduce the lines of advance to comparatively few well-known routes, with dangerous peat bogs and forest strongholds scattered over the upland, and with the slopes of the trench and its floor sufficiently exposed to give artillery and machine-gun fire full play, the barrier is doubtless correctly considered as practically impregnable. The massive Vosges barrier with its deep valley trenches made a fitting basis for the southern element in the Verdun-Toul plus Épinal-Belfort scheme of defense for France's eastern frontier.

A noteworthy feature of the crystalline Vosges is the large number of valley trenches and ridges which run parallel to the crest of the range and hence at right angles to the general courses of the transverse Meurthe and Moselle. It is not necessary to enumerate these valleys, which will appear on any good detailed map, nor to discuss individually their military value. Suffice it to say that this systematic parallelism of surface form, which probably has its origin in parallel folding or fracturing of the crystalline mass, and which affects both the continuity of the crest line and the character of certain passes, results in a succession of trenches and walls too difficult to be crossed save along the courses of a few transverse valleys.

Passes of the Vosges

The crest of the Vosges Mountains is not composed of a single continuous ridge, but of a series of three main ridges and possibly one subordinate crest arranged *en échelon* (Fig. 98).[8] The Franco-German frontier of 1871, following the crest of the range northward to the Donon, was forced to jump from one ridge to another, thus acquiring its peculiar offsets to the west. It further results from this pattern of the topography that the passes by which the range may be crossed are of two principal

[8] Barré, L'architecture du sol de la France, pp. 113–114.

types: those leading directly over one of the main crest ridges, and those which insinuate themselves, so to speak, between two of the ridges where they overlap. To the former class belong the pass of Bussang (731 meters above sea level; the highest point of the range is 1,424 meters above sea level) and the high pass of the Schlucht (1,139 meters), saddles in the first main crest giving access to the Moselle headwaters; the passes of Bonhomme (951 meters) and Ste. Marie aux Mines (753 meters), the latter utilized by the French armies during the wars of the end of the seventeenth century, both saddles in the second crest giving access to the headwaters of the Meurthe; and finally the pass of Donon (750 meters) across the third ridge, also giving access from Germany to the upper Meurthe. There are seven more of these passes crossed by minor roads; but not until we reach the low trough of Saverne (404 meters), at the northern limit of the High Vosges, do we find a pass carrying a railway. Prior to 1914 the main Vosges barrier had never been breached by a line of rails. As a rule the passes of this first class are difficult of ascent, even for a highway, especially on the steep eastern side, where the roads first follow the short valleys till near their heads (Fig. 116), then zigzag up some ridge to gain the crest. Traffic along the perilous trails climbing the eastern scarp was not infrequently punctuated with disaster, many men and animals having lost their lives when the Rotenbach and other passes little used today were crossed by caravans carrying commerce between Alsace and the transmontane regions. The Schlucht Pass was rendered really practicable only in 1860 by skillful engineering work which perched an excellent highway in a niche cut along the face of sheer granite walls and yawning precipices and made it secure from the falling rock slides which imperiled the poor trail of earlier days. It is easy to see how a small body of defenders, installed in good positions on the crests commanding such a road, could effectively block the pass against greatly superior forces restricted to this one line of advance. After severe fighting in 1914 the French held the Schlucht Pass for the remainder of the war; and the sentry in his sandbag shelter where the road over the pass intersected the new military road along the ridge, during most of this time experienced no greater excitement than the daily "strafing" of the crossroads by the German guns down in the forested ravines of the eastern slope.

The second class of passes, those connecting longitudinal valleys between the main crest ridges where these overlap, includes but two; the relatively unimportant Louchpach (Luschbach) Pass, between the first and second ridges, and the Saales Pass (560 meters) between the second and third ridges. The Saales Pass is the lowest and most important gap in the entire range from the Belfort gateway to the trough of Saverne. The second and third crest ridges are farther apart than the first and second, and the col between them correspondingly deeper. On the north the longitudinal valley of the Bruche (Breusch) heads against the low col and flows northward between the overlapping ridges until it can turn around the northern end of the eastern ridge to reach Strassburg. Southward a longitudinal branch of the Meurthe leads down to St. Dié. Here, then, is the one really accessible gateway through the High Vosges from the Rhine valley to the plain of Lorraine; or, in 1914, from Germany into France.

That the Saales Pass should have played an important rôle in directing movements across the Vosges from the earliest times can easily be comprehended. Before the Roman period, invading hordes from the east poured through it to spread over the plain of Lorraine, and the Roman conquerors later built a road across it. In 1870 it was the only pass through the High Vosges utilized by the Germans. When after that war the territory to the north and east became German, the German government constructed a railway up the valley of the Bruche to the pass, by which they could in "the next war" transport armies to the very threshold of this entrance into France. On the other hand, the French government, acting on the advice of the Minister of War, refused permission to build a line on the French side of the pass linking the German railway with the French system. Now that Alsace-Lorraine has returned to the mother country, a railway through the pass will doubtless soon be completed for commercial, political, and strategic reasons.

The "Ballons"

The crest of the Vosges rises now and again in the broad grassy domes, or *ballons*, between which are the saddles forming the transverse passes of the first class described above. It is very striking to behold these "balds" rising above the dark mantle of the Vosges forest, in a range which nowhere attains an altitude of 5,000 feet. But there is, despite the low elevation for this latitude, a distinct tree line at about the 4,000 foot contour. This seems to result from the severe climate of the higher Vosges, which are covered with a deep mantle of snow late into the spring, while in favored localities in the highest valleys or glacial cirques patches of snow sometimes last through the summer. As a consequence of these climatic conditions enormous kennels of dogs were maintained by the armies on the Vosges front for sledge transport during a large part of the year.

Because of the superior heights of the *ballons*, guarding the passes on either side and dominating the approaches to them through the valleys far below, they possess some military value. Their smooth surfaces were fortified with field works (Fig. 105) and utilized as observation stations when not too far from the battle front. From the Ballon d'Alsace one has a wonderful view in all directions: westward the even-crested upland slopes gently down to the plain of Lorraine, with the trench of the upper Moselle visible for a long distance; northward is the asymmetrical crest bearing other domes which merge on the west with the even upland sky line but drop off abruptly into the ravines scouring the steeper eastern scarp; eastward one gazes down into the ravine heads which seem to undermine the very feet of the observer and out upon the plain of the Rhine far beyond, where passing trains can be seen with the naked eye; southward the Belfort gateway and the town itself are in plain view. The Ballon de Servance likewise gives commanding observation in every direction, and the fortress upon it, the highest on the eastern frontier of France north of the Swiss boundary, dominates the valley routes below and connects the fortified line of the upper Moselle with the entrenched camp of Belfort.

East of the main crest is a short minor ridge (Fig. 98) carrying the highest dome of all, the Ballon de Guebwiller. Its summit, 1,424 meters or about 4,700 feet above sea level, is the culminating point of the Vosges and gains in impressiveness because of its advanced position, close to the low plain of the Rhine valley. From its higher slopes a broad area of the plain from Mülhausen to Colmar may be kept under easy observation and much of it dominated by artillery fire. This is why so many bitter conflicts raged for the possession of a knob on the eastern slope of the *ballon*, known as Hartmannsweilerkopf. Loss of this knob by the Germans meant that the enemy would gain excellent views far behind their front, as well as direct observation into ravines concealing the big guns supporting the German line, and would enfilade their trenches for a long distance to the south, compelling a retreat into the exposed plain, where hidden gun positions would no longer be available. Only when the consequences of yielding such a position were considered could one who looked upon the splintered trees and shattered rocks of Hartmannsweilerkopf, with shells bursting around its crest and fountains of rock and earth vomiting on all sides, imagine why so many lives should be sacrificed to retain one particular height amidst a maze of peaks and ridges.

East of the main crest of the Vosges the short valleys which drop so abruptly near their heads, but more gently farther down, have the forests cleared from their floors (Fig. 116), so that enemy movements by the only practicable routes are exposed to observation and fire from above. The valley walls are often so steep as to be left with their cover of trees, although ridge tops are in places cleared for pastures. To penetrate the eastern valleys of the Vosges under enemy observation and fire, to traverse exposed valley roads enfiladed by artillery posted on the heights, to negotiate tortuous zigzag roads scaling precipices and steep rock slides, with sheer granite walls on one side, yawning chasms on the other, and a determined enemy advantageously posted in front, is obviously a military operation of such extreme difficulty that only in case no easier approach to the enemy was available would the terrain of the Vosges become the scene of major operations. Throughout most of the war this portion of the front was lightly held while both sides sought victory on fields better adapted to military maneuvers. "This sector in the beautiful wooded hills and mountains of the Vosges was quiet in that it was practically free from danger of a great German offensive, as there was nothing to gain, while enormous losses were certain in trying to march armies over the mountains. The French and Germans had used the sector as a position where divisions worn out with fighting elsewhere in the line could break in their recruits or 'replacements'. The shell fire was consequently held down to a minimum by both sides so that the much-needed rest could be obtained before the call came for the division to go to an active front."[9] Into this sector one division after another of American troops were brought to give the men their first acquaintance with modern warfare, until the Vosges became one of the chief training grounds of America's new armies. Those compelled to leave for other fronts "still speak of the happy days in the Vosges."

[9] Shipley Thomas: The History of the A. E. F., New York, 1920, p. 62.

CHAPTER X

MILITARY OPERATIONS ON THE BATTLEFIELD OF LORRAINE

First Advance into Alsace

Scarcely had France recovered from the first shock of the declaration of war, when her general staff launched an offensive into Alsace. This operation was designed, among other things, to render more secure the right flank of the armies defending the vital Charmes Trough and soon to advance into the Lorraine gateway; to establish the right flank of the Allied line firmly on the strong Rhine barrier; and to assure the possibility of debouching from the Vosges passes into the Rhine plain on a wide front. It was also doubtless intended by this movement into a "lost province" to stimulate the patriotic ardor of the whole French people and to arouse the Alsatians to fight against their oppressors. If successful in pushing down the Rhine valley for any considerable distance, the advancing force would threaten the flank and rear of the German armies massing on the threshold of the Lorraine gateway for the impending irruption along the Trough of Charmes.

The first effort of the French was executed with a number of troops utterly insufficient for so important a task and with a lack of prevision which was swiftly punished. Debouching through the Belfort gateway on August 7, 1914, the light covering forces of the Germans were so rapidly hurled back that in a few hours the Altkirch position on the upper Ill was forced and the next day the soldiers of France joyously entered Mulhouse (Mülhausen). Before the echoes of Joffre's proclamation of deliverance had died away, however, the feeble French columns were violently assaulted by superior enemy forces, and three days after they had quitted Belfort they were again back under its protecting guns.

Immediately a new army of Alsace under General Pau was formed to undertake the task with larger forces and better appreciation of the difficulties to be overcome. It was determined after driving the enemy back behind the flat-bottomed trench of the Ill River to breach that barrier and throw him beyond the formidable obstacle of the Rhine, where he could be held in check by a limited number of troops while the advance down the Rhine valley toward the north continued. To aid this operation other French forces would descend the steep eastern face of the Vosges along the short valleys eroded in the escarpment, clearing the Germans from the mountain stronghold. Pau's army could then move north with its right flank protected by a practically impassable river and its left by an impregnable mountain buttress.

The capture of the Vosges passes was a necessary preliminary to this operation. The French withdrawal for a distance of eight kilometers (elsewhere ten) from the frontier in order to avoid border incidents and to demonstrate to the world that the war was one of German aggression, coupled with the perfection of Germany's preparation for launching the blow at the moment chosen by her as most favorable, gave to the enemy a great initial advantage. In the Vosges, however, more than anywhere else, this advantage was offset by natural disadvantages which no military skill could completely overcome. The Germans had to supply their forces by few and perilous routes clinging precariously to granite walls along the steep mountain face; whereas the French, maneuvering on the gentle backslope of the range, found the movement of troops, the handling of supplies and munitions, the bringing up of big guns, and many other necessary operations much less difficult. The natural features of the terrain told in favor of the French, and at one pass after another the enemy was pushed over the crest and back down the eastern slope. The recapture of the Bussang and Schlucht passes was found to be especially facilitated by the fact that German artillery under the steep eastern scarp could lend very poor support to their infantry on the upland, whereas the French artillery operated easily and effectively on the much more favorable terrain west of these passes. Before a week had elapsed every pass in the range between the Ballon d'Alsace in the south and the imposing Mont Donon in the north was in French hands.

With possession of the Vosges passes assured to the French, and the Germans clinging with difficulty to the steep eastern face of the great mountain barrier, the situation was ripe on August 14 for the new offensive in Alsace. General Pau's army debouched through the Belfort gateway into the Rhine plain, crushed the German resistance along the river Ill, captured Mulhouse, and threw the enemy back on the Rhine. His left wing, pushing down hill from the passes, drove the Germans from the heights into the lower valleys, captured Thann near the mouth of the Thur valley, and much of the dominating mass of the Ballon de Guebwiller farther north. By the 21st Colmar was closely threatened, and the possibility of an attack on the communications of the German armies in the Lorraine gateway seemed on the point of changing into a probability.

Dubail's First Army, on the left of Pau's army, having secured the northern passes of the range, was likewise pushing toward the plain. Through the vital Saales Pass especially his forces were pouring down the Bruche valley toward Strassburg, protected by the dominating height of the Donon, towering above them to the west. From one end of the Vosges to the other the armies of France, allied with the force of gravity, were beating down upon the retreating enemy, with visions of a decisive success beckoning them onward, when failure on other battlefields wrecked the Alsatian campaign. The retreat to the Marne had become necessary, and every ounce of strength must be conserved for the impending struggle farther west. At the same time the defeat of the French in the Battle of Sarrebourg-Morhange (Saarburg-Mörchingen) was entailing the retirement of Dubail's left wing from the northern passes of the Vosges. The armies in Alsace were therefore ordered to fall back upon the impregnable position of the Vosges crest south of the Bonhomme Pass.

The Battle of Sarrebourg-Morhange

On August 14, simultaneously with the invasion of Alsace by the armies of General Pau and of General Dubail's right wing through the Belfort gateway and the Vosges passes as detailed above, Dubail's left wing and the French Second Army under General de Castelnau invaded Germany through the Lorraine gateway. Their advance was directed toward the line of the Sarre River, which together with its outlying lake-forest-marsh barrier on the southeast and the hills rimming the Sarre Basin on the northwest, constituted a formidable obstacle across the Lorraine plain, connecting the Vosges Mountains with the mountains of western Germany. Between the French and their objective the German armies were holding an exceedingly strong position protecting the Metz-Strassburg strategic railroad through the Lorraine gateway, a position based on the northward extension of the Saffais plateau and the lake-forest-marsh region of the plain next to the east. General de Castelnau, whose intimate knowledge of the terrain, keen appreciation of the importance of topographic barriers in military operations, and calm judgment made him one of the most trusted advisers of the French high command throughout the war, was not unaware of the tremendous difficulties which must be encountered in any attempt to dislodge the Germans from the terrain on which they had elected to receive the French attack. Indeed, there is good reason to believe that this unfortunate offensive was undertaken in obedience to orders which his judgment did not approve.

The German right wing rested on the commanding ridge of the Côte de Delme (Fig. 113) crowning the northern extension of the Saffais upland and dominating the exposed Seille lowland from the east. The center lay across the double bastion of Morhange-Dieuze (Fig. 98), each of the two walls of this projecting plateau mass being held by strong forces entrenched on the upland, often concealed under forest cover and sweeping the intervening valley, as well as the plain to the north and south, with their fire. At Dieuze the marshy valley of the Seille links up with a series of marshes, lakes, and forests stretching northeast, east, and southeast to the great belt of lakes and forests paralleling the Sarre valley on the west, and forming thus a triangular mass (Dieuze-Gondrexange-Mittersheim) of difficult country, the northern exits from which were held by the German right wing standing behind the upper Seille barrier. The wooded heights of the bastion north of Dieuze dominated this humid lowland for many miles to the east and south. East of the marshes ran the natural trench of the Sarre River, on the farther bank of which were strong German forces. Altogether the enemy's position, while less formidable than the high plateau of the Moselle which the Germans were later to attack, was exceedingly strong. It had been skillfully organized with trenches, barbed-wire entanglements, concealed machine-gun nests, and other devices of modern defensive warfare soon to become a familiar feature of the struggle, but which were then sufficiently novel to baffle and disorganize the most determined assaults, even those delivered by greatly superior forces.

Against the difficult terrain thus strongly defended, along roads accurately registered by the German artillery, through a country teeming with spies who by innumerable devices kept the enemy well posted as to the movements of the attacking forces, Dubail's left wing and De Castelnau's Second Army advanced to the assault. The former, debouching from behind the protecting trench of the Meurthe, pushed forward toward the line of the upper Sarre along the easier pathway between the savage country of the sandstone Vosges on the right, where his right wing was delayed by the difficulty of dislodging the enemy from mountain strongholds, and the lake-forest barrier on the left. Dubail thus sought to turn the marshy triangle by advancing down the valley of the Sarre River. When the upper Vezouse valley was encountered, the light German forces fighting a delaying action from beyond Domèvre on the west to Cirey on the east were overthrown after a short struggle, and the line of the Sarre from Sarrebourg up to the base of the Vosges gained without serious resistance. On the 20th, however, the real battle was precipitated when the French sought to cross the Sarre trench. Crushed by sheer weight of the artillery storm let loose by the defenders, the attacking forces, unable to cross the obstacle in front of them and hemmed in on the west by the wilderness of lakes and woods, were forced to retreat. Sarrebourg, which had been seized, had to be evacuated, and heavy losses in men added to the costliness of the check. But on the right the Germans hurled themselves in vain against the strong mountain positions from the border of the plain to the Donon, and Dubail might have resumed the offensive next day had not the forces of De Castelnau been constrained to an extensive retirement.

On Dubail's left the Second Army attacked the main German positions by four principal routes. One column was directed toward the marshy triangle Dieuze-Gondrexange-Mittersheim, to enter and clear that stronghold of enemy detachments and debouch if possible from the northern exits as the forces farther east advanced parallel with them down the Sarre valley. A second column advancing across the Arracourt curtain toward Dieuze was to attack and dislodge the enemy from the wooded heights of the Dieuze ridge, while a third performed a similar function for the northern, or Morhange, wall of the double-walled bastion. The fourth column on the extreme left was merely to keep the enemy pinned to his positions on the Côte de Delme and so protect the left flank of De Castelnau's army from attack. Once the Morhange wall was in French possession, the Côte de Delme position at the western base of the Nied valley curtain would be outflanked.

As before Dubail's army, so also before De Castelnau's the Germans fell back without offering any opposition more serious than stiff rearguard combats at favorable points on the terrain, until the strong positions described above were reached. The first column cleared the marsh-girt triangle but encountered great difficulty in debouching from the northern exits and forcing the marshy Seille barrier. At Dieuze the second column, caught in the marshes at the base of the ridge under heavy fire and finding the wooded heights practically impregnable, could make little progress despite heavy sacrifices. Farther north the steep southeastern face of the Morhange wall, repeatedly assaulted with indomitable bravery on August 20, exacted so heavy a toll from the attacking troops as to break the force of the French offensive. Exposed on the smooth open slopes to machine-gun fire from well-chosen positions, the French infantry was literally "mowed down in swathes" and finally broke in confusion. All along the line the skillfully defended natural obstacles had proved insurmountable, and General de Castelnau ordered his troops to fall back on their strong plateau positions. The Battle of Morhange, like the Battle of Sarrebourg, had ended in a German victory; and the victors now followed hard upon the retreating Frenchmen, eager to launch their projected offensive along the vital Trough of Charmes.

The Battle of the Trough of Charmes

While the forces under De Castelnau took their stand upon the Grand Couronné north of Nancy (Pl. VI) and the Saffais plateau to the south, Dubail's troops retired across the Vezouse and Meurthe trenches, along each of which delaying actions were fought, until his east-west front linked up with the Second Army's north-south front on the strong Saffais upland (prolonged as the Essey bastion) on the west and was defended on the right by the wooded labyrinth of the sandstone Vosges. In this latter sector the troops which had reluctantly yielded the high buttress of the Donon and the northern passes of the Vosges brought the invaders to a halt on the sandstone escarpment at the Col de la Chipote (p. 465). With the enemy checked at this almost impregnable barrier, Dubail's forces stretching across the floor of the Charmes trough were reasonably secure against the danger of a turning movement from the east, a security repeated on the west by the strong position on the plateau scarps. To meet the shock of the enemy's assault all available artillery was concentrated in the hidden ravines dissecting the Essey bastion, particularly on the Borville plateau near its apex, from which the plain beyond toward Rambervillers and all approaches to the scarp could be held under fire. In case of emergency, the defenders still had the strong line of the Moselle to fall back upon.

On August 24 it became clear that the German high command was indeed going to attempt the logical drive through the Trough of Charmes, so long foretold by French military students. Large enemy forces were reported moving southwest along the depression. Apparently they were ignoring for the time the French units which from the Côte de Delme and Morhange had retired to the vicinity of Nancy. The strong position of the Grand Couronné could be effectively turned from the south if the Trough of Charmes were breached. The first act in the great struggle to break and roll up the right wing of the Allied front developed rapidly. De Castelnau, taking prompt advantage of the enemy's mistake in exposing his right flank to attack, directed a part of his forces to move forward to the crest of the Saffais plateau north of the River Sanon called the "Rembétant" and out across the Manonviller bastion to the south of the river, and from these commanding positions to harass the flank and rear of the southward moving German columns so as to break the speed and force of their assault along the line of the Mortagne. The following morning the French loosed an offensive which wrecked the whole German plan of campaign. From the strong Essey bastion and Saffais heights, dominating the Mortagne moat on the southwest, a terrific fire was poured upon the front and flank of the attacking forces. Simultaneously the troops of De Castelnau's Second Army, debouching from the plateau strongholds farther north, pushed home the attack against the enemy's flank and rear.

Staggered by the suddenness and the fury of an offensive delivered by troops seriously defeated with heavy losses only a few days before, unable to overcome the admirable advantages of position which the French enjoyed on the Saffais upland, and badly worried by the flank attack coming from the plateau farther north, the enemy beat a hasty retreat. Efforts to pass by the right flank of the First Army in the Vosges near St. Dié were also frustrated by forces basing their defense on that difficult terrain. There was nothing to minimize the decisiveness of the enemy's defeat.

The German high command now saw that until De Castelnau's army, firmly entrenched on the plateau heights dominating the German right flank, was disposed of, any further attempt to drive forward along the Trough of Charmes was doomed to end in failure, if not in disaster. The Battle cf the Trough of Charmes, sharp but short, had ended in an Allied victory of the highest importance. Admirable use of a terrain peculiarly adapted to a defensive-offensive had defeated the first great attempt of the enemy to turn the right wing of the Allied line. The second and supreme attempt was to begin without delay.

The Battle of the Grand Couronné

A successful attack on the plateau crests to the north and south of Nancy was now essential if the German scheme of enveloping the Allied armies from both wings was not to end in failure thus early in the campaign. Something much more important than entering the chief city of France's eastern frontier was involved on the German side. If the Moselle plateau barrier was conquered and the French expelled from their natural bridgehead fortress east of the Moselle River, one of the two chief obstacles to the isolation of Verdun would have disappeared, the German flank would be relieved of a dangerous menace, the Trough of Charmes opened, and the envelopment of the Allied right wing rendered possible. When to all this was added the moral effect of a spectacular entrance of the Kaiser into Nancy, both in stimulating German ardor for the war and in depressing French public opinion, the fruits of the hoped-for victory are seen to have been of sufficient magnitude to justify a tremendous effort on the part of the German high command.

To appreciate fully the whole significance of the Battle of the Grand Couronné, often called the Battle of Nancy, one must review certain interesting chapters in the development of French military opinion. The question of the fortification and defense of Nancy had long been vigorously debated. In the judgment of many experts Nancy could not effectively be defended. It was believed that Germany's political constitution must enable her to mobilize and concentrate her troops for attack much faster than France, and therefore any attempt to gather forces of defense east of Nancy would only expose them to a disastrous onslaught before their organization had been perfected. This difficulty might have been obviated by turning Nancy into an entrenched camp like Verdun, surrounded by a circle of modern forts and furnished with a garrison strong enough to hold an enemy at bay until French mobilization was completed in good order. But this proposal always raised the objection that the fortification of Nancy might be regarded by Germany as a *casus belli* and so precipitate the very disaster it was intended to avert. There even grew up in certain circles the legend that a codicil or appendix to the Treaty of Frankfort forbade the fortification of the city—a result, it is said, of some statement in diplomatic correspondence to the effect that Germany would regard such fortification as an affront. In any case it was a favorite theory among French military authorities that the first great battle for the defense of eastern France must take place with the French army standing on the western bank of the Moselle, and hence with Nancy unprotected.

It is largely due to General de Castelnau that a contrary opinion ultimately prevailed. When he became chief of staff he took up this question with characteristic vigor and demonstrated that the peculiar topography of the Nancy region made it both possible and desirable to defend the city and deliver the first battle east of the Moselle. His intimate knowledge of the terrain and keen appreciation of its possibilities awakened confidence in his contention that the natural bridgehead of the Grand Couronné projecting east of the Moselle trench (Pl. VI) could be turned into an impregnable stronghold and must be held to permit the French armies to debouch from behind the river barrier at will. Failure to do this would not only condemn the French to a purely defensive campaign and prevent them from following up effectively any reverse they might inflict upon enemy forces trying in vain to breach the river barrier; it might also lead to the enemy's outflanking the French position from the north, owing to the fact that the westward slope of the plateau would permit enemy forces established on the heights of the Grand Couronné to dominate the lower western wall of the Moselle trench and so to force a crossing. On the other hand, with the Grand Couronné securely held by the French, a flank attack against enemy forces seeking to force the Moselle barrier farther south and to penetrate along the Trough of Charmes could always be counted upon to counteract that danger. The detailed forms of the terrain were so favorable to the defense that even moderate forces properly established on the dominating heights could be relied upon to hold in check greatly superior numbers of the enemy. Finally, the pronounced bastion-and-curtain pattern assumed by the plateau margin would both simplify the defense and make it difficult, if not impossible, for the enemy to penetrate as far as Nancy. By utilizing a combination of the Amance bastion and Nancy curtain of the Moselle plateau, and the Manonviller bastion, Lunéville curtain, and Essey bastion of the Saffais plateau (Fig. 118), a combination rendered desirable by the direction of the German attack and the change of course of the Moselle scarp south of Nancy, the city could be rendered inapproachable either from the northeast into the Nancy curtain

FIG. 118—Natural bastions and curtains of the Nancy region formed by salients and re-entrants of the Moselle and Saffais plateaus.

or from the southeast into the Lunéville curtain, by flanking fire from the bastion walls.

General de Castelnau made his views prevail to such good purpose that the fortification of the Nancy region was finally authorized, and the work began a few months before war was declared. Unfortunately, he could derive little benefit from this fact in the battle which raged from August 28 to September 12, 1914, so little was the important task advanced. But the masterly way in which the natural fortifications were employed vindicated the general's opinion as to the exceptional defensive value of the terrain.

We may consider that the Battle of the Grand Couronné opened with that portion of the enemy's offensive in the Battle of the Trough of Charmes which was directed westward into the Lunéville curtain against the Saffais escarpment southward from Rosières. In this initial blow against the natural fortifications of Nancy the strength of the French position immediately revealed itself. Every approach to the escarpment was dominated by artillery concealed in the ravines dissecting the plateau and directed from excellent observation posts along the crest (Figs. 102 and 115). "On the 24th a fierce onslaught was made against the right of the French position about Saffais, but the rush of the attack spent itself, and during the 25th and 26th the Germans were forced back step by step in spite of most determined efforts to hold their ground." The topographic advantages were telling heavily in favor of the defenders, and, when the flank attack from the western part of the Manonviller bastion and the Saffais plateau farther north threatened to turn defeat into disaster, the Germans massed their forces against that part of the front. Early in September the enemy returned to the assault against the Saffais positions but again met a decisive check along the plateau wall.

The next blow was aimed at the Nancy curtain and the Amance bastion. Against this impregnable stronghold the Kaiser's trained legions were flung in wave after wave, day after day, in numbers greatly superior to those of the defenders, who were constantly being weakened by heavy withdrawals of troops to stem the German onrush farther west. Under cover of darkness the gray-clad invaders would debouch from the concealing shades of Champenoux Forest and press forward into the Nancy curtain (Fig. 114). But, when the light of day revealed the whole situation to the defenders on the heights, a terrible flanking fire would pour upon the Germans from the Amance bastion on the north, where batteries were concealed on the reverse slopes of the Mont d'Amance mesa, the Pain de Sucre butte, and in the ravines and valleys of the main upland spur (Figs. 106 and 110); and to some extent from the heights of the Saffais plateau above Dombasle and the hills between Dombasle and Lunéville, forming the western base of the Manonviller bastion. Under the murderous fire, directed with great accuracy because of the exceptionally fine observation enjoyed by the defenders, the Germans would fall slowly back. Thus on the plain in the curtain the battle front ebbed and flowed with the alternation of day and night, but every attempt to make permanent gains of ground toward Nancy was decisively beaten.

Against the heights themselves the enemy assaults broke even more hopelessly. Concentrations for the attack and columns debouching from the cover of the woods were decimated by artillery fire directed from above, while a hail of machine-gun fire sweeping parallel with the smooth, open slopes below the wooded crest mowed the assailants down in rows. A violent assault on the Mont Ste. Geneviève sector of the bastion had developed early in the struggle. If the Ste. Geneviève-Mont Toulon ridge (Fig. 106) of the plateau could be stormed, the way would be opened for an advance up the Moselle to Nancy, turning the Mont d'Amance position from the rear and shattering the whole French scheme of defense based on the Moselle plateau. The first attempt against the formidable position (Fig. 108) having proved its exceptional strength, the enemy paused to deliver a heavy bombardment of two days' duration. The infantry then advanced to the attack, but was caught in a storm of shrapnel from the French 75's so accurately served that the ranks melted rapidly under the cruel fire. Machine-gun fire sweeping the slopes completed the slaughter. Demoralized by the fearful losses inflicted upon them, the assailants withdrew into the plain.

The prize for which the German high command was bidding was so precious that the reverses already suffered were insufficient to discourage new efforts. A more violent attack against the Mont d'Amance position was in preparation, and about September 1 began a violent bombardment by heavy guns lasting for several days. Then came the determined assaults by which the Germans debouching from the village and forest of Champenoux sought to storm the plateau bastion which formed the key to the natural defenses of Nancy. It is reported that the capture of the position and the entrance into Nancy were considered so important by the German high command that the Kaiser himself came to the front to lend his troops the inspiration of his presence, and issued the order that Nancy was to be taken at all costs and under his own eye. On the 6th and again on the 7th the struggle raged with undiminished fury both around Mont Ste. Geneviève and the Mont d'Amance. The crest of the former plateau spur was won by the enemy on the night of the 7th, but his precarious hold on this one advanced spur of the bastion was quickly loosened.

Finally on the 8th came the supreme effort. Sweeping over the plain and up the slopes, one gray wave after another melted under the accurate French fire. Six times the desperate assaults on the impregnable bastion were repeated, and six times the best Bavarian troops were hurled reeling back into the plain. "At some places the bodies were piled up five or six feet high." But the wholesale sacrifice availed nothing against the unshakable walls of the plateau. The supreme effort to break through the right wing of the Allied front was a costly, a ghastly failure, and the beaten armies of Prussian militarism fell back close to the German frontier. Both the Battle of the Trough of Charmes and the Battle of the Grand Couronné had been won by the French, in large measure through a masterly use of the formidable terrain provided by the Moselle and Saffais plateaus.

Second Advance into Alsace

After the victories of the Marne, the Trough of Charmes, and the Grand Couronné, the depleted French forces in the difficult country of the Vosges, opposed by likewise depleted enemy forces, were able to resume their advance down the steep eastern face of the mountain range and through the gateway of Belfort. But on both sides other battlefields had assumed such importance that henceforth only local operations by limited forces would disturb the relative calm of this sector of the front. By the end of October the French had descended the short eastern valleys as far as the Ballon de Guebwiller and Thann in the southern area and had pushed through the Belfort gateway to the line of the Ill near Altkirch. The only part of enemy territory held by the French throughout the war was the steep eastern side of the southern Vosges, where Nature favored the French to an unusual degree. Northward the front climbed obliquely across ridges and valleys to the crest near the Pass of Bonhomme, then descended the western slope, where the Germans, profiting by their railway to the low Saales Pass, were able to keep forces well supplied to a certain distance into French territory. Bitter struggles for Hartmannsweilerkopf and other dominating points would ensue in the following months; but to all appearances the war ended in a stalemate on the Battlefield of Lorraine. The great offensive which was to have been launched from the plateau strongholds on November 14, 1918 (pp. 412, 444), as the knock-out blow to bring the German colossus to its knees, was forestalled by the armistice.

CHAPTER XI

THE BATTLEFIELD OF THE TRENTINO: THE ALPINE-AND-PIEDMONT BATTLEFIELD

The Trentine Alps consist of complex masses of rocks intensely folded and broken and deeply dissected by stream erosion. Because the rock masses were raised more than 10,000 feet above sea level in their higher parts, streams cut remarkably deep canyons, giving a wild, mountainous country most difficult to traverse. The difficulties were increased many fold when the ice streams of the glacial period flowed down the valleys, cutting them much deeper, steepening the valley walls into rocky precipices, leaving the side valleys hanging hundreds of feet above the floors of the overdeepened main valleys, sharpening the intervalley ridges into knife-edge arêtes, and carving the dome-shaped peaks into jagged needles and horns. The resulting topography is one of indescribable ruggedness (Pl. VII), in which precipitous cliffs, inaccessible peaks, steep-sided divides, and hanging valleys present to the engineer almost insurmountable obstacles. Even in time of peace man makes his way across this mighty Alpine barrier only by means of highways which zigzag dizzily over lofty ridges and railroads which rise by devious looped detours into side valleys or through tunnels cut in the solid rock, to reach the few practicable passes.

It requires no great imaginative power to realize that to make war in such a region must tax to the utmost the ingenuity of man. And yet for a variety of reasons, both political and military, the Trentine Alps were destined to become one of the two main battlefields of the Austro-Italian front. The Italian

NOTE—For Chapters XI and XII the reader should constantly consult the sketch maps (Figs. 119 and 120) and, in the pocket, the block diagram of the battlefield (Pl. VII).

army engineers, worthy successors of the Romans in the art of road construction, built magnificent highways to apparently inaccessible points, over which autocamions could speed swiftly to the battle front on the peaks, bearing their burdens of men, munitions, and material (Fig. 121). If a precipice barred the path, the indomitable workmen blasted their way into it and by a system of tunnels scaled the barrier (Fig. 123). All along the Alpine battle line there soon appeared a system of these remarkable roads which made it possible to supply armies and carry on campaigns in a region seemingly designed by Nature to separate peoples by an impassable wall (Fig. 122).

In regions where not even the genius of the Italian engineer could carry roads, the *teleferica* (Fig. 124), or aërial tram, raised men and provisions, guns and munitions, to battle grounds among the clouds and lowered the wounded to hospitals in the valleys far below. On the lofty Adamello six successive hoists by as many *teleferica* lines raised the Italian soldier between 3,000 and 4,000 feet to the reserve positions behind the main front line; and, when Monte Grappa became the mountain buttress of the Piave front, great numbers of these aërial cables were used to aid the roads in supplying troops on that important height. Between *teleferica* stations tramways drawn by mules sometimes traversed the high mountain valleys, and, where snow fields and glaciers lay along the spectacular battle ground, dog sleds (Fig. 126) formed the line of communications, either across the surface or through tunnels several miles long in the ice, according to the degree of exposure to enemy fire. Assuredly this is the terrain *par excellence* of "mountain warfare."

Nor would the story of the adaptation of methods of warfare to this mountainous country be complete, even had we covered fully the question of communication lines. Big guns with specially devised wheels which facilitated the ascent of very steep slopes; sandbag breastworks or stone walls replacing trenches where impossible to excavate the latter in the solid rock; white uniforms for portions of the front on snow fields (Fig. 127);

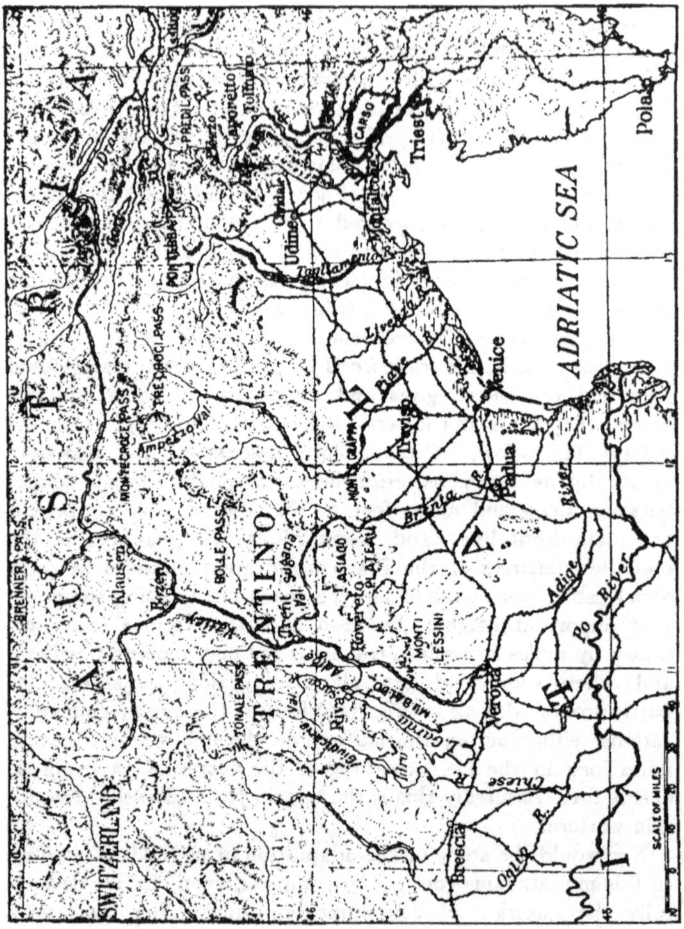

FIG. 119—General map of the Italian theater of war, showing the principal mountain and plateau barriers, river trenches and marshes, affecting the Italian campaigns.

Alpine troops on skis carrying ice axes and alpenstocks as essential parts of their equipment; electric motor lorries with trolleys driven up the steepest road by power derived from local mountain torrents—these are but a few of the responses which the contestants on one side or the other made to the special physical conditions which confronted them on the Alpine front. The flying corps had to contend with vast stretches of country affording no suitable landing place and with the treacherous air currents peculiarly frequent above such rugged terrains. The supplying of water to the troops engaged on high mountain ridges and plateaus of fissured limestone presented a series of complex problems which became acute whenever an advance was attempted. These problems were solved by the construction of great aqueducts and pumping plants, by the laying of miles of pipe lines leading to the stationary sectors, and by the organization of trains of pack animals to distribute the precious fluid to the most inaccessible portions of the field as rapidly as the front was pushed forward in any given sector. Men and mules were lost in crevasses in the glaciers. Avalanches, some starting from natural causes, many dislodged by heavy artillery fire, swept down upon helpless troops, a single battalion advancing to attack losing forty of its number in this manner.

Such was the warfare along the northern Italian frontier, and such the terrain upon which it was waged. An observer looking over the wild waste of Alpine crags from some lofty summit or gazing down upon the white-capped waves of the frozen granite sea from some soaring airplane (Fig. 128) would scarcely imagine that armies could find footing in the wilderness of ice and rock, much less fight battles there. Yet through the maze of peaks and ridges there pass narrow runways which were threaded by migrating hordes before the dawn of history, corridors along which invading armies advanced and retreated as the walls echoed the din of early battles. Roman legions conquered the mountain fastnesses and passed beyond to subdue the lands to the north. In later centuries rude invaders from the north repeatedly reached the sunny south through these same mountains. Napoleon's armies marched and countermarched from victory to victory in this forbidding realm, as well as on the open plains of other lands. Clearly there must be, either in the terrain itself or in its position with respect to other lands, or perhaps in both, something which made of it one of the predestined battlefields of Europe.

Stretching southward from the southern foothills of the Alps is the alluvial plain of the Piedmont,[1] formed of rock débris eroded from the mountains and spread out in great fans along the mountain base. The rivers which made this alluvial deposit flow southeastward down its slope to the Po or to the Adriatic (Fig. 119). Near the sea the plain is very low and marshy (Fig. 129) and is characterized by extensive lagoons back of narrow barrier beaches. This lagoon-and-marsh belt varies in width from ten to thirty miles or more and interposes a serious obstacle to the movement of troops. But between the marshes and the mountains only the transverse rivers oppose the advance of armies over the level surface of the plain. Some of these rivers, particularly the Tagliamento and the Piave, have an extraordinarily braided pattern. The interlacing network of channels crossed by good bridges only at infrequent intervals, the level sand bars exposing troops to the murderous hail of a grazing machine-gun and rifle fire (Fig. 130), and the good defensive positions formed by dikes built to retain flood waters (Fig. 131) make of these rivers strong military barriers, despite the absence of the steep, high banks which help to make the mountain valleys difficult to cross.

Strategic Position of the Trentine Alps

Among the circumstances which determined that the Trentino should become an important battlefield in the World War was the fact that it constituted one of the "unredeemed provinces" of Italy. This was, indeed, one of the principal factors determining Italy's entrance into the struggle on the side of the Entente, despite her alliance with the Central Powers. It is

[1] The word is here used in the physiographic and not in the political sense. The Italian province of the Piedmont, whose name is, of course, derived from its physical character, takes in only the westernmost section of the foothill plain. The portion of the plain here under discussion is politically part of Lombardy-Venice.

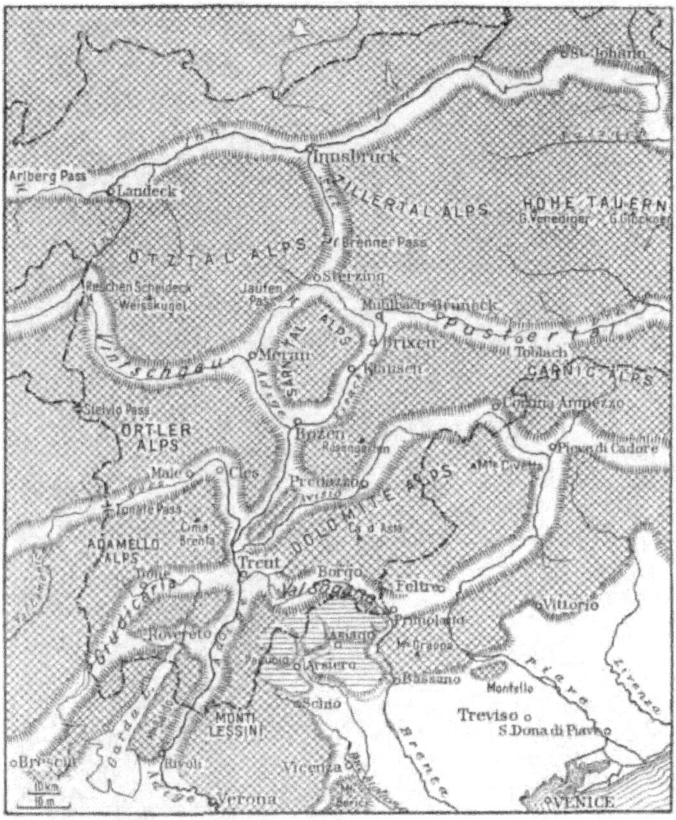

FIG. 120—Generalized sketch map of the Battlefield of the Trentino. White areas represent lowlands or valley trenches, ruled areas plateaus, cross-ruled areas mountains. See also the general map of the Italian theater of war (Fig. 119) and the block diagram of the Trentino (Pl. VII).

true that Italy demanded much more territory than she could justly claim on the basis of nationality or the right of peoples to decide their own destiny. While the southern part of the

Fig. 121—New Italian military road ascending the steep southern slope of the Monte Grappa massif, from the Piedmont plain to the Italian front on the crest. Observe the steepness of the mountain wall and the effectiveness with which it dominates the plain.

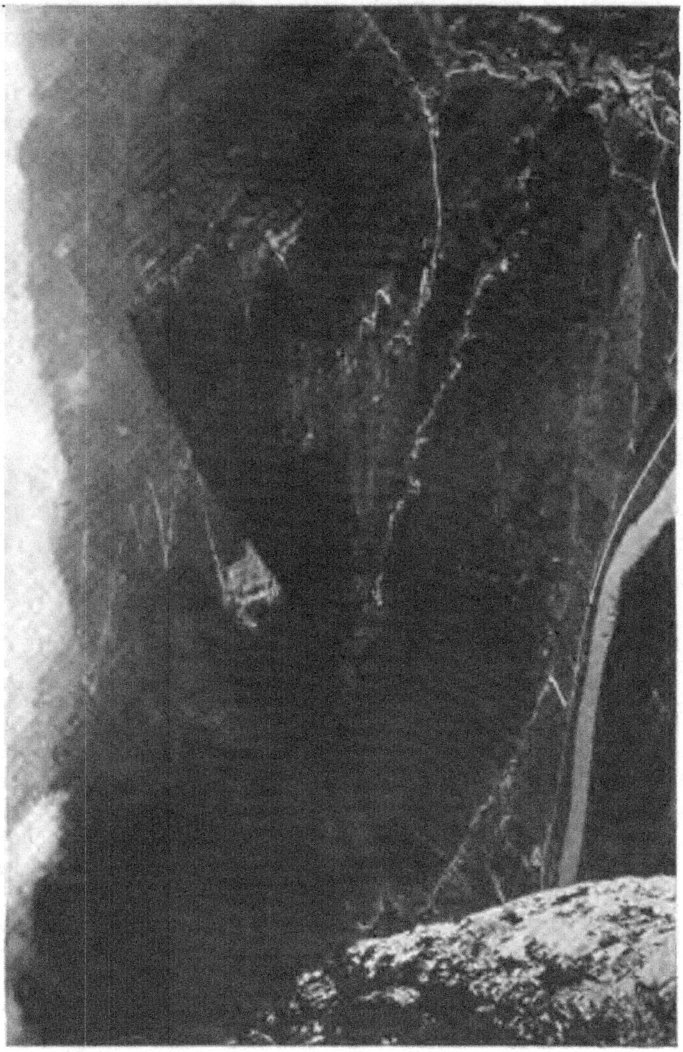

Fig. 122—The trench of the Adige in the foreground, with the high ridge of Coni Zugna, 6 miles south of Rovereto, in the distance. Note the military roads zigzagging up the steep mountain slopes to reach the Italian front on the lofty ranges of the Trentine Alps.

Trentino is overwhelmingly Italian, the northern part, or "Alto Adige," as the Italians call the upper basin of the Adige (Etsch) River, is as overwhelmingly German, and has been so for centuries. The fundamental reasons for Italy's claim to the lands of Andreas Hofer and the Tyrolese patriots were purely strategic, and the historical and other arguments offi-

FIG. 123—New Italian military road zigzagging up a steep precipice on the slope of Monte Baldo. Several tunnels may be noted within the space of a few hundred yards.

cially advanced in support of the claim were obviously weak. But the claim to the southern Trentino on racial grounds was valid, and strong strategic arguments could be adduced to support demands for further territory beyond the racial boundary. This was sufficient to bring the Trentino into the field as a prize of war, as well as a field of maneuver for hostile armies; and, in accordance with the rule "if you want a piece of ground you must sit on it," it was considered highly important that Italian armies should affirm the right of the kingdom to the territory in question by doing all that was practicable to conquer it. That Austria should dispute the attempted conquest was inevitable, and hence the two contestants were sure to clash upon the difficult terrain of the Trentine Alps.

The Battlefield of the Trentino is of peculiar interest also because it lay across the shortest and easiest route from the land of Italy's strongest foe, Germany, to the Italian frontier.

FIG. 124—A *teleferica* on the slope of the Adamello group.

The Brenner Pass, only 4,495 feet high, is not only one of the lowest and most accessible passes in the Alps but is the only one by which the central part of the massive barrier can be crossed without traversing other additional passes. Here, and here only, valleys cut clear through to the central axis on either side, and a single low gap divides the two transverse corridors. This was the direct route from Berlin to Rome in time of peace, and from time immemorial the chief route for invading Italy from the north. The natural defenses against such invasion are found on the Battlefield of the Trentino.

The Form of Italy

It was not merely the fact that Italy was in danger of attack from the north which made the Trentine Alps so vital from the military standpoint. The geographic form of Italy and the location of her great industrial regions both conferred upon this northern barrier a degree of importance which it might not have possessed under other conditions. The form of the Italian peninsula renders the rapid mobilization of Italy's man power peculiarly difficult. The peninsula is long and narrow (for nearly 600 miles of its length it averages little more than 100 miles in breadth), and the danger zone is at one extremity. Imagine California to be about half its present width, surrounded by water on both sides, and threatened at the northern end by an enemy able to concentrate quickly from many directions (Fig. 125). Such is Italy's situation. Over 55 per cent of her man power comes from south of the constriction near the latitude of Bologna and must journey to and through that constriction on four main railway lines, of which three must traverse the Apennine mountain barrier and two can be destroyed from the sea. Italy's geographic form thus makes her peculiarly dependent upon the natural defensive qualities of her northern frontier, which a fraction of her man power must be able to defend successfully against greatly superior numbers until the whole can be mobilized. The labyrinth of ridges and peaks forming the Battlefield of the Trentino is admirably suited to perform this function. When we add the fact that Italy's vital industrial regions are in the north, centering about Milan, and de-

FIG. 125—Comparison of Italy and California, same scale and orientation, to illustrate the difficulty of mobilization to meet an attack along the northern frontier, owing to the long, narrow form of the Italian peninsula. Italy has, however, the advantage of more railways.

pend for their protection on the defense of the Alpine barrier, the strategic significance of this northern battlefield can readily be appreciated.

Relation of the Trentino and Isonzo Battlefields

The full importance of the Battlefield of the Trentino does not become apparent, however, until we consider it in relation

FIG. 126—Crossing the glaciers of the high Adamello group by dog sleds. This was the usual route to this part of the Italian front in the difficult terrain between the Noce-Avisio and Giudicaria-Val Sugana corridors.

to another battlefield, that of the Isonzo (p. 541), on the eastern frontier. The Piedmont plain may be attacked either from the north or from the east. Armies defending the eastern frontier necessarily depend upon supply lines which traverse the plain for 150 miles in sight of an enemy advancing over the northern mountains. Hence the eastern armies must always fight under the menace of a disaster which is inevitable if the enemy on the north succeeds in reaching the plain and cutting their communications. In the World War Cadorna's eastern operations came to an abrupt halt in May, 1916, when he was compelled to transfer large forces westward to check the dangerous Austrian advance across the Asiago plateau almost to the edge of the plains. Irretrievable disaster to the eastern armies was nar-

FIG. 127—Special costumes worn by the "Alpini" in mountain warfare on snow-covered terrain. (Italian official photograph.)

rowly averted. The magnitude of the Caporetto disaster, consequent upon the Teutonic armies breaking through to the plains near the extreme eastern end of the northern frontier, enables one to picture the far more serious consequences which must have ensued if the northern mountain barrier had been breached farther west and the communications of the eastern armies destroyed 150 miles in their rear. Some idea of the wonderful command of the Italian plain possessed by an enemy on the dominating mountains may be secured from Figure 132, from a photograph taken looking southward from Monte Grappa, upon the northern end of which the Austrians maintained a firm hold for many months.

It is evident that the Teutonic attack condemned the Italians to conduct two principal campaigns, a northern and an eastern. Since their military forces would not admit of two offensive

FIG. 128—The Monte Pasubio massif as seen from an airplane, showing the spectacular surroundings of the battle front on the Alpine heights. (Italian official photograph.)

campaigns against so powerful an enemy, at least one of these campaigns had to be defensive. Topographic conditions dictated that the defensive campaign should be the northern one, for a successful offensive across the main Alpine barrier, supported by but one through railway line, had less chance of success than an offensive in the east, where the terrain was less difficult, railways were more numerous, and support by sea was possible. The Battlefield of the Trentino, with its defensive rôle, thus entered most vitally into the whole scheme of Italian military operations.

Position of River Defenses

Should Italy's eastern defenses be breached, the natural defensive lines which would remain to the Italian armies on that front are the rivers traversing the gently sloping Piedmont plain from the mountains to the sea. After the Isonzo the Tagliamento, then the Livenza, Piave, Brenta, Bacchiglione, Adige, and finally the Po, in succession offer natural obstacles of no mean value at which enemy pursuit might be checked. Following the Caporetto disaster it was on the Piave that the fleeing Italian armies finally made a successful stand (Fig. 133). Back of this barrier other lines of defense were then prepared or reconnoitred along the Brenta, the Bacchiglione, the Adige, and the Lake Garda-Mincio River-Po system. It will be observed that every one of these defensive lines rested its right upon the coastal marshes and the sea, its left upon the Alps. The Battlefield of the Trentino thus became the critical field of operations, upon the successful defense of which the safety of the Italian armies depended. Should the mountain barrier be breached, the defensive lines in question would be taken in the flank and rear (Fig. 119). It was apparently because the Adige could be outflanked only at two points which were capable of being effectively blocked (the Lake Garda and Chiese River gateways) as well as because the river was "broad, swift and deep, never fordable" that Napoleon considered it the best defensive barrier among these streams.[1]

It will be pertinent now to inquire what facilities for defense this terrain offers to the army which controls it.

ELEMENTS OF A STRONG DEFENSIVE TERRAIN

The extreme ruggedness of the Trentine mountains, with their lofty ridges and glacier-clad peaks, their towering precipices and yawning abysses, has been sufficiently emphasized on earlier pages. Our attention may now be directed to that system

[1] Napoléon Buonaparte: Mémoires pour servir à l'histoire de France, Paris, 1823, Vol. 3, p. 126.

observable in the mountainous highlands and valley depressions which both reinforces and makes available the defensive possibilities of this majestic region. In pursuing our investigation of the Trentine terrain it is desirable to keep in mind certain generalizations which our previous study of the battlefields of

FIG. 129—Coastal marshes bordering the Adriatic at the foot of the Piedmont plain. These marshes assured protection to the Italian right wing on the Piave against a flanking movement by the enemy. (Italian official photograph.)

France would appear to justify. These relate to the elements of terrain which might be looked for in the ideal battlefield most perfectly adapted to defense, yet permitting easy transition to the offensive as occasion may warrant. Briefly stated, these generalizations might be formulated in some such terms as the following:

The terrain should constitute a natural topographic barrier, impassable except at a few points, where the direction and nature of the enemy attack are restricted and known to the defenders. A lofty Alpine watershed, crossed by a limited number of narrow passes, is ideal in this respect.

In general the defensive front should be the shortest possible strategic line (although not the shortest line tactically, as explained below), in order to reduce to a minimum the forces required to hold it. An exception to the above generalization occurs where a longer front forms a salient projecting into the enemy's territory and is so situated topographically as to be easily defended while at the same time offering good interior lines for use in offensive operations.

The defensive front should not, in detail, be a straight line, but should form a series of bastions and curtains, providing ample opportunity for flanking and enfilading fire.

The topography of the battlefield should provide three further elements of natural fortifications: rivers, lakes, or marshes forming natural moats; exposed valley sides or mountain slopes forming natural glacis; and, back of these, steep and impassable walls or scarps. Rivers, lakes, and marshes, always serious obstacles to the rapid advance and effective supply of troops, have, as already noted, assumed an increased importance since tanks have become such a valuable instrument of offensive warfare; for they effectively check tanks and prevent them from destroying protective wire entanglements. Valley walls and hill or mountain slopes increase the physical labor of the attacking force, tend to impose upon the troops below the conviction that they are at a disadvantage and to give to the defenders above a moral feeling of superiority and the vital advantage of better observation. The slopes of the natural glacis should be of moderate inclination, open and smooth, affording a good field of fire. The summit portion or wall should be not only steep and inaccessible but also high enough to command a good view of all approaches to the moat.

There should be, back of the fighting front and parallel to it, a valley or glacial trench forming a lateral corridor through which troops and supplies may be rapidly and safely shifted from one point to another along the battle line. If close to the front, the corridor must be so deep and steep-sided, or the barrier between it and the enemy so high, that the floor is in a dead angle, even

FIG. 130—Multiple channels and sand bars of the Piave River, showing enemy troops slain in attempting to cross the obstacle. It was the Piave barrier which finally halted the Austrian advance after the disaster of Caporetto. (Italian official photograph.)

FIG. 131—Machine-gun positions along a dike bordering the Piave River. The stream itself forms a natural moat in front of the dike. (Italian official photograph.)

during high-angle artillery fire. Protection is thus also afforded for large masses of reserve troops and supplies, both on the corridor floor and on the protected wall.

There must be not one but a series of parallel defensive positions of the type described in the foregoing paragraphs, so that, if one is breached by enemy attacks, the others will check his advance. There should also be a number of easily defended transverse lines connecting each main position with the next parallel position behind it, so that an enemy success resulting in the breaching of one defensive position may be contained and thus localized to a small sector.

There should be no river, marsh, or lake barrier behind the defensive positions of sufficient magnitude or so insufficiently bridged as to imperil the retreat of the defenders in case retreat becomes necessary. Since an embarrassing river obstacle in one of the lateral corridors behind a first defensive position may become a valuable protective moat capable of checking an enemy who has broken through that first position, it is evident that the advantages and disadvantages of such a river position must be weighed against each other. If bridges are numerous and well protected from enemy fire, the river will not greatly embarrass the retreat; but after the bridges are destroyed it will effectively check enemy pursuit.

The topography should provide numerous high points from which excellent visibility of enemy back areas may be assured. This is essential to the most effective control of artillery fire and is important as a means of maintaining continuous watch over all the enemy's movements behind his front.

The defensive positions should be served by a good network of railways and roads. This is one of the elements most essential to a strong defense. Even the poorest natural positions may be well defended by armies having the advantage of excellent railway communications, as was amply demonstrated in the early part of the World War by the success of the German armies against the Russians whenever the former operated in contact with their strategic railway system along the otherwise weak Russo-German frontier.

The defensive terrain should include either bridgehead positions defended by a favorable topography, or positions from which it will be comparatively easy to flank the enemy out of territory which may be made to serve as a bridgehead, on the farther side of the natural defensive moat formed by river valley, lake, or marsh. Otherwise, as was made clear in the study of the Battlefield of Lorraine, the army of defense may, after repulsing enemy attacks, find it impossible to debouch from behind the barrier, and may thus be condemned to a wholly defensive campaign when offensive operations are essential to a military decision.

With the elements of a strong defensive terrain in mind, let us examine the military geography of the Trentino region.

Surface Features of the Battlefield of the Trentino

Through the maze of rugged mountains forming the central Alps there runs, in a general east-west direction, a series of parallel glacial trenches of the greatest strategic importance, forming four principal corridors (Fig. 120) which control movements east and west parallel to the Alpine crest. Between the corridors are high mountain walls, but at intervals a deep transverse trench connects one corridor with its neighbor to the north or south, or a fairly accessible pass across the intervening barrier permits somewhat more difficult communication. The most noteworthy of these transverse connections is a continuous north-south trench formed by the valleys of two streams flowing north and south from the Brenner Pass. This is the great trench carrying the principal railway and road connecting Italy with Germany, the main north-south route of commerce and travel already mentioned. It crosses and connects all of the east-west lateral corridors.

THE INN CORRIDOR

If we examine the lateral corridors more closely, we note that the northern one is wholly in Austrian territory and is the only

FIG. 132—The Piave River crossing the Piedmont plain, as seen from Monte Grappa. Austrian troops on the mountain enjoyed a similar view over the back areas behind the Italians defending the river barrier. The low ridge in the middle distance, around the eastern end of which the river bends, is the Montello. (Italian official photograph.)

one of the four provided with railway communication throughout its entire length. It is a very striking topographic feature (Fig. 134), and, as it is drained by the Inn River for most of its length, it may be called the Inn corridor. Near its western end the corridor is interrupted by a low divide at Arlberg,* but the Arlberg tunnel carries the railway through into the valley of the Ill, a tributary of the upper Rhine, whence communication on westward is very easy. This corridor is, in fact, part of the highly important route nearly 400 miles long connecting Vienna, through the Alps, with Bregenz on Lake Constance.

THE VINTSCHGAU-PUSTERTAL CORRIDOR

South of the Inn corridor and roughly parallel to it is the long trench formed by the Vintschgau, Pustertal, and connecting valleys (Fig. 120). This may be called the Vintschgau-Pustertal corridor. Its western end (Engadine) is drained by the upper Inn River and lies in Switzerland; the remainder lay wholly in Austrian territory before the war but, except for the portion of the Pustertal east of the low divide at Toblach, was drained by branches of the Adige River southward into Italy. The main corridor passes south of the Sarntal Alps, by Bozen, but there is an alternative passage around the northern side of the Sarntal mass, by Sterzing. This northern passage is completely interrupted at one point by a high divide, the Jaufen Pass (6,990 feet), which is, however, crossed by a practicable road, much used in ancient and medieval times. The main corridor is provided with railway communication, except near its western end. Some conception of its usually open character may be gained from Figure 135, but it narrows to gorgelike dimensions at Finstermünz (Fig. 136), just north of the Reschen Scheideck Pass, and at Klausen and Mühlbach. Like the Inn corridor, it is part of a vitally important through route, 330 miles long, connecting Marburg on the eastern edge of the Alps with the Swiss

* Unless otherwise stated, places mentioned in Chapters XI and XII may be located on Fig. 120 or Pl. VII.

frontier. In the region under consideration the only connection between the Inn corridor and the Vintschgau-Pustertal corridor is by road and railway over the Brenner Pass and by carriage road (possibly now by rail) up the Inn valley from Landeck to the Reschen-Scheideck Pass.

THE NOCE-AVISIO CORRIDOR

The third corridor is drained by the Noce and Avisio Rivers in its central portion, by the upper Oglio and the Adda west of Tonale Pass, and by the upper Piave and Tagliamento east of Cortina d'Ampezzo (Fig. 120). It will be called the Noce-Avisio corridor, from its more important central portion. In general its floor is less open and less continuous than is the case of the two corridors previously described, and it carries railways only for short distances in its central portion and in its western end; but it is provided with roads throughout its length. Several passes interrupt the trench floor, one of the best known being Tonale Pass; but even here the continuity of the trench walls is clearly apparent. As in the case of the second corridor, the third is provided with alternative passageways near its central portion: a northern one carrying a good road over a low pass and reaching the Adige valley at Egna, and a southern one down the gorgelike lower Avisio valley, traversed by a practicable road. Both the eastern and western ends of the third corridor were in Italian territory before the war; but the Italians could not make effective use of the ends so long as the whole central portion, from Tonale Pass to Cortina d'Ampezzo, lay in the southward projecting triangle of the Austrian Trentino. Communication between the Vintschgau-Pustertal corridor and the Noce-Avisio corridor is easy in the central zone by means of the Brenner railway and carriage road through the main north-south Adige trench, by local mountain railway and mountain road from Bozen to Cles, and by a good road from Bozen southeastward over the Rosengarten mountains. To the west a single carriage road, the highest in the Alps (over 9,000 feet), across the lofty and difficult Stelvio Pass connects the Vintschgau with the third corridor, west of Tonale Pass; and to the east the connection is effected by three carriage roads leading from the Pustertal southward across the mountain barrier.

THE GIUDICARIA-VAL SUGANA CORRIDOR

The fourth and southernmost corridor is composed in part of the Giudicaria and Val Sugana trenches but is continued north-

FIG. 133—Italian first-line trenches behind the defensive barrier of the Piave. (Italian official photograph.)

eastward by the upper Piave valley until it coalesces with the third corridor near Pieve di Cadore (Fig. 120). We may call this southern trench system the Giudicaria-Val Sugana corridor, after its two best known elements. A good road traverses its entire length, and both the Val Sugana and the Piave valley sectors carry railways; but the railways in these two sectors are as yet unconnected, and there is no railway in the western third (Giudicaria sector). At Trent the corridor divides into two branches, after the manner of the two next north, the northern branch con-

FIG. 134—The Inn corridor in the vicinity of Innsbruck. Note the breadth of the flat valley floor and the steepness of the walls leading up to the high mountain barriers.

FIG. 135—The flat-floored trench of the Vintschgau, forming part of the Vintschgau-Pustertal corridor in the Trentino. In the distance are seen the glacier-clad peaks of the Ötztal Alps.

tinuing west and then southwest past Tione, the southern branch turning south to Rovereto and then west past Riva. Near Brescia this corridor opens upon the Piedmont plain. Just as in the case of the third corridor, Italy formerly held only the eastern and western extremities of the fourth. The vital central sector, from Lake Idro to the eastern end of the Val Sugana, was included in the southward projecting peninsula of Austrian territory. At Trent the Giudicaria-Val Sugana corridor and the Noce-Avisio corridor next north are almost in contact and are connected by railway and carriage road through the north-south Adige trench. Eastward as far as the junction of the third and fourth corridors the only connection between the two is one carriage road across the broad intervening mountain barrier, from Predazzo to Feltre. West of Trent there is likewise but one fair carriage road crossing the barrier from Male in the Noce valley to Tione in the Giudicaria; but a railway running from Edolo west of Tonale Pass in the third corridor, southward through the glacial trench of the Val Camonica, issues upon the plain only a few miles west of the western end of the fourth corridor.

THE MOUNTAIN BARRIERS

No detailed description of the mountain barriers separating the parallel lateral corridors is necessary. In every case the barrier consists of high, inaccessible, and often glacier-clad peaks and ridges, which, as we have already seen, are crossed in but few places by practicable passes or connecting trenches. The barrier which is on the whole the broadest, highest, most extensively glacier-clad and least broken by cross passages is that between the Inn and Vintschgau-Pustertal corridors. The Ötztal group (Fig. 135) alone has more than 100 peaks approaching or exceeding 10,000 feet in altitude and carries 229 separate glaciers. Assuredly the best strategic frontier for northern Italy lay along this watershed barrier. Between the Vintschgau-Pustertal and the Noce-Avisio corridors we have the high, glacier-clad Ortler group (Fig. 137) in the west and the formidable Carnic Alps in the east. The central portion of this barrier is less imposing

FIG. 136—The wild mountain gorge forming the Finstermünz Pass, through which traffic over the Reschen-Scheideck col on the "Brenner watershed" must cross.

but nevertheless is a natural obstacle of the first importance. The frontier which Italy was willing to accept from Austria, rather than enter the war against her, lay within this barrier belt, except that it crossed to the Sarntal Alps between the alternative passageways north of Bozen. The barrier between the Noce-Avisio and the Giudicaria-Val Sugana corridors includes within its mass the high, inaccessible Adamello group (Fig. 126) and the Brenta mountains northwest of Trent, while to the northeast lie the Dolomite Alps, a broad zone of wild, mountainous terrain, including the Marmolada and Civetta groups.

South of the Giudicaria-Val Sugana corridor and separating it from the Piedmont plain is a mountain barrier of rather formidable proportions, despite the numerous gateways which open through it to the plains. Three railways and half a dozen good roads traverse it. In this mass are found the high ridge of Monte Baldo, Monte Pasubio, and the maze of peaks surrounding Monte Grappa (Fig. 128 and Pl. XI, *B*). Between the latter two massifs lies the high Asiago plateau, or plateau of the Sette Comuni (Pl. X, *B*), broken by a steep escarpment into two benches or steps, the lower of which is from 4,000 to 4,500 feet above sea level, the upper being about 1,000 feet higher (Fig. 138). Through this plateau the Astico and Assa Rivers, after heading near the Adige trench, cut deep pathways from the higher to the lower bench and, in the case of the Astico, through the lower bench to the Piedmont plain; while the Brenta River on leaving the Val Sugana corridor turns sharply southward through a wild gorge between the Asiago plateau and the Grappa massif to reach the plain at Bassano. These were dangerous gateways from Austria into Italy, the last-mentioned being guarded by the famous fortress of Primolano. It was the Italian positions on the Monte Baldo-Pasubio-Grappa mountain barrier which protected the Piedmont plain to the south from invasion during the World War. In respect to this barrier the plain with its railway system takes the place of a natural corridor supplying lateral communications.

THE ADIGE TRENCH

All of the mountain barriers, as well as all of the lateral corridors, are cut transversely near the central zone by the great north-south Adige trench (Fig. 122 and Pl. XI, *B*) and its continuation as the valley of the Sill north of the Brenner Pass. Through this main north-south trench runs the Brenner railway, giving

FIG. 137—The Königspitze, part of the glacier-clad Ortler group forming the lofty mountain barrier between the Vintschgau and Noce corridors in the Trentino.

access to, and connecting with most of the railways in, the lateral corridors. Manifestly the three vitally important elements in the military geography of this section of the Alps are the four east-west mountain barriers, the four east-west lateral corridors, and the one main north-south transverse trench. Of secondary importance are the minor north-south connecting trenches.

MILITARY STRENGTH OF THE TRENTINE BARRIERS

To a remarkable degree the Battlefield of the Trentine Alps fulfills the requirements of the ideal defensive terrain as summa-

FIG. 138—The upper bench of the Asiago plateau, with the dominating peak of Campolongo at the right, the Astico valley at the left. The massif of Campomolon is across the valley, beyond the left margin of the view. (Italian official photograph.)

rized on earlier pages. It constitutes a most formidable barrier which armies can attack at comparatively few points. The successive positions form strategically short lines, although tactically of such form as occasionally to offer salients and reentrants capable of being effectively defended, as for example the horseshoe-shaped bastion of the Sarntal Alps, permitting enfilading fire along both the Vintschgau and Pustertal trenches; and the Bozen basin, commanded by the heights to the east and west. Each mountain barrier is fronted by a natural moat in the form of one of the glacial trenches through which flow one or more rivers, and has behind it a deep, lateral corridor carrying roads and railways. The river is not always of sufficient volume to be a serious obstacle in itself, but the exposed valley floor is not only difficult to traverse under fire but usually contains smoothly sloping alluvial fans forming a natural glacis easy to sweep with grazing machine-gun fire. As each fan is convex upward in cross profile and rises many feet above the valley floor, the crest offers a commanding position enfilading the length of the valley, the half toward the enemy forming a smooth glacis and the other half an area where secondary defensive positions hidden from hostile view may be organized. Trenches on the fans may be concealed behind existing terraces and stone walls, the stream channel across each one serves either as fosse or communication trench, and the gorge in the mountain wall at the head of the fan provides shelter for reserve troops. A significant proportion of the defensive positions in the mountain trenches were accordingly based on the alluvial fans or cones. The four parallel trenches assure as many successive lines of defense, and, as they are connected by easily defended cross trenches or passes, the problem of localizing an enemy irruption into a given portion of any trench is much simplified. Visibility of enemy back areas is difficult in a mountainous terrain, and for this reason the Italians suffered more than one unpleasant surprise attack; nor is the railway system serving the defensive positions what one could desire. Yet despite these deficiencies, the favorable elements of the terrain are so preponderant that it is no exaggeration to say that the Battlefield of the Trentino presents one of the strongest natural defensive systems to be found anywhere in the world.

POSITION OF THE AUSTRIAN FRONTIER

In 1914 Italy suffered from the fact that the Austrian frontier, drawn so as to impose on the southern kingdom a fatal strategic disadvantage, accomplished in advance of hostilities precisely what a hostile army would seek to do with any series of defensive positions, natural or artificial. It drove a wedge clear through the successive mountain barriers and valley trenches to the last crest overlooking the Piedmont plain (Fig. 120). It was possible, in fact, for Austrian armies standing on the frontier, or even well north of it, to see far across the plain and observe at leisure trains passing on the vital railways supplying the Italian armies both in the Trentino and 150 miles away on the eastern frontier. Italian armies in the plain, on looking up to the crests, saw the enemy frontier passing from peak to peak and dominating them from its lofty position. This frontier, furthermore, systematically beheaded even small valleys descending into the Italian plain so as to secure to Austria strategic control of a large majority of the practicable passes. The Italian people could not rest secure in their own house so long as an hereditary enemy perched threateningly on the threshold. To remove this intolerable frontier to a safe position farther back in the mountains was a perfectly understandable and wholly legitimate aim of Italian policy. Whether it was wise for Italy to demand and for the Allies to consent that the frontier should be pushed far into German territory is a political question which need not here be discussed.

In full possession of this strategic salient for many long years, Austria had so fortified it as to render it practically impregnable, an easy thing to do in a terrain so favorable to defensive warfare. Mountain peaks were honeycombed with tunnels, gun chambers, ammunition storage vaults, quarters for officers and men, and thus transformed into rocky strongholds from which hidden artillery could pour a destructive fire on any attacking force (Fig.

139). From a central position like Trent any threatened portion of the long frontier rimming the salient on the west, south, and east could promptly be strengthened by reserves, or an offensive could be launched with overwhelming suddenness against the Italians, who, even if well aware of the concentration for the

FIG. 139—Gun position at exit of a tunnel excavated in the solid rock of an Alpine mountain. The sides of certain mountains in the Trentino facing the direction of enemy attack were pierced by great numbers of such openings, which transformed the natural obstacles into almost impregnable fortresses. (Italian official photograph.)

attack, could not foretell against what part of the long line the blow would be hurled at the last moment. Thus were they confronted with the almost impossible task of holding the whole line in strength sufficient to contain the maximum force the Teutons might suddenly concentrate against any part of it. To shift troops quickly from one part of this front to the other was impossible, because the wedge of hostile territory driven through the whole defensive system of the Trentino blocked the corridors along which alone could rapid transfers be made. Italian units operating in the western and eastern ends of the same corridor were completely isolated from each other and denied any possibility of effective co-operation in their campaigns. If one desired to send troops to support the other, a long delay was necessary while the troops to be shifted were moved down out of the mountains on one side of the salient, across the plain, and back up into the mountains on the opposite side. With difficult routes and limited railway facilities, the transfer of any large quantity of men, equipment, and supplies was so slow a process that an enemy offensive could make dangerous progress before the defense was effectively reinforced. At the very beginning of the Austro-Italian campaign, therefore, the Italians found themselves largely deprived of, and their enemies largely possessed of, the strong defensive features of the Battlefield of the Trentino.

CHAPTER XII

MILITARY OPERATIONS ON THE BATTLEFIELD OF THE TRENTINO

The Blocking of the Passes

On May 24, 1915, immediately following the Italian declaration of war against Austria, Cadorna despatched Italian forces to seize the gateways through which Austrian armies might seek to debouch from the Trentine mountains into the Italian plain. The Italian Commander-in-Chief wisely decided upon a defensive campaign on this difficult terrain and concentrated his principal forces for an offensive on the Battlefield of the Isonzo. But since this latter offensive could proceed with safety only in case the Austrian armies threatening the Italian communications from the dangerous salient of the Trentino were securely bottled up, it was essential to push across the frontier and farther into the mountains, both to seize and fortify the northern approaches of the gateways to the plains and to place a greater depth of the strong defensive terrain back of the Italian front for its better support.

In the southern part of the Trentino the great Giudicaria-Val Sugana lateral corridor was the first objective. If it could be taken, the greatest danger to Italy would be removed. So long as it remained in Austrian hands an enemy concentration about the military base and fortress of Trent could without warning launch an attack eastward into Italy along the Val Sugana end of the corridor, southwestward along the Giudicaria end, or southward along the north-south Adige trench. Trent could also serve as the base for an attack directed westward along the Noce-Avisio corridor into Italy through the Tonale Pass, or eastward through the Cortina d'Ampezzo gate. If, therefore, the southern corridor containing the most important fortified military base in the Trentino could be captured, a great advance would have been made toward freeing the northern plain from the danger of hostile invasion.

Italian troops converged upon Trent by each of the routes predetermined by Nature (Fig. 120), repeating on the same terrain some of the maneuvers executed by the Napoleonic armies in 1796 and 1797 and by the Italian armies in 1866.[1] One unit pushed northward along the Adige trench (Pl. XI, *B*), supported by a vigorous attack along the high Monte Baldo ridge, which separates the Lake Garda trench (a parallel water route of some importance) from the Adige and dominates military operations in both; but it was stopped at the Austrian main line of defense on the southern side of the lateral corridor, in front of the outlying fortress of Rovereto. It had at least secured a firm hold on the approaches to the famous Chiusa di Verona, the narrow limestone gorge at the exit of the trench which placed the name of near-by Rivoli on many a page of military history, and which Bourcet in his rare treatise on "Principes de la Guerre des Montagnes" cites as an example of a natural position capable of being held by small forces against a whole army. Another force pushing northeast up the Chiese River through the Giudicaria valley was halted a short distance inside the Austrian frontier. The third column made good progress westward along the Val Sugana, capturing Borgo, nearly halfway to Trent, before it was halted. A fourth advance along the road zigzagging over the high shoulder of Monte Pasubio (Fig. 128) captured that rugged massif and the Fugazze Pass by which the road crosses the divide just south, but was stopped short of Rovereto. Thus on all sides the Italians were checked a short distance within the frontier by the obstinate defense of a formidable terrain which could be traversed only at the cost of terrible sacrifices. The portion of the Giudicaria-Val Sugana, or southernmost, corridor held by the Austrians was reduced at both ends and its southern wall reached but not crossed at a number of

[1] Napoléon Buonaparte: Mémoires pour servir à l'histoire de France, Paris, 1823, Vol. 3, Chapters X and XIV; Bolton King: History of Italian Unity, London, 1899, Vol. 2, p. 297.

points. In other words, the danger of an enemy irruption from the Trentino into the Italian plain had been minimized but not removed, while the Austrian defensive system had been little impaired. The southern enemy front still lay along the southernmost trench, and back of it were three other trenches still able to serve as lateral corridors of communication.

The Noce-Avisio corridor next north was attacked at both ends. On the west the Italians pushed a short distance through the Tonale Pass, on the east they captured the gateway of Cortina d'Ampezzo. But the Austrian hold on this corridor was very little reduced. A push north toward the vital Pustertal corridor was quickly stopped, although the Italians were able to hold the crest of some of the passes giving access from that trench to the headwaters of the Piave and Tagliamento Rivers. Other minor passes were seized and strong positions established on the eastern and western ends of the mountain barriers between the three southern corridors, as on the Cividale and Adamello (Fig. 126) glacier-clad massifs.

For more than three years the Italian armies were to make no further appreciable progress in the Trentino. An offensive in the following October made little impression against the massive stone walls of the mighty natural fortress. Some months later, indeed, the Italians were to be pushed back to the brink of the Asiago plateau overlooking the plain, and then to regain a part only of the lost ground. Aside from this nothing but local changes of no great consequence marked the struggle in the Trentine Alps until the final collapse of the Austrian army in the fall of 1918 made possible the rapid Italian occupation of the whole region.

The Battle of the Asiago Plateau

It was in the month of May, 1916, one year after Italy's entrance into the war, that the Austrians made their first great attempt to break through to the plain and by reaching the sea completely to surround and annihilate the main Italian armies, busy pushing their offensive on the Battlefield of the

Isonzo. Heavy reinforcements of men and enormous quantities of supplies and munitions were brought down the Adige valley and concentrated about Trent. The process of accumulation was necessarily slow, for through the single narrow trough of the Adige all the men and supplies, big guns and munitions, withdrawn for the purpose from the Russian and Balkan fronts, had to pass to their destination. New roads and railways were constructed and provision made to distribute water over the difficult terrain to the great armies as they advanced. The high ridges and deep valleys afforded some measure of concealment for these preparations, which, even if known, would not reveal the direction in which the Austrians, profiting by interior lines, would strike their chief blow.

When the accumulated troops exceeded a third of a million men, and the massed batteries numbered two thousand guns of all calibers, the great offensive was launched on May 15 following a terrific bombardment. The front of attack extended from west of the Adige trench to north of the Val Sugana corridor, the enemy seeking to push south and southeast into the plain through the time-honored routes of invasion formed by the Adige trench and the Val Sugana corridor, by the road crossing the shoulder of Monte Pasubio via the Fugazze Pass, and by the roads crossing from northwest to southeast over the deeply trenched Asiago plateau and descending the Assa, Astico, and Posina valleys (Fig. 138 and Pl. X, *B*). If the invading forces could master the Asiago plateau, then the deep gorge of the Brenta, carrying road and railway from the Val Sugana corridor through the barrier to the plain, and the strong defenses of Primolano, guarding the northern entrance into that gorge, would be dominated from the plateau rim and another route of invasion thus opened up.

The high, rocky ridge of Coni Zugna (Fig. 122), towering aloft between the Adige trench on the west and the Vallarsa carrying the road to Fugazze Pass on the east, transformed by the Italians into a strong point supplied by new roads traversing its crest and zigzagging up its precipitous sides, brought the Austrian advance to a sharp halt on that sector of the front. The natural fortress of the superb Pasubio massif (Fig. 128) proved an impassable barrier farther east and checked the attempted irruption through Fugazze Pass. In vain the Austrians poured a terrific fire from heavy howitzers upon the heights and assaulted it again and again for several weeks with forces vastly superior numerically to those of the defenders. The rock-ribbed mass remained an impregnable buttress against which the Austrian storm lashed in impotent fury. In the Val Sugana corridor the enemy soon found progress arrested by strong Italian positions on the two walls. But across the less difficult terrain of the Asiago plateau (Pl. X, *B*) the advance soon assumed highly dangerous proportions. The upper bench of the plateau (Fig. 138), including the strong massif of Campomolon south of the Astico valley, was quickly lost when enemy columns filtered down the valleys dissecting it. The victors then pushed slowly out over the lower bench toward the margin of the tableland looking down upon the plain. For a week, for two weeks, the furious battle raged. The height of Pria Fora, dominating the southwestern part of the lower bench, and the town of Arsiero at its base, where several of the plateau roads unite in the Astico valley, fell to the enemy, as well as the town of Asiago, the center of roads on the northern part of the lower bench. The Austrians were within three or four miles of the open plain, and the third week found them driving slowly forward.

The situation was critical in the extreme. Not merely the Trentino campaign, but the mass of the Italian armies on the Isonzo front, and the very life of Italy, were in jeopardy. Under the terrific pounding of heavy artillery the Italians, fighting doggedly to maintain their precarious foothold on the edge of the mountains, were being slowly pushed back toward the brink of the abyss. Just behind the Italians, on the plain below, the Austrians could see the railway arteries upon which the eastern armies depended for their existence. A few more hard pushes and the invaders could sweep down into the plain like an avalanche, cut the vital arteries, take the Isonzo armies in the rear, and overwhelm them with what would be almost sure to prove irretrievable disaster.

But the Austrian machine was beginning to lag. Its advance over the difficult terrain in the face of a stubborn resistance was destroying its momentum. Lines of supply were lengthening, and food, water, and munitions had to be brought farther and farther over wild mountain trails. Nevertheless, with the advantage of a downhill push against the Italians, the Austrians were likely at any moment to precipitate the dreaded catastrophe. Support from without and from within was necessary to restore the situation. Appeals were made to the Czar to advance the date of a contemplated Russian offensive in Galicia, in order to relieve the pressure on the Italian front. This was done, and in the first week of June Austria found her eastern front violently assailed. Soon some of her best divisions had to be recalled from the Trentino to meet the menace in Galicia. At the same time Cadorna had massed reserves behind the Italian front and was successfully appealing to his men on the battle line to defend to the death their last hold on the mountain flank. With splendid determination the soldiers of Italy, their backs to the brink, withstood one violent assault after another which the enemy hurled with desperation against the Arsiero-Asiago front. Their very proximity to the plain with its excellent communications gave the Italians some advantage in the ready supply of men and munitions to their front, despite the interference of enemy fire directed against roads and railways.

By the latter part of June Cadorna began to react with a counteroffensive which drove the Austrians first from one commanding hill, then from another. It was clear that the Austrian drive was over. Weakened by the heavy losses inevitable in an advance across such a terrain as that of the Trentine Alps when well defended, and by withdrawals of troops for service on the Galician front, the Austrian command abandoned the great prize almost within its reach and began to evacuate some of the ground gained. Falling back upon the solid defensive position provided by the higher bench of the Asiago plateau, the enemy brought the Italian counter-offensive to a halt before an almost impregnable mountain scarp. The Battle of the Asiago Plateau had netted the invader a small section of Italian territory and a final position nearer the plain. But, despite the strong dissatisfaction which the loss of ground created throughout Italy, the result was in effect an important Italian victory, since the Austrian offensive failed completely of its objective and only purchased minor advantages at the cost of enormous sacrifices.

The First Battle of the Piave

After the disaster of Caporetto in late October, 1917, to be discussed later (Chapter XIV), the fleeing Italian armies were finally able to stand behind the natural barrier formed by the Piave River from the mountains to the sea and, with the aid of French and British divisions hurried to their assistance, to hold their pursuers at bay.

Few military critics believed that the Italians would maintain the Piave front, and a further retirement was predicted; but the position was much stronger than it appeared. Along the sea its right rested on a vast expanse of marshland threaded by water channels on which barges camouflaged with bushes and bearing big guns could lie concealed in the reeds, with the power of shifting to new positions as rapidly as enemy fire located them (Fig. 129). Such a terrain, defended by heavy artillery fire, was practically impassable, and assured the security of the Italian right wing. On the left the line rested against the Alps. Unfortunately, the Italians had been forced to evacuate the whole sweep of the high ranges from the Isonzo to the southern tip of the Trentino. The rugged Carnic Alps, the second and third corridors, and the massive barriers between them had been lost, together with all northeastern Italy. There remained to the Italians east of the Adige only a portion of the last mountain wall south of the last lateral trench, the merest fringe of the Alps, consisting of the Monti Lessini, the Pasubio massif, the

Asiago plateau, and the Monte Grappa group. It was the fact that a short push from these heights into the plain would outflank the whole Piave front which made the latter appear untenable.

Yet this last mountain wall, which had broken the great Austrian offensive on the Asiago plateau, was in itself a formidable buttress. With rare skill the Italian engineers had constructed a series of magnificent military roads by which motor transport could deliver men and munitions to the crest of the barrier with a speed which the Austrians could not equal (Fig. 121). And while the Austrians on this part of the front were reduced to dependence upon a few main lines of communication, chiefly the one following the Adige trench, immediately back of the Italians lay the whole network of lines on the Piedmont plain. To capitalize this advantage further than roads alone would permit, scores of *teleferica* lines were erected to each important massif. Thus was assured an abundant flow of supplies and munitions to the Italian front, which of itself was sufficient to give the defenders marked advantage over their assailants. Add the fact that the mountains had been honeycombed with tunnels, galleries, invisible observation posts, concealed gun chambers (Fig. 139), and shelters for troops, and so transformed into vast fortresses upon which the heaviest enemy fire could produce little or no effect, and the security of the left end of the Piave line will be seen to have been far greater than appeared to casual observation. Through the heart of the Monte Grappa massif, upon which the Piave front rested most directly, one could traverse miles of subterranean galleries connecting with scores of openings on the northern side, from which guns of medium and large caliber hurled volcanic eruptions of fire and steel against the Austrian positions.

A further element of strength on the Italian left was the curious limestone arch of the Montello (Fig. 132), rising out of the plain a little south of the main mountain wall, and along the eastern end and northern side of which (the only parts exposed to enemy attack) the Piave River forms a natural moat. This plateau belongs to the karst type of landscape, for its surface is deeply pitted with innumerable sink holes, chasms, and caverns, forming a terrain almost impossible to traverse in the face of a vigorous defense. It is the *carso verde*, or green karst, of the Italians, for vegetation covers much of its surface and conceals from view the countless depressions offering strong defensive positions. Big guns in the larger sink holes were completely hidden in a natural camouflage of leafy branches, and machine-gun nests were as effectively masked by bushes and other smaller growth. The solid buttress of the Montello, impregnable itself and dominating the level plain for many miles in every direction from its superior height of some 1,200 feet (nearly 900 feet above the plain), was one day to prove the salvation of the Piave front.

The river itself was a barrier of no mean importance. Aside from its numerous channels, sometimes swollen with flood waters, its open expanses of sand bar flats (Fig. 130), and its marshes, the dikes served the purposes of the defenders (Figs. 131 and 133). Even when not holding waters in check they formed strong lines of defense ready at hand and, like the railway dike from Dixmude to Nieuport on the plain of Flanders, were strongly organized with machine-gun emplacements, both near the crest and in tunnels cut through to the outer base. The low plain on either side of the river is traversed by minor streams and canals, which together with their containing dikes served as secondary defensive positions. They also divide the plain into compartments in one or more of which an enemy irruption could be contained without serious damage to the position as a whole, and which could be flooded without deluging other portions of the plain.

The intense cultivation of the plain, however, seriously embarrassed the defense. As on the Flanders plain, the difficulty of observing an approaching enemy and controlling artillery fire upon him was very great. Roads are camouflaged by close-set bushes and trees, fields are divided by high hedges and rows of trees, orchards are numerous, high corn and other crops add their concealment in season. Thus the defense has no adequate field of fire. An enemy rushing from hedge to hedge filters into the defensive positions; the line becomes confused and irregular; and field artillery, compelled to keep on the front itself in order to distinguish friend from foe, is liable to capture. The Italians lost whole batteries during the Piave battles on this account. Effective indirect artillery fire is hindered because neither the foe nor the results of the fire can properly be observed. Such are some of the difficulties incident to the defense of an intensively cultivated plain.

It was in the early part of November that the Italian front began to stabilize itself behind the Piave barrier. Cadorna had been retired while the retreat was in progress, and General Diaz made Commander-in-Chief. Under the new leadership vigorous efforts were undertaken to check the enemy advance. Dikes were opened to flood areas the Austrians would have to cross. Bridges were destroyed, and the passages of the multiple channels of the river, as well as the exposed sand bar flats (Fig. 130), were swept with machine-gun fire from well-chosen positions. With these defensive measures were coupled vigorous counterattacks which threw the pursuers into confusion and reduced the violence of their impact against the obstacle which barred their way. It soon became evident that the Italian high command had decided to fight something more serious than a rearguard delaying action along the river barrier. The First Battle of the Piave was in progress.

All along the line from the sea to the mountains the Austrians tried in vain to push forward. Everywhere their efforts encountered an obstinate and effective resistance. Once more the Italian troops, who had a few weeks before fallen victim to the disloyal divisions which had undermined the patriotism of the nation behind the front and to the wiles of enemy propaganda pushed with skill and energy, were fighting with a spirit worthy of the best Italian traditions. Here and there the enemy might force a temporary crossing, but it was usually to be thrown back with heavy loss. At Zenson alone, where, five miles northwest of San Dona di Piave, the river makes a bend hard to hold, did he succeed in maintaining a bridgehead of any importance, and this he lost some weeks later. The Piave line could be broken, if at all, only after systematic preparation.

Attempts to outflank the line of the Piave by breaking through the mountain wall on the north were equally fruitless. Attacks on the Asiago plateau and the Grappa massif failed of their objective, and the Austrian soldiers were halted with the fertile plain of the Piedmont spread green and enticing in full view of their hungry eyes (Fig. 132). In December the attack on the mountain wall was vigorously renewed. Important parts of the Monte Grappa massif, including the peaks of Asolone and Tomba, fell into enemy hands but were later recovered. The main mountain fortress held firm against all attacks, and the First Battle of the Piave slowly died out. Comparative calm reigned along the front for some months.

Back of the Piave front the Italians now reconnoitered and partially organized additional main defensive positions, in order not to repeat the unpardonable neglect which had left the Carso front practically unsupported. The Brenta River would serve as an intermediate position, but the first strong defensive line was based on a series of hills stretching southeastward into the plains—the easternmost ridge of the finger-like Monti Lessini from near Pasubio to Vicenza, then the Monti Berici with their imposing eastern face rising like a stone wall from the level floor of the plain, and finally the Euganei Hills, just to the southeast. Along the eastern base of the Berici and Euganei massifs, forming a natural moat connecting the hill buttresses into a single defensive system, runs the Bacchiglione River, which continues the defensive barrier to the lower Brenta and the coastal marshes and lagoons. The Adige line next to the southwest is in turn followed by the very strong line of Lake Garda-Mincio River-Po River. Henceforth the Italian armies fighting on the Piave knew that should they fail to halt the enemy there, other strong defensive positions awaited them a few miles to the rear.

The Second Battle of the Piave

In the middle of June, 1918, while Ludendorff was hurling his masses forward in one great drive after another in the desperate effort to win a decision before America should assemble overwhelming forces against him, the Austrian command launched a violent offensive designed to bring Italy to her knees. Massing against the Piave front and against the Asiago-Grappa mountain wall large forces set free by Russia's betrayal of the cause of democracy, the Austrians sought at a single blow to breach the Piave barrier by frontal attack and to take its defending armies in the rear by debouching from the mountains into the plain farther west.

The outlook for the Italians was none too reassuring. From the crest of that last mountain wall the enemy could survey the whole plain to the border of the sea (Fig. 132). On a clear day Venice was in view, and trains crossing the level surface could easily be discerned. The interlacing channels of the Piave and Brenta stretched far away to the southward amid green fields and orchards. To encourage his soldiers the Austrian general pointed to the promised land which lay so close below, where they could live in the midst of plenty after throwing the enemy from his last foothold on the heights. Whatever the result of the fighting along the Piave, victory final and complete would crown their irruption into the plain.

After a brief but violent artillery preparation the whole Piave front was furiously assaulted by the forces of General Boroevic, the major efforts being directed against the weakest part of the front between the Montello and the northern edge of the coastal marshes. Undaunted by the murderous fire which swept stream crossings and sand flats, the avalanche of attackers swept over the river and submerged the first Italian positions. Near San Dona di Piave, on the edge of the marshes, they pushed several miles beyond the barrier toward Venice. But now the Austrians were fighting with a river behind them, across which reserves, supplies, munitions, and guns moved slowly and irregularly over crossings deluged with Italian artillery fire. The attack lost its "punch," and a bitter struggle began for each compartment of the terrain, divided off by canals and dikes from behind which an obstinate defense was maintained. Advanced enemy units found themselves surrounded on three sides by solidly entrenched Italian forces and cut off from effective support by inundated areas and the river in their rear. Their condition became critical when the Piave suddenly rose in flood on June 18, swept away the temporary bridges and boats used for the crossings and even barges loaded with men and materials of war. Isolated from all help, subjected to a murderous fire directed in part from guns on monitors in the river and on barges in the marshes, and attacked with great vigor by General Diaz's reserves, the Austrian units west of the Piave finally fled in confusion to their old positions on the eastern bank, leaving the river choked in places with the corpses of their slain and drowned comrades. The barrier had proved impregnable against direct frontal attack.

Meanwhile the key position of the Montello (Fig. 132) was being vigorously assaulted. The river was crossed at the cost of heavy losses, and the conquest of the formidable terrain of the limestone arch was pushed with great energy. If this commanding height could be carried, the Italian positions behind the Piave farther south would be dominated and turned, and the whole front would crumble. Foot by foot the invaders forced their way up its eastern end, inch by inch they edged westward along the crest. But now they were face to face with the same appalling difficulties which the eastern Carso had earlier opposed to the advance of the Italians. From the labyrinth of sink holes, fissures, and caverns, camouflaged by the natural vegetation, a deadly fire from rifles, machine guns, and artillery poured upon them and exacted a heavy toll for every bit of ground wrested from the determined defenders. Slowly the offensive was worn down by the insuperable difficulties of a terrain admirably adapted to defense. Efforts to push around the ends of the arch and encircle it were crushed by fire accurately directed from the upland. Neither by direct assault nor by flank attack could the Montello, "the central pivot of the whole Italian line," be carried.

Completely beaten, the Austrians abandoned the ground gained and fell back on their old lines east of the Piave.

Neither the failure to breach the river barrier along its central portion nor the failure to conquer the Montello buttress would have wrecked the great Austrian offensive had the enemy forces farther west been able to take these strong positions in the rear by debouching from the mountain wall into the plain. Simultaneously with the attacks just described a desperate effort was made to break through this latter barrier. From the Asiago plateau to the Monte Grappa massif the soldiers of Field Marshal von Hoetzendorf flung themselves against the last giant dike which held back the Teutonic flood from the low Piedmont. On the plateau the assaults broke down before strong positions defended by French and British troops. On the natural fortress of Monte Grappa the Italians soon halted the enemy's advance. The third great effort to break through the mountain wall was a costly failure.

From the sea to the Trentino natural barriers ably defended by the Italian armies had resisted the supreme effort of the Austrians. That effort had cost the assailants in men more than 20,000 prisoners and between 150,000 and 200,000 casualties, and had gained them absolutely nothing. "They had failed grotesquely, and their offensive power was at an end. Their morale was hopelessly lowered, and domestic revolt threatened."[1] The way was being prepared for the complete collapse of the whole Austrian military machine in the following October.

The Battle of Vittorio Veneto

On October 24 General Diaz delivered a final blow against the crumbling Austrian armies. The strategic plan of the battle[2] "was to separate the Austrian mass in the Trentino from that on the Piave by a decisive break-through and then, by an enveloping action, to bring about the fall of the whole mountain front, which

[1] C. J. H. Hayes: A Brief History of the Great War, New York, 1920, p. 319.
[2] Royal Italian Army: Report by the Comando Supremo on the Battle of Vittorio Veneto, Rome, 1919, p. 11.

would cause, necessarily, the complete yielding of the enemy's front in the plain." This plan was to be realized by piercing the Austrian front where the lines on the plain met those in the mountains, then advancing to Vittorio in the foothills northeast of the Piave, whence vital communication lines in the mountains behind the Austrian front could be reached.

Both in the mountains and along the Piave front the Italians struck with vigor in order to give the *coup de grâce* to their tottering enemy. In the mountains, where the terrain was so formidable that even a few loyal troops could hold large numbers at bay for a time, the Italians encountered much resistance at first and suffered heavy casualties. But the defense was short-lived. The Piave was forced, the Grappa front collapsed next, and the victorious armies of General Diaz were soon sweeping forward along the whole line, gathering in the demoralized enemy hordes which threw down their arms and surrendered wholesale. "In ten days the Austrians lost an immense quantity of material of all kinds, nearly all their stores and depots, and left in Italian hands some 300,000 prisoners and not fewer than 5,000 guns." It was a rout in which the débris of the Austrian armies was more intent on safety than on defending any position, however strong.

To understand the ease with which the Italian armies now swept over a mountainous terrain which had earlier defied their own best efforts as well as those of the enemy, we must recall the general situation of the Austrian armies in late October, 1918. For three months they had been looking on with fear and consternation while the great German colossus, their only hope of ultimate victory, was being beaten to its knees by the legions of France, England, and America. Not a ray of light relieved the black certainty that Austria's partner in the conspiracy of 1914 would be finally and decisively crushed. Utter hopelessness, the surest solvent of armies, had seized the Austrian military forces, from the highest officers to the men in the ranks, who could not now be deceived as to the actual state of affairs. Under the circumstances no longer could the notoriously poor fighting qualities of the Austrians be stiffened by a leaven of German officers and

German units. Indeed, Ludendorff was urging his little-respected ally to send him troops to help stem the tide of disaster on the main western front. For the Austrians there remained nothing but the certainty of ultimate defeat, and the only question was the time and manner of its arrival.

Another factor of vital importance had completed the disintegration of the Austrian armies. When the amazing Treaty of London was published, the Austrians were able, by spreading its terms broadcast among their regiments of Yugoslav troops, to show that Italy was fighting the war only to subject vast areas of their lands and a large part of their population to Italian domination. This galvanized the Slavonic elements, strongly disaffected toward their Austrian oppressor, into renewed hostility against the Italians and made it possible later for the Italians to say that the Yugoslavs had fought against them as bitterly as had other Austrian troops. Intelligent Italians soon realized, however, that Italy had only herself to thank for this condition of affairs, and set about the task of removing the obstacles which prevented two enemies of Austria from working together for the defeat of their common oppressor. At the Rome conference of April 10, 1918, Italians and Yugoslavs came to an agreement that their respective territorial claims should be settled "on the basis of the principles of nationality and the right of peoples to decide their own fate." Yugoslavs and Italians now united in proclaiming this just and friendly understanding to the Slavonic troops of Austria and in carrying on a gigantic propaganda designed to win them to the Allied cause. So successful was this effort that by August there were observed large desertions of Yugoslav troops, the withdrawal of Slav units from the Austrian front line trenches because of their unreliability, and the disintegration of the Austrian armies to the extent that further offensive operations by them were practically impossible. This was, moreover, only one phase of a disintegration in which Poles, Czechs, and other subject peoples of the Dual Monarchy were daily becoming more outspoken regarding their intention to throw off the Austrian yoke and more determined not to fight against the armies of the

Entente. As early as the fall of 1917, according to Ludendorff, "the Austro-Hungarian army was worn out; it had lost 1,800,000 prisoners; it was short of recruits. Its fighting value was slight."[4] When Diaz's blow fell Austria's "remaining armed forces were so honeycombed with disaffection and sedition that they were incapable of making even a defensive stand."[5]

A third circumstance of great moment rendered any effective resistance by the Austrian armies impossible. It will be remembered that in the summer of 1918 the Austro-Italian front ran from the Swiss frontier to the mouth of the Piave River. Then came a long stretch of sea, to where the next battle front began in Albania and stretched across southern Macedonia and northern Greece to reach the sea once more east of Saloniki. Thus on the south the dam which held back the Allied flood and prevented it from sweeping into the plain of Austria-Hungary behind the armies facing Italy was a comparatively short front across the narrow southern part of the Balkan Peninsula. It was, furthermore, a dam composed largely of Bulgarian troops, the Austrians holding but a small part of the extreme western end of the line, in Albania. Now this dam was simply annihilated in the middle of September, 1918, when the entire Bulgarian army was routed and forced to abject surrender. Immediately the Austrian armies faced the alternative of stretching a new front across the broad base of the Balkan Peninsula to check the northward rush of the Allied armies, or of seeing their armies on the Italian front cut off from their bases and forced to capitulate. In their weakened condition it was difficult, even with German and Bulgarian aid, for the Austrians to hold the shorter fronts assigned to them. To increase enormously that frontage at the moment when German and Bulgarian aid failed was hopelessly impossible. There was nothing possible for the Austrian forces on the Italian front save to look on in despair while the Allies pushed swiftly northward

[4] Erich von Ludendorff: Ludendorff's Own Story, August, 1914–November, 1918: The Great War from the Siege of Liège to the Signing of the Armistice As Viewed from the Grand Headquarters of the German Army, 2 vols., New York, 1919; reference in Vol. 2, p. 161.

[5] Hayes, A Brief History of the Great War, p. 352.

through Serbia to take them in the rear, hoping against hope that diplomacy might avert the inevitable disaster. The Austrians on the Piave-Trentino front knew they were beaten the moment the Bulgarian front collapsed.

On October 6, while the Allied armies were advancing rapidly along the great Morava-Vardar corridor to the Danube plain, Austria asked the American government to intervene with the European Allies for an armistice. October 12 the Serbians re-entered their ancient capital at Nish, and on October 19 Allied troops stood on the Danube in northwestern Bulgaria. The hour of destiny was already sounding in their rear when on October 24 the Hapsburg armies received the Italian blow. Despairing of German victory, demoralized by internal disintegration, and doomed to defeat, whatever happened, because of the foe advancing from behind, the Austrians staggered under the impact, gave ground slowly at first, then incontinently fled. Five days later the Austrian government was begging for a separate armistice and separate peace, and at the end of a week the readiness of its armies to capitulate was announced.

It is thus quite evident that the failure of the Austrians to hold the Italians at bay along the many natural obstacles which this difficult terrain offered for prolonged rearguard actions is in no wise a measure of the defensive value of those obstacles. As in the case of the German retirement without adequate defense of the topographic barriers in France and Belgium during the last few weeks of the war, so in the case of the much more rapid Austrian retirement over a much more formidable terrain, the internal condition of the retiring forces rendered the strongest natural positions of no avail. In the words of Marshal Foch: "All terrains are passable if not defended by rifle shots, that is to say, by valiant and active men." [6]

[6] Ferdinand Foch: Des principes de la guerre, Paris, 1917, p. 29.

CHAPTER XIII

THE BATTLEFIELD OF THE ISONZO: THE KARST PLATEAU BATTLEFIELD

Standing on the old international frontier southeast of Udine, one has before him a broad panorama of plain and mountain. In front stretches the smooth Piedmont sloping gently toward the sea. Down this slope from north to south the Isonzo River takes its sluggish way, choked with sand bars and wandering through marshy flats. Low dikes hold the waters in check lest they flow too far over the adjacent lowland. Beyond the river rises a steep mountain wall, even-crested toward the south (Pl. X, *A*) but merging into the complex mass of Alpine ranges farther north, where lofty ridges and inaccessible peaks remind one of the savage terrain of the Trentino. Directly opposite, the towers of Gorizia are visible in a wide recess in the mountain wall formed by the broad Wippach valley. South of the city the even crest of the wall is formed by the top of the far-famed Carso plateau; while on the north stretches the Bainsizza plateau. West of the river the Alps send a long spur southward to end in the hills of Podgora, grim sentinels before the city gates.

It is in this interesting region that the east-west ridges of the main Alpine chain shoot off a spur to the southeast called the Julian Alps, which rapidly broaden into the limestone hills and plateaus of the Dinaric Alps bordering the eastern coast of the Adriatic. Already the barren wastes of the desolate *karst*, that belt of arid limestone uplands which forever condemns to poverty the western margin of the Balkan Peninsula, begin to assert themselves in the Carso (Karst) and Bainsizza plateaus (Fig. 141). The northwest-southeast structure of the rocks,

NOTE—For Chapters XIII and XIV the reader should constantly consult the sketch maps (Figs. 119, 140, and 142) and, in the pocket, the block diagram of the battlefield (Pl. VIII). Unless otherwise located, places named in the text can most readily be found on Figs. 140 or 142.

folded on axes trending in this same direction, is clearly manifest in the course of the Carso and of the lowlands bordering it on either side, and in the parallel ridges and depressions of the Istrian peninsula (Pl. VIII). Farther north the east-west structure of the greater Alps reveals itself in the lofty Carnic range, continued east of Tarvis in the Karawanken Mountains, as well as in the ridges and valleys which west of the Isonzo carry the Alpine highlands down to the plains. Thus the Battlefield of the Isonzo lies at the critical junction of two distinctly different mountain systems (Fig. 140).

Strategic Position of the Isonzo Battlefield

Just as the Trentine Alps form the natural protective barrier between the Italians and the Teutons, so the Julian Alps stand guard between the Italians and the Slavs. In both cases the Italian side of the main divide is trespassed upon by the neighboring race, which in the Isonzo region reaches far down into the plain. But notwithstanding this failure of ethnic boundaries to respect the topographic frontiers designed by Nature, it remains true that the Julian Alps form the logical and only effective protection for the Italian plain on the east, just as they form the similar western protection for the Slavonic lands of the Save lowland. In the World War it was as inevitable that the Italians should seek to breach this barrier and burst into the Slav provinces of Austria beyond, as it was that Austria should strenuously defend this back door to Vienna.

From time immemorial one of the principal gateways into Italy has been the Pear Tree Pass,[1] a low gap in the limestone plateau near Adelsberg by which passage from the Save basin to the Piedmont plain was easily effected. A well-known route of early migrations, it was later crossed by a Roman road connecting the Adriatic Sea with the navigable Save-Danube system. This "most harmful door left open by Nature to chastise the faults of Italy" offered the shortest and lowest transit

[1] E. C. Semple: The Barrier Boundary of the Mediterranean Basin and Its Northern Breaches As Factors in History, *Annals Assoc. Amer. Geogrs.*, Vol. 5, 1915, pp. 27–59.

FIG. 140—Generalized sketch map of the Battlefield of the Isonzo. See also the general map of the Italian theater of war (Fig. 119), the sketch map of the Isonzo River front (Fig. 142), and the block diagram of the Isonzo battlefield (Pl. VIII).

route from the interior to the Mediterranean basin to be found in the whole 1,300 mile stretch of mountains between the Bosporus and the Rhône valley, and lay between two natural thoroughfares, the Piedmont plain of Italy and the level expanse of the Danube plain. It is not strange, therefore, that barbarian hordes from the Danube basin poured through it repeatedly to overrun the level lands of sunny Italy. In our day road and railway from both Trieste and Fiume on the Adriatic cross the Pear Tree Pass to Laibach on the Save River. Farther north another railway from Trieste ascends the Isonzo valley to the vicinity of Tolmino, then follows up a branch valley to cross by tunnel under the Bacha Pass (Fig. 140) into the upper Save basin, there to reach the vital junction point of Assling. Southward, the mountains are for the last time crossed by a standard-gauge railway where they are narrowest, in the latitude of Fiume, both road and railway connecting Fiume with the lower basin of the Save. The approaches to all three of these strategic gateways lie within the broader limits of the Battlefield of the Isonzo and were important objectives of the Italian eastern campaign. Beyond them lay the Austrian military bases of Klagenfurt and Laibach, as well as Agram some seventy-five miles southeast of Laibach on the Save River, the capture of which would enormously cripple Austria's further military efforts.

Within the limits of the battlefield were areas of predominant Italian population constituting, like the Trentino, an "unredeemed province" of Italy. "Trent and Trieste" was a battle cry which aroused the patriotic ardor of the Italian masses and, independently of purely military considerations, created a political necessity for an advance into the Isonzo region. As in the case of the Trentino, however, Italy demanded far more than racial, economic, geographic, or historical grounds could justify. The real basis of her claim was strategic. If she could have assigned to her the three strong bridgeheads on the eastern coast of the Adriatic (in Albania, Dalmatia, and the Julian Alps) by which according to the Treaty of London her entrance into the war on the side of the Entente was to be recompensed, her military supremacy in the western Balkans would be assured.

Most valuable among these bridgeheads was the northern one, which would carry the Italian frontier clear across the Julian Alps barrier to the heights dominating the Laibach basin and place the Adelsberg gateway well within Italian territory. Demands supplementary to the Treaty of London, but pushed with great insistence at the Peace Conference, were designed to place the Bacha Pass tunnel south of Assling and the vital port of

FIG. 141—The barren upland of the Carso in the vicinity of Oppachiasella. (Italian official photograph.)

Fiume within Italy's boundaries. Thus all three strategic gateways would pass wholly under Italian control and all northwestern Yugoslavia lie open to easy invasion. Possession of Fiume in particular would prove of the highest strategic value to Italy, because the geographic character of the eastern coast of the Adriatic makes of this port the only practicable economic outlet for Yugoslavia, thus assuring to the power controlling the port a potential control of the economic life of the Yugoslav people. As in the case of the Trentino, the racial argument was emphasized to stimulate the support of the Italian people and to affect world opinion; but the annexing to Italy of half a million Slavs against their violent protest, which would be necessary in order to save the few thousand Italians constituting at Fiume a tiny Latin island in the midst of a Slavonic sea, could be justified, if at all, only on military grounds.

That the real objective of Italian policy was to secure ascendancy in the Balkans by controlling strategic bridgeheads of the highest possible importance was clearly avowed in the defense of the Orlando-Sonnino government published by the *Giornale d'Italia*, generally regarded as Baron Sonnino's organ, after the fall of that government from power. After rebuking certain influential Italian elements for refusing to admit "the strategic, political, and economic reasons for which Italy must set foot in Dalmatia, thereby constituting with Istria and Albania the triple bridgehead for expansion in the Danubian and Balkan system, which expansion is feared by others and is the true motive of the resistance offered to our Adriatic claims," the defense enumerates the following objects which Orlando and Sonnino had in view in consenting to discuss at Paris the so-called "Tardieu project" for a free state of Fiume:

(1) Annexing to Italy the whole of Istria, even including that part of eastern Istria which Wilson denies us.

(2) Giving to the small free state of Fiume such a statute as would have effectively placed it for fifteen years under our government through the long arm of a local government faithful to us, pending its eventual annexation to Italy.

(3) Saving in Dalmatia the harmonious system of Zara-Sebenico-the islands, while leaving to Yugoslavia a part of the interior; but thus establishing an adequate political, economic, and military bridgehead, together with a substantial guarantee of the *italianità* of Dalmatia and full security against any future contingency.

(4) Having Albania under mandate and thus insuring through its integrity and national independence our influence, our expansion, and our strategic safety.[1]

It is clear that with such far-reaching strategic plans in view the Italian high command must have concentrated its chief offensive energies on the eastern front, even had not the nature of the Trentine terrain destined it for a merely defensive rôle. East and southeast of the Isonzo River lay the keys to the most im-

[1] *Giornale d'Italia*, July 4, 1919.

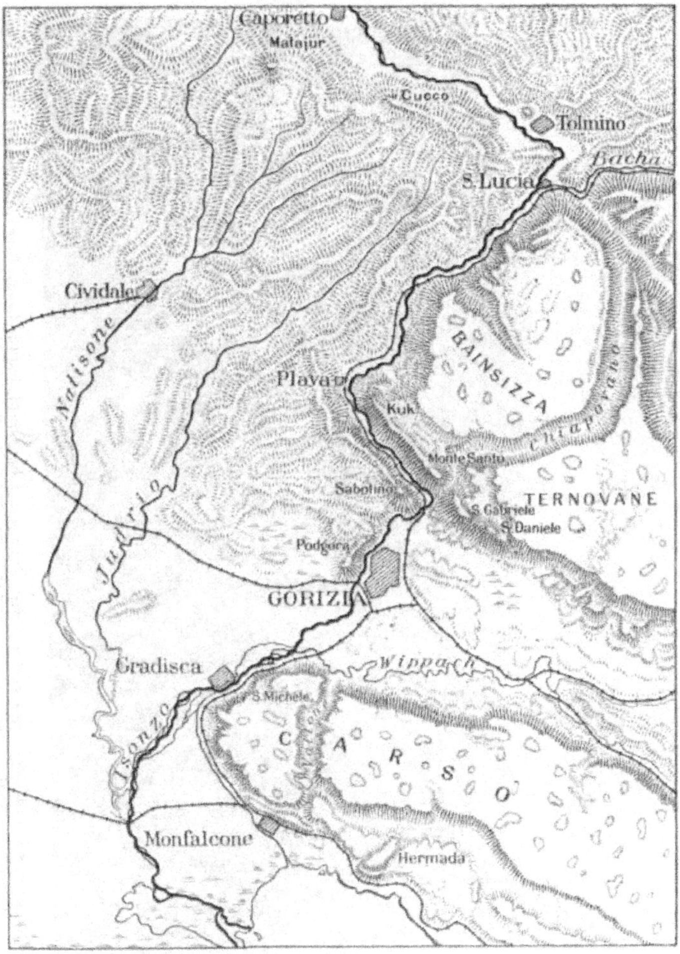

FIG. 142—Map of the Isonzo front, showing the Carso and Bainsizza plateaus and the mountain peaks which formed the principal Austrian strongholds. Scale, 6.7 miles to the inch.

FIG. 143—Parallel ridges of the Alps in the northern part of the Isonzo battlefield, and the Isonzo River at Caporetto (just to the left of the center of the view). In the left foreground one of the important passes breaching the mountain wall to reach the Piedmont plain is visible; in the right background the Monte Nero range. The Battle of Caporetto began with a German attack which broke the Italian line along the [illegible]

portant bridgehead of the three and to the only trade route possession of which would at a single stroke render practically impossible any effective resistance to possible future Italian expansion in the Balkans. The Battlefield of the Isonzo was certain, therefore, to witness the main efforts of the Italian armies, until the disaster of Caporetto should throw them back on the Piave.

The only two ports by which Austria-Hungary and Germany could communicate effectively with the Austrian fleet in the Adriatic were Trieste and Fiume. The railways leading to these ports and to the great Austrian naval base at Pola (Fig. 119) could be cut by a moderate eastward advance, the ports themselves isolated and captured, and the Central Powers cut off from all communication with their Adriatic fleet save by the roundabout and difficult route via narrow-guage railway across the barren karst highlands to several minor ports farther south. Thus the position of the Battlefield of the Isonzo made it an attractive field of operations designed in part to assure full Allied control of the Adriatic Sea.

Surface Features of the Isonzo Battlefield

The battlefield to which the valley of the Isonzo is the threshold, and which we have chosen to designate by the name of that river, is characterized by three striking topographic elements which deserve some further notice (Fig. 142). These are the Alpine ridges at the north, the karst plateaus at the south, and the Isonzo River and its tributaries draining parts of both these regions. A full description of even these three elements of the terrain would carry us further than the fate of the operations on this eastern front would seem to warrant; for despite valiant endeavors the Italians failed to progress appreciably beyond the margin of the field, and their moderate gains were completely wiped out by the great Teutonic victory of Caporetto.

THE ALPINE RIDGES

The northern ridges belonging to the main Alpine system resemble the Alps of the Trentino in more respects than one (Fig.

FIG. 144.—The Matajur-Cucco-ridge viewed from the plain. Along the crest ran the old Austro-Italian frontier, dominating the plain from its advantageous position. (Italian official photograph.)

143 and Pl. VIII). They are eroded from the same complexly folded and broken system of rocks of varied types and show the same ragged crests and peaks separated by the same deep, glacier-scoured valleys. It is not necessary to repeat in detail the effects of such a terrain upon methods of warfare, already described at some length in connection with the Battlefield of the Trentino. It will be sufficient to modify the account there given by noting the fact that no peak in the Isonzo region rises sufficiently high to have glaciers still persisting on its flanks, and so to eliminate from our present consideration such spectacular features of Alpine warfare as depend upon the presence of permanent ice fields.

As in the Trentine mountains, so in those of the northern Isonzo region, movement is limited to a few important passes. Among these the Pontebba, or Pontafel, Pass formerly crossed by a Roman road, now carries a railway through the upper Tagliamento valley and across a low col at Tarvis into the Drave valley by a transverse gorge cut clear through the Carnic-Karawanken ridge; while the difficult Predil Pass at the head of the Isonzo is crossed by a highway which ascends that valley to reach the Drave by the same strategic gateway at Tarvis. The only other important crossing of the northernmost ridges is the tunnel already referred to through which the railway from Trieste crosses into the Save valley south of Assling.

A few of the minor passes across the more southerly members of the Alpine ridges have considerable strategic importance. It will be observed that the upper Isonzo River cuts a zigzag trench through the east-west ridges, the general course of the trench being southward from the Predil Pass (Fig. 140). Across the high Monte Matajur-Monte Cucco ridge, steepest on the northeast toward the Tolmino-Caporetto sector of the Isonzo trench and sloping more gently southwestward to the plain around Cividale and Udine, are cut a series of low gaps by which troops may easily pass from the plain into the trench or debouch from the trench into the plain. The lowest of these is the wild gorge of the Natisone (Fig. 142) which from the floor of the trench just west of Caporetto (Karfreit) turns abruptly southward and cuts through the mountains to the plain. A good highway runs from Caporetto along the narrow floor of the Natisone gorge to Cividale. Two or three miles southeast of Caporetto a branch of the Natisone heads on a low col (Fig. 143) and flows southwest to join the main river before Cividale is reached. By this col and branch valley a second highway crosses the high Matajur-Cucco ridge to connect

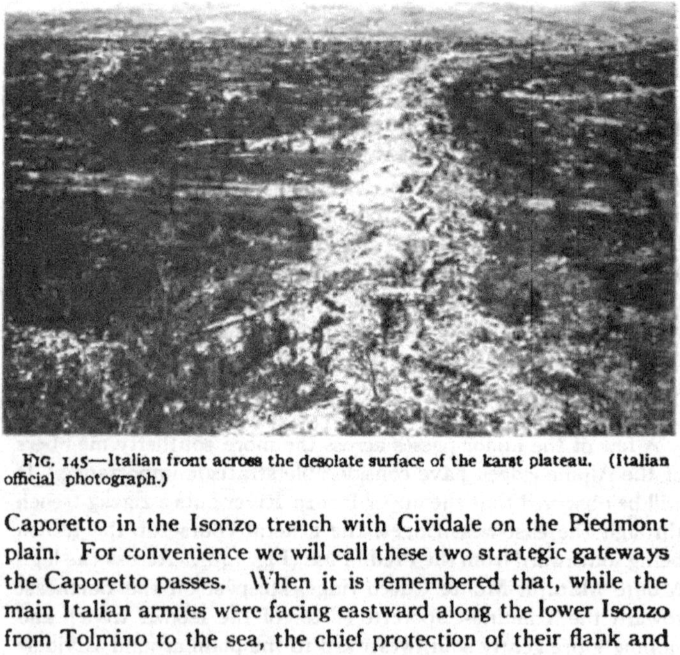

FIG. 145—Italian front across the desolate surface of the karst plateau. (Italian official photograph.)

Caporetto in the Isonzo trench with Cividale on the Piedmont plain. For convenience we will call these two strategic gateways the Caporetto passes. When it is remembered that, while the main Italian armies were facing eastward along the lower Isonzo from Tolmino to the sea, the chief protection of their flank and rear in this region was formed by the Matajur-Cucco ridge, the vital importance of the Caporetto passes can easily be imagined.

The old Austro-Italian frontier across the Battlefield of the Isonzo traversed the Pontebba Pass and ran close to the Predil Pass on the northernmost ridges, then followed southward parallel to and west of the Isonzo trench to reach the Matajur-Cucco ridge, along the crest of which it pursued an easterly course to a point near Tolmino, whence it ran southwestward to the open plain along a minor stream valley just west of the Isonzo. On the Matajur-Cucco ridge the frontier for a distance of 15 miles or more overlooked the plain at close range (Fig. 144). As in the

FIG. 146—Sink hole in the limestone of the barren karst country. (Italian official photograph.)

Trentino, Austria had kept for herself the great mass of the Alpine barrier, and all along the frontier from the sea nearly to the Predil Pass the Italian plain lay open to hostile invasion. It was a frontier imposing fatal strategic disadvantages on Italy, and one which Italian statesmen could not be expected to tolerate when once the World War opened the way to a possible revision. The "American line" which the Peace Conference offered to Italy but which Italy refused to accept, carried her frontier eastward to the main crest of the Julian Alps far into Slavonic territory, blocked the gateways of possible invasion by giving her the southern half of the tunnel south of Assling, the Birnbaum plateau and other highlands dominating the Adelsberg region and the Pear Tree Pass, and the great Monte Maggiore ridge commanding the exits from the Fiume basin, thus assuring reasonable strategic security to Italy, something which the old frontier had denied.

THE KARST PLATEAUS

Southward from the latitude of the Bainsizza plateau the northwest-southeast system of ridges and plateaus prevails (Pl. VIII). Irregular sandstone hills, valleys eroded on shales and other weaker formations, and sharp ridges where the folded rocks are tilted up on edge vary the landscape. But it is the flat-topped limestone plateaus which give the dominant character to the country (Fig. 145). Of these the great Carso plateau, lying between the broad Wippach valley on the northeast and the narrow corridor leading along the sea to Trieste on the southwest, projects westward as a formidable bastion dominating the plain in all directions. In a great arc around its western end sweeps the natural moat of the Isonzo River and its marshes (Fig. 142).

It is not easy adequately to conceive the stupendous difficulties of the Carso terrain. The plateau is in fact a flat-topped mountain from four to six miles broad (Pl. X, *A*). Its sides are precipitous, and, as it rises from three or four hundred to more than a thousand feet above the surrounding lowlands, it constitutes a gigantic rock-walled fortress whose guns control with ease the city of Gorizia, the crossings of the Isonzo moat, and the roads and railways running to Trieste through the Wippach valley and the coastal corridor. Its lower slopes are partly wooded, but higher up the steep rimming scarp is bare and rocky. Merely to scale the walls is therefore a military feat of no small magnitude.

The upland surface is a dry, thirsty, wind-swept desert (Fig. 147). The rock is limestone, and its ready solution permits the development of underground channels by which all moisture quickly escapes to great depths. Scanty vegetation relieves but slightly the barren aspect of this desolate land without rivers, brooks, or springs. It is the typical rocky desert of the karst type, for which its name, Carso, is the Italian equivalent. Like other karst lands the surface is excessively irregular, pitted with sink holes without number and undermined by subterranean caverns. The sink holes, or "dolinen," are funnel-shaped depressions from fifty to many hundred feet in diameter and fifty, a hundred, or even several hundred feet in depth (Fig. 146).

FIG. 147—Defensive position on western margin of the Carso north of Monfalcone. The barren aspect of the limestone country is apparent, and stone breastworks and shelters replace trenches and underground dugouts. (Italian official photograph.)

They may end in passageways connecting with the vast labyrinth of underground caves and galleries. Over the surface rise many low hills and an occasional mass of more imposing dimensions, like the Monte San Michele (Pl. X, *A*). Nature thus provided ready to hand innumerable concealed sites for heavy artillery, machine-gun emplacements, observation stations, and secure underground retreats for vast numbers of troops. And what Nature offered, the Austrians had accepted and improved by long years of elaborate fortification. Trenches had been cut in the solid rock, elaborate systems of galleries and tunnels had been excavated, gun emplacements had been prepared in pits quarried for the purpose, and the whole system connected by covered

FIG. 148.—The Hermada hills on the southern border of the Carso, which blocked the Italian advance along the coastal corridor toward Trieste. (Italian official photograph.)

communication trenches and supplied by water pumped up to the thirsty surface and distributed by pipe lines.

The western end of the plateau is cut off from the main Carso by a deep trench, plainly visible on the map, called the Vallone (Fig. 142). Thus even should the Italians cross the Isonzo barrier, scale the rocky walls of the plateau and conquer Monte San Michele and the rest of the western upland (sometimes called the Doberdo plateau), their further progress would be disputed at the deep, wide trench just described.

South of the formidable Carso bastion is the coastal corridor leading to Trieste. So narrow is the depression that it could hardly be called a "curtain" in the terminology of fortifications, although it can be effectively swept by flanking fire from the bastion walls. These are broken and terraced on the southwestern side, and an outlying strip carries the heights of Hermada, which rise even higher than the adjacent surface of the main plateau and effectively block the route to Trieste (Fig. 148). To the north the broad Wippach valley constitutes a natural curtain, upon the floor of which good defensive positions are to be found in a maze of low hills and on detached remnants of the plateau of greater elevation. It is within the shelter of this re-entrant that the city of Gorizia nestles under the protection of the bastion walls.

North of the Wippach curtain is the massive Bainsizza bastion, repeating in essentials the features of the Carso bastion. Again we have the high limestone plateau, with a labyrinth of hills and sink holes rendering its arid surface almost impassable, and steep rimming walls making it almost inaccessible. Again there are bordering valley lowlands on either side; again the western end of the plateau is cut off from its eastward continuation (the Ternovane plateau) by a deep, dry valley, this time known as the Chiapovano valley; and again the Isonzo River forms a natural moat around the projecting apex of the bastion. And just as the outlying highland strip carrying the Hermada strengthens the southwestern wall of the Carso bastion, so the southwestern wall of the Bainsizza bastion is reinforced by the detached ridge carrying Monte Santo, San Gabriele, and San

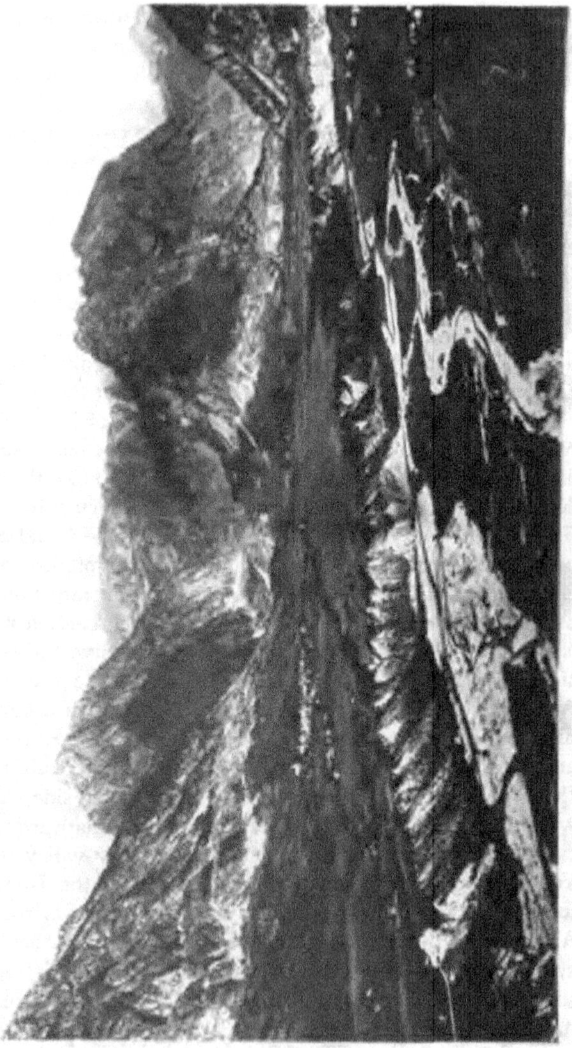

FIG. 149—The Isonzo River issuing from the gorge at the right to flow through the open Plezzo basin in the foreground. (Italian official photograph.)

Daniele (Fig. 142). A new feature appears in the strong outlying defenses to the west formed by the fading spurs of the Alps, which terminate southward in Monte Sabotino and the lower hills of Podgora.

THE ISONZO VALLEY

The Isonzo River moat, which defends the approaches to both bastions, the Wippach curtain, and the Trieste corridor, is in its northern part a deep, steep-walled glacial trough, with a fairly open floor (Fig. 149), much of the way from its head to Tolmino. Here the isolated hills of Santa Lucia and other heights rising from the floor defend the important Tolmino basin, where streams from three directions join the Isonzo and where the Predil Pass, Assling, and Idria mining region routes meet on their way to Gorizia. From Tolmino southward the valley is a wild and picturesque gorge over 2,000 feet deep, with scarcely room in its narrow bottom for the two roads and railway which traverse it. It is this portion of its course which forms the deep moat contouring the western end of the Bainsizza bastion. At Gorizia the river passes from the mountains to the plain and immediately takes on the braided pattern so characteristic of the other rivers of the Piedmont. Across the broad entrance to the Wippach curtain it stretches its multiple, shifting channels and sand bars, a barrier not to be despised by any army. On reaching the Carso bastion it swings southwest, then south, and finally southeast around its western end, sometimes close to the limestone wall as in the vicinity of Gradisca (Fig. 150); sometimes farther out, as in its course over its own marshy delta southwest of Monfalcone.

Like the rivers on the Flanders plain the course of the Isonzo across the Piedmont has to be diked to prevent flooding broad areas of the adjacent country. Especially when melting snows or heavy rains in the mountains add their unusual volume to the stream does the protection become imperative. Then indeed the full channel with its muddy flood becomes a military obstacle most difficult to traverse under fire from the bastion walls. Should it fail, however, to check a persevering enemy, the defense may be strengthened, as in the case of the Yser, by breaking the dikes and flooding the bordering lowland. During one of the Italian attacks on the Carso, where the natural moat lies close at its base, the Austrians availed themselves of this method of defense, and the river flood, then unusually high, deluged the surrounding lands to a depth of six feet in places.

THE BORÀ

Even the winds of this forbidding region oppose themselves to an army attacking from the west. The *borà*, a violent wind which renders so many channels and harbors along the eastern coast of the Adriatic unsafe, pours like a furious, invisible flood over the margins of the karst plateaus, sweeping clouds of dust or storms of rain, sleet, or snow into the faces of all who may try to scale the slopes. Hodgkin relates that, when the Romans in the Wippach valley tried to stem the tide of invasion under Theodosius the Great pouring through the Pear Tree Pass in 394 A.D., they were thrown into confusion by the clouds of dust which, sweeping down upon them from the heights, destroyed their aim, and by the fact that the violent *borà* turned their darts back upon themselves. It was the *borà* which "decided the battle of the Frigidus near fifteen centuries ago and gave the whole Roman world to the family of Theodosius and the dominion of the Catholic faith."[3] Iron railings are now fastened to the rocks at some of the karst passes to prevent travelers from being blown over the cliffs by a wind against which neither man nor beast can stand unaided.

Such are the natural barriers which blocked the eastward advance of the Italian armies. The steep-walled bastions of the Carso and Bainsizza plateaus, defended by the Isonzo moat, were rightly regarded by the Austrians as almost impregnable. An advance to Trieste along the coastal corridor was impossible so long as the Austrians held the Carso bastion and the high hills of Hermada. To reach that port by way of the Wippach curtain was impossible until both the Carso and Bainsizza bastions completely dominating it were cleared of enemy forces.

[3] Thomas Hodgkin: Italy and Her Invaders, 8 vols., Oxford, 1880–1899; reference in Vol. I, pp. 575–577.

Farther north an advance would merely carry one ever deeper into the maze of difficult Alpine ridges. There was no alternative to a frontal attack on the two bastions and their protecting moat.

MILITARY VALUE OF THE NATURAL BARRIERS

A better appreciation of the formidable character of the natural obstacles which confronted Italy upon her entrance into

FIG. 150—The Isonzo River at the base of the Carso plateau near Gradisca. The bridge crossing the river has been partially destroyed to prevent easy passage of the barrier. (Italian official photograph.)

the war would have served to mollify the criticisms directed against her slow progress across the Battlefield of the Isonzo. The fact that she was confronted by troops of poor fighting quality, troops which were regularly beaten by Serbs, Rumanians, or Russians except when stiffened by German units and directed by German officers, was more than offset by the nature of the terrain. Let the reader imagine himself west of the Isonzo, attempting the task of the Italian army. Gorizia, an important field base and a prize of considerable political importance, is one of the first objectives. There it lies in full view just beyond the river, nestling securely in the re-entrant angle of the Wippach curtain. You approach the treacherous Isonzo and find the bridges wholly or partially destroyed. The guns of Monte Sabotino and Podgora on the Alpine spurs fire upon you from the north; the guns of Monte Santo, San Gabriele, and San Daniele on the outer wall of the Bainsizza bastion threaten you from the northeast; the guns of Monte San Michele on the crest of the Carso bastion belch steel upon you from the south. Boats and pontoon bridges are destroyed in this hurricane of cross fire, and it would almost appear that you cannot cross the river moat until you have captured the protecting bastions, and cannot reach the bastions until you have crossed the protecting moat. At last you can appreciate the full significance of one phrase in the memorandum said to have been issued to the Austrian soldiers at the beginning of the struggle on the Isonzo front: "We have but to retain possession of a terrain fortified by Nature."

CHAPTER XIV

MILITARY OPERATIONS ON THE BATTLEFIELD OF THE ISONZO

The First Battle of the Isonzo

War was declared by Italy on the evening of May 23, 1915. Cadorna struck without a moment's delay, and by morning his armies were sweeping across the level plain toward the Isonzo barrier. The soldiers of Italy were now to attempt the same maneuvers which Napoleon's legions had repeatedly performed upon this identical terrain in the years 1797 to 1809. Like Napoleon, Cadorna directed one column against the Pontebba Pass to close that gateway to the plain and also to cut off Austrian forces which might attempt to descend the Isonzo via the Tarvis gateway and Predil Pass. Like Napoleon, he directed his main offensive against the Isonzo barrier and the karst plateaus beyond, in order to seize the gateways through the mountain wall and to debouch into the Austrian lands beyond. If successful, he might, like Napoleon, re-unite his columns after their passage through the gateways, and prepare for an advance on Vienna.

But the power of the modern defensive has restored to mountain ranges all of that importance as military barriers which they possessed in the early days of warfare and which they seemed in some measure to lose when the genius of Napoleon, adopting novel methods of campaigning which his slow-witted opponents could neither understand nor imitate, forced one mountain pass after another with disconcerting ease. Where Napoleon found a mere stage upon which to execute his maneuvers, Cadorna discovered an impregnable fortress belching death from myriad unseen caverns, camouflaged sink holes, and concealed tunnel exits. Long-range heavy artillery fire, the deadly machine gun, and barbed-wire entanglements had turned rivers into almost impassable moats and long hill slopes into fatal glacis.

The Isonzo River was the first serious barrier encountered by the advancing Italians. Along its lower course a crossing was effected by means of pontoon bridges and boats under heavy fire and with great difficulty. It is related that the advancing troops, inspired by high patriotic fervor and the lofty ambition to free their oppressed brethren of Gorizia and Trieste, flung themselves into the flood to gain the farther bank. Gradisca and Monfalcone at the base of the Carso were captured, but the Italian right wing hurled itself in vain against the walls of that great bastion (Pl. X, *A*). In front of the Wippach curtain the river moat, swept by fire from the bastion walls, proved too difficult an obstacle to negotiate, and the hills of Podgora and Monte Sabotino, outworks of the Bainsizza bastion, resisted capture. Farther north the Isonzo moat before the main Bainsizza bastion was reached, but its passage was too heavy a task for the forces engaged. Still farther northward, in the Caporetto sector (Fig. 143), heavy fighting closed the passages from the trench into the plain, carried the Italians across the river and up the slopes of the ridge beyond, where the bare rock precipices of Monte Nero dominated the whole valley from Tolmino past Caporetto to Plezzo (Flitsch). The bridgehead gained here on the northeast side of the great Isonzo trench was of prime importance, for it threw the enemy back from the dangerous Caporetto passes opening through the Matajur-Cucco ridge southwest of the river. Should the enemy ever debouch through those passes, the remaining armies along the Isonzo all the way to the sea would be threatened from the rear and constrained to a precipitate retreat.

All summer long the Italian armies struggled in vain to breach the formidable barriers which opposed them. Their most violent assaults made little impression on the frowning walls of the Carso and Bainsizza bastions, and much of the Isonzo moat still lay between them and the enemy. A heroic attack on the Carso in the latter part of July netted small gains; and, when the fires of the battle burst forth again in October and the whole

Isonzo front was assaulted in order to relieve pressure upon the Serbian armies, the massive natural fortress remained unshaken. The long and bitter First Battle of the Isonzo had done little more than secure bridgeheads east of the Isonzo below Gradisca and above Tolmino, with a smaller foothold about Plava at the western apex of the Bainsizza bastion.

The Second Battle of the Isonzo

The summer's experience had demonstrated that the fortified natural bastions on the Isonzo front could be carried only by a slow siege warfare which should prepare the way for a grand assault many months later. To this laborious work the Italian armies now addressed themselves. Day after day, week after week, they burrowed up the barren slopes of the Carso, driving deep trenches in the solid rock farther and farther toward the crest, boring great tunnels hundreds of yards in length which should ultimately open within a few feet of the Austrian main defenses. Spade and entrenching tool were useless here, and the miner's pick, sledge-hammer, drill, and dynamite were called into service. Sandbags, brought up from the plain below, were built into breastworks which afforded the toilers some protection from the deadly flying splinters of rock dislodged by the enemy's constant artillery bombardment. Farther north the same methodical preparations were being carried out against enemy strongholds on the advanced spurs of the Alps west of the Bainsizza bastion. Rock trenches were being driven toward the works crowning the hills north of Podgora while the almost inaccessible positions on Monte Sabotino were secretly being reached by a tunnel thousands of feet in length. Week after week, month after month, the laborious task was pursued. What surface topography made impossible must be achieved by an advance underground.

At last, on August 4, 1916, more than a year after the Italian armies had first flung themselves at the Isonzo barriers, the second great blow against the natural fortifications was struck. After a terrific bombardment "the Italian infantry, believing that the hour of victory had at last arrived, charged with unexampled impetuosity," while a fleet held the coastal corridor under fire to prevent reinforcements coming from Trieste. The advantages gained by the long siege operations were now apparent. Leaping from their protected advanced positions the assaulting forces overwhelmed the astonished enemy before he could take effective countermeasures. The walls of the Carso bastion were scaled, the cruel guns of Monte San Michele silenced, the western end of the Carso forming the Doberdo plateau overrun, and even the deep trench of the Vallone crossed after a bitter struggle. Here, however, the difficulties opposed by Nature wore down the offensive so effectually that further progress on the main Carso was soon reduced to minor gains purchased at a heartbreaking price.

Northwest of Gorizia the brilliant assault was equally successful. The hills of Podgora and the supposedly impregnable stronghold of Monte Sabotino were carried by storm. Thus the last of the natural outworks of the Bainsizza bastion were in Italian hands. With the enemy pushed farther back toward the east on both protecting bastions, the Wippach curtain was less secure, and Italian troops crossed the river, captured Gorizia, and established themselves on the hills east of the city. But here again the offensive was soon halted. Further advance into the curtain was impossible so long as the enemy still poured a murderous fire into it from the bastion walls on both sides, just as the advance on Trieste by the coastal corridor must await the clearing of the Carso and the outlying hills of the Hermada. In September, in October, and again in November, the bloody struggle was continued with admirable vigor and determination, and some ground was added to that previously gained. But the formidable obstacles mentioned above were beyond the power of the now exhausted Italian armies to conquer.

The Second Battle of the Isonzo was for the Italians only a partial success, since it failed to break through the system of natural barriers on which the Austrian defense was based. It was, nevertheless, a magnificent exploit, the full significance of which can only be appreciated by those who know what a maze of military difficulties is ever associated with the barren limestone plateaus of the savage karst.

The Third Battle of the Isonzo

In the middle of May, 1917, Cadorna began another desperate effort to smash through the rock wall barriers which had so long defied the assaults of the Italian armies. The first rampart of the Bainsizza bastion carrying the peaks of Kuk, Monte Santo, San Gabriele, and San Daniele, the first sweeping some miles of the Isonzo trench with fire, the latter two dominating the Wippach curtain with their big guns, was one of the chief objectives. At the same time the frowning mass of the Hermada hills, blocking the coastal corridor route to Trieste, was to be assaulted and the enemy lines on the adjacent Carso to be pushed back toward the southeast. From the 12th to the 14th the whole front was deluged with heavy artillery fire, and on the latter date the infantry attacks were launched.

The first success was scored in the north, where the important height of Kuk was quickly carried by troops scaling the eastern wall of the Isonzo moat. When the Austrians were fully engaged with the task of meeting the menace to this part of their front, Cadorna struck in the south. Here the Austrian first lines were overrun and several villages captured. But against the Hermada bulwark (Fig. 148) the Italians beat in vain, and the force of their blow expended itself in destroying the advanced defenses organized on an almost impregnable terrain, and in checking violent Austrian counterattacks launched from strong positions. In the north little progress could be made from the peak on the rampart seized at Kuk. For a third time the natural barriers of the Isonzo had resisted the most violent assaults.

The Fourth Battle of the Isonzo

While pushing his operations at various points of the line Cadorna was organizing a new thrust on a scale which entitles it to rank as a separate great battle. Indeed, this action has been called the "Eleventh Battle of the Isonzo" by the Austrians and Germans, who separate the different phases of the offensives here grouped as three main battles. A crushing weight of Italian, French, and British artillery was massed in a gigantic attempt to smash the enemy's resistance, while squadrons of bombing airplanes carried consternation into the Austrians' back areas and disorganized their communications in the hope of overcoming the great natural advantages which the defense enjoyed on this difficult terrain.

On August 19 the attacking forces, crossing the Isonzo by a dozen or more temporary bridges built under fire, swarmed up the steep walls of the first rampart of the Bainsizza plateau and farther south hurled themselves once more against the natural fortress of the Hermada. For nearly a week a fierce struggle raged on the Bainsizza rampart before Monte Santo passed wholly into Italian hands. Along the wall to the southeast, and separated from Monte Santo by a deep notch cut by a transverse stream, San Gabriele resisted much longer, changing hands repeatedly before the attackers could call it securely their own. The southeastern end of the rampart still remained to be conquered. On the main plateau to the north the Italians pushed slowly forward to the deep Chiapovano trench, paying heavily for every foot of advance. Against the Hermada the offensive once more broke in vain.

At the cost of repeated offensives entailing heavy losses, the formidable terrain of the Isonzo battlefield was being conquered bit by bit. There still remained a long and difficult succession of strong defensive positions to be taken, but sufficient advance had been made to cause the Austrian command great anxiety and to lead it to demand that stiffening of the mediocre Austrian troops by German units which had already so often proved necessary. In the words of Ludendorff, "The Austro-Hungarian armies had indeed held their ground, but their losses in the Carso mountains had been so heavy and they were so shaken that the responsible military and political authorities of the Dual Monarchy were convinced that they would not be able to stand a continuation of the battle and a twelfth attack on the Isonzo. The Austrian army on the Italian front needed stiffening by German troops."[1]

The Battle of Caporetto

The German high command now planned with admirable skill a stroke designed to shatter the whole eastern Italian front and restore the morale of the Austrian armies. A repetition of the attempt to debouch from the Trentino and take the mass of the Italian armies in the rear was considered, but was rejected on the ground that the task was beyond the power of the forces Germany could spare from other fronts so long as the pressure on those fronts was perilously heavy. The last mountain wall defending the plain had already proved a barrier which could be overcome only with greatly superior forces and at a heavy cost. There was, however, another terrain which offered promise of a strategic success only second in magnitude to an irruption from the Trentino. On the upper Isonzo the Italian front turned from its south-to-north course and from near Tolmino ran northwestward along the Monte Nero ridge (Fig. 143) past Plezzo to gain the main ridge of the Carnic Alps. If this sector of the front could be broken between Tolmino and Plezzo, the Isonzo crossed east and west of Caporetto, and the passes through the Matajur-Cucco ridge on the opposite side of the valley captured, the assaulting forces could "roll like an avalanche" down the last slope of the Alps to the plain at Cividale, thus outflanking the whole Isonzo front.

It is true the terrain presented by the Isonzo moat and the mountain rampart beyond was most forbidding. In Ludendorff's opinion "the difficulties of the ground seemed almost insurmountable, and the communications on the Austrian side were as bad as could be imagined." Mountain artillery and pack transport had to be provided and the German artillerymen given

[1] Erich von Ludendorff: Ludendorff's Own Story, August, 1914–November, 1918: the Great War from the Siege of Liège to the Signing of the Armistice As Viewed from the Grand Headquarters of the German Army, 2 vols., New York, 1919; reference in Vol. 2, p. 94.

a long course of special training in shooting in mountainous districts. Alpine troops experienced in mountain warfare had to be chosen, and specially equipped and trained. For the concentration on the front of attack two mountain roads were available, so narrow in places that movement in one direction only was possible; hence the concentration and deployment of the troops would require a long time, increasing the danger of detection by the Italians.[2]

But there were compensating features. In the first place the Italians, paying insufficient heed to the maxim that "all terrains are passable if not defended . . . by active and valiant men," had presumed upon the natural strength of the position by entrusting its defense to weak forces of Territorials of poor quality. These troops were further weakened by a war weariness aggravated by Italy's long, costly, and unsuccessful efforts to break through the mountain barriers of the Trentino and Isonzo. Their morale had been dangerously lowered by the deplorable divisions which rent the Italian people behind the front, by the Italian government's neglect of the welfare of its troops, and by Cadorna's failure to check an enemy propaganda which fostered sedition and treachery within the ranks under his command. Furthermore, the mountainous terrain made observation of enemy back areas difficult, and in the deep valleys the German concentration could be more or less effectively concealed. Ludendorff states that the Italians discovered the movement, but that Cadorna apparently took no steps to meet the menace. Probably the extent of the danger was not easily observable.

On October 24, 1917, after a few hours' bombardment, the Austrian spear, tipped with its German point, was thrust deep into the Italian flank. The stroke was eminently successful. The corrupted and demoralized troops to which Cadorna rashly entrusted the safety of his Isonzo armies threw down their arms and fled or surrendered. With little opposition the Teutons poured through the Caporetto passes and down into the plain.

[2] Ludendorff, Vol. 2, pp. 98–113.

Threatened in its rear, the whole Isonzo front collapsed and the Isonzo armies began a hurried retreat.

It now began to appear that the Italian command had taken wholly inadequate measures to organize and fortify the natural defensive positions in the rear of the Isonzo. Not even the Tagliamento, that historic barrier against invasion from the east, seems to have been placed in a state of defense. With only brief checks at the successive obstacles which are provided by the parallel rivers flowing from the mountains to the plain and along which many Italian units fought magnificently to save the débris of their defeated armies, the enemy swept on, while forces in the Italian rear worked feverishly to prepare behind the Piave a position on which the retreat might be halted. When that line was reached in November the Austrians had taken over 200,000 prisoners and much of the Italian artillery and supplies; Cadorna had been retired; and the Italian nation had been shaken to its foundations. Only the recuperative powers of her armies, the patriotic devotion of her people, and the triumph of those among her leaders whose faith in ultimate Allied victory was still unimpaired, coupled with the loyal support of her Allies, saved Italy from the worst consequences of the disaster of Caporetto and enabled her to take ample revenge on another battlefield precisely one year later (see Chapter XII).

CHAPTER XV

THE BATTLEFIELD OF THE BALKANS: THE RANGE-AND-BASIN BATTLEFIELD

Near the head of the Adriatic there rise several small streams whose waters flow almost due eastward through the Save and Danube Rivers, to empty into the Black Sea. South of this west-to-east river trench, and separated by it from the open plains of Hungary and Rumania, lies the rudely triangular mass of complex mountainous country known as the Balkan Peninsula (Fig. 151). The geological history responsible for this mountain terrain is strikingly reflected in the topography of the country. A great series of rocks, including massive beds of limestone, were folded along northwest-southeast axes, and these folds deeply dissected by erosion—in places, indeed, worn down to an undulating surface of low elevation. Later the region was rent by great fractures, many of which trended more or less nearly in the same direction as the folds, and the blocks between the fractures were displaced and tilted, some being raised into mountain ranges, others dropped to form basin lowlands. Continued erosion dissected the mountains into wild and rugged forms (Fig. 153), while the eroded débris was spread out in the basins in the form of sloping alluvial fans (Fig. 154). Where the basins have not yet been sufficiently filled with débris, lakes remain or vast marshes in which the malaria-carrying mosquitoes breed.

Because the folding trended northwest-southeast, the mountain ranges eroded on the folds show a prevailing trend in that direction. Occasional cross folding gives cross ridges, but the general rule holds. Because many of the basins are produced by

NOTE—For Chapters XV and XVI the reader should constantly consult the general map of the Balkan Peninsula (Fig. 151), the sketch map of the Balkan battlefield (Fig. 152), and the special maps, Figs. 159, 160, and 165. Unless otherwise located, places mentioned in the text will be found on Fig. 151 or Fig. 152.

fracturing of the earth's crust, their borders and those of the rimming mountains are often remarkably straight and abrupt. The western margin of the great Monastir basin (Fig. 158) and the imposing straight front of the Belashitsa Range (Pl. XI, *A*, and Fig. 159) show types of topography repeated throughout the Balkans and with which the American becomes familiar in the Great Basin region of the United States. It is this "range-and-basin" topography which is the dominant element in the surface form of the country and which most affects the life of the people and the prosecution of military operations in the peninsula. The ranges are savage and inaccessible, often forested, and frequently the homes of wild mountain tribes whose poverty and independence are alike traceable to their mountain environment. It is in the basins, where broad alluvial plains offer fertile lands for agriculture (Fig. 155), that the people find an ampler return for their toil and form more prosperous communities.

"The mountain massifs, as well as the larger number of the great depressions which they shut in, naturally divide into a great number of little isolated units, into 'geographic regions,' or *zhupa*."[1] This isolation of the different basins by mountain barriers often prevents their serving as effective routes of travel, and only the fact that they are often connected by stream gorges or low passes cut through the rimming ranges gives them any considerable value in this respect. Thanks to such connecting links a series of basins are occasionally strung together to form an important lowland route across the mountainous land. Thus from the Skoplye (Üsküb) basin containing the Vardar to the basin of the lower Struma (Fig. 154) and the Gulf of Orfano there is a series of basins and passes which, before the construction of the railway down the lower Vardar valley, formed the great route of travel between the central and southern parts of the peninsula, between Serbia and the Byzantine countries.[2] It was only when the railroad made it easier to traverse the gorges

[1] Jovan Cvijić: La Péninsule Balkanique: Géographie humaine, Paris, 1918, p. 17.
[2] Cvijić, p. 21.

FIG. 150. Map showing the battle areas and trenches of the Balkan Peninsula.

connecting several basins on the lower Vardar, as well as the gorge between the Leskovats and Kumanovo basins, that the great Morava-Vardar route could be strictly adhered to throughout its whole length.

Today the famous Morava-Vardar corridor uniting Belgrade and Saloniki is not only the vital artery it has always been throughout history, but can easily be traversed without even local detours. A similar series of basins with connecting gorges and passes makes up the historic Morava-Maritsa corridor uniting Belgrade and Constantinople. These two great pathways constitute the only important lines of communication across the Balkan barrier and form the key to the military geography of the region. Our emphasis will therefore be placed upon these vital arteries. For them the bounding ranges on either side serve as natural protective barriers through which the more isolated basins offer indifferent passage.

Strategic Position of the Balkan Battlefield

It is obvious that a terrain which forms a mountain barrier just at the doorway from one continent to another, and which at the same time includes within its limits two of the chief pathways offering possibility of movement between the two land masses, must have a peculiar strategic importance. Here indeed is one of those vital spots on the earth's surface where to block the pathways and defend the barriers may be to turn aside the whole course of history.

Importance of the Morava-Maritsa Corridor

The full significance of the Morava-Maritsa corridor can be appreciated only in case we recall the important rôle it has always played in the history of the Nearer East. From all parts of Europe highways of travel converge southeastward toward the points where Occident and Orient touch hands at the Bosporus. Whether coming from the plains of the Po over the Pear Tree Pass, from western and central Europe along the upper Danube, or from farther north through the Moravian and other gaps to the Vienna gateway, travelers find the mass of the Balkans blocking the path to Constantinople and the East; just as in other days the hosts which invaded Europe from the lands of Asia Minor found in this same barrier an impediment to progress toward the northwest. Under these conditions it was inevitable that a continuous corridor cutting clear through the barrier from the plains of Hungary to the shores of the Bosporus should become a topographic feature of commanding historical importance.

Long before the time of the Romans the Morava-Maritsa depression had become a highway for peoples migrating east or west through the mountainous Balkan lands. In a later day one of the principal Roman military roads led from Belgrade through the trench to Constantinople. The great Slavonic flood which issued from the plains of northeastern Europe through the Moravian and Vienna gateways entered the Morava valley and, in the seventh century of our era, was flowing through the trench to surge about the walls of Adrianople. A few centuries more, and the mountain sides were echoing the shouts of the Crusaders who toiled along the same pathway to fight for the Holy Sepulcher. Back through the same defile came those hordes of conquering Turks who pushed the limits of their misrule to the very gates of Vienna. In our day a double line of steel rails has succeeded trail and military road, and the smoke of the Orient Express hangs low in the very valley where, centuries ago, dust clouds were raised by the passing of Roman legions, Crusading knights, or Turkish infantry. Here is the vital link in the great Berlin-to-Bagdad railway route, the channel through which German ambition hoped to reach the Far East, and the path by which the Teutonic powers had to send men and munitions to the hard-pressed Turks and bring back food to their own hungry people.

Importance of the Morava-Vardar Corridor

Second only to the Morava-Maritsa corridor in importance is the deep trench which cuts through the Balkans from north

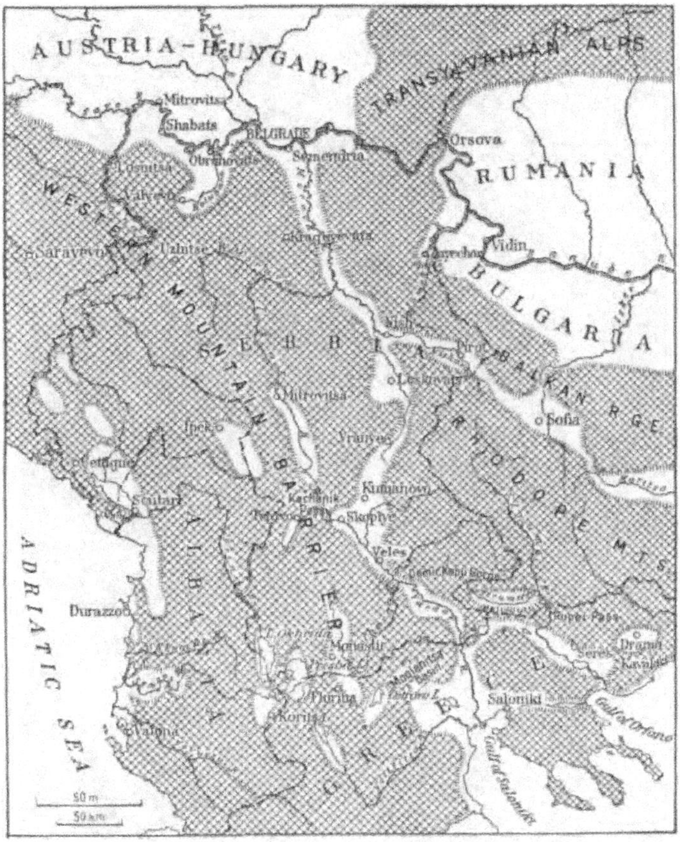

FIG. 152—Generalized sketch map of the Battlefield of the Balkans. Cross-ruled areas represent mountainous country, white areas the principal basins and valley trenches. See also the maps of the barriers and trenches (Fig. 151), of the Saloniki region (Fig. 159), of northwestern Serbia (Fig. 160), and of the Moglenitsa region (Fig. 165).

to south, connecting Belgrade with Saloniki. The Morava-Vardar depression does not lead to the land bridge uniting Europe with Asia Minor, but it does serve as a most important

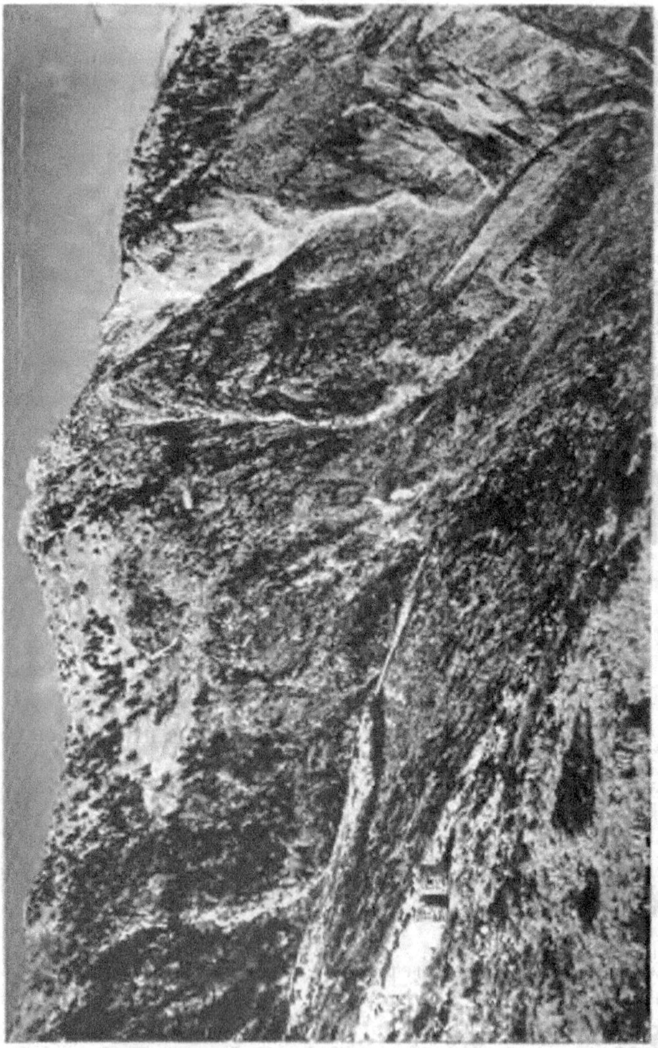

FIG. 153—The rugged limestone mountains of southern Albania, showing the steep descent into the Chimara basin south of Valona. Note the stone-wall defenses of the pass, which must serve in lieu of trenches on this rocky terrain.

outlet channel from the plains of Hungary to the Mediterranean Sea, and is one of the shortest routes from Central Europe to the Suez Canal. From southern Germany and the eastern Alps, the Carpathian foothills and the Alps of Transylvania, and from all of the great Hungarian basin, the valley routes lead straight to Belgrade, whence the Morava-Vardar valley cleaves a way through the mountains to the open waters beyond.

It is not without reason that the Morava-Vardar corridor has been called the key to the history of the Balkan Peninsula. Through it ebbed and flowed the tides of repeated invasions from the dawn of history. Under Roman dominion most of it was occupied by an important military road. Through it the Ostrogoths entered northern Greece in the fifth century of our era, while names still found on the map of Greece bear witness to the great Slav flood which, two centuries later, flowed through the corridor and overwhelmed the Greek peninsula. The story of the Serb race is largely the story of a struggle for control of this vital artery of communication. Austria's ambition to seize for her own uses a channel to the sea which should not open on the inclosed Adriatic was the mainspring of her reactionary policy in Balkan affairs. Bulgaria, realizing that the nation which dominates the Morava-Vardar depression must ultimately dominate the politics of the peninsula, precipitated the Second Balkan War in order to make good by force of arms her claim to a section of the corridor; and the same incentive played an important part in determining Bulgaria's alliance with the Teutonic powers in the World War.

The Corridors As Objects of Enemy Strategy

The immediate object of the Balkan campaign of 1915 was to secure for Germany complete control of the Morava-Maritsa corridor and the Orient railway which runs through it from Belgrade to Constantinople (Fig. 151). Roughly, one-third of the depression was in Turkish territory and therefore already subject to German supervision; one-third was in Bulgaria; and the remaining third in Serbia. German diplomacy set itself

FIG. 154—The flat floor of the Struma basin, exposed to fire controlled from the dominating heights of the Krusha Balkan, from the northern slope of which the view was taken; and from the more imposing wall of the Belashitsa Range, seen bounding the basin on the north. Note that the basin floor merges into sloping alluvial fans which lead gently upward to the base of the mountain wall. (Italian official photograph.)

the task of inducing Bulgaria to become an ally of the Central Powers, in order that the middle third of the Morava-Maritsa corridor might pass under German control without a contest and in order, further, that Bulgarian troops might bear the brunt of the fighting necessary to capture the remaining third from Serbian hands.

This was truly an ambitious plan, but certain considerations having a geographic basis made it possible for Germany to crown the program with success, and that with slight cost and incalculable profit to herself. The close of the Second Balkan War found Bulgaria not only bitter from the disastrous defeat with which her treachery to her allies had been punished, but suffering serious geographical disadvantages from the illogical boundaries forced upon her. Rumania's appropriation of the Dobrudja brought hostile territory close to Bulgaria's chief seaport, Varna, and also menaced the safety of the railway connecting with the port, since this line lies parallel to the new boundary and close to the frontier. The natural outlet for all central Bulgaria is to the Mediterranean by way of the lower Maritsa River; but the reconquest of Adrianople by the Turks led to a division of territory which forced Bulgarian goods en route downstream to the Bulgarian port of Dedeagach to cross through a small section of Turkey. The only other natural channel to the Mediterranean from Bulgarian lands was down the Struma valley to the port of Kavala; but Greece in her turn had insisted on a boundary which should leave the lower course of the river and the port in her hands, thus compelling Bulgarian commerce by this route to pass through Greek territory. Finally, Serbia obtained possession of that section of the Morava-Vardar corridor which Bulgaria had coveted, leaving to the latter no part of the key to future power in the Balkans. The opening of the present war thus found Bulgaria with a serious geographical grievance against every one of her neighbors. With coast lines bordering on two seas, every bit of her commerce, save only that with Russia, was forced to pass through hostile lands.

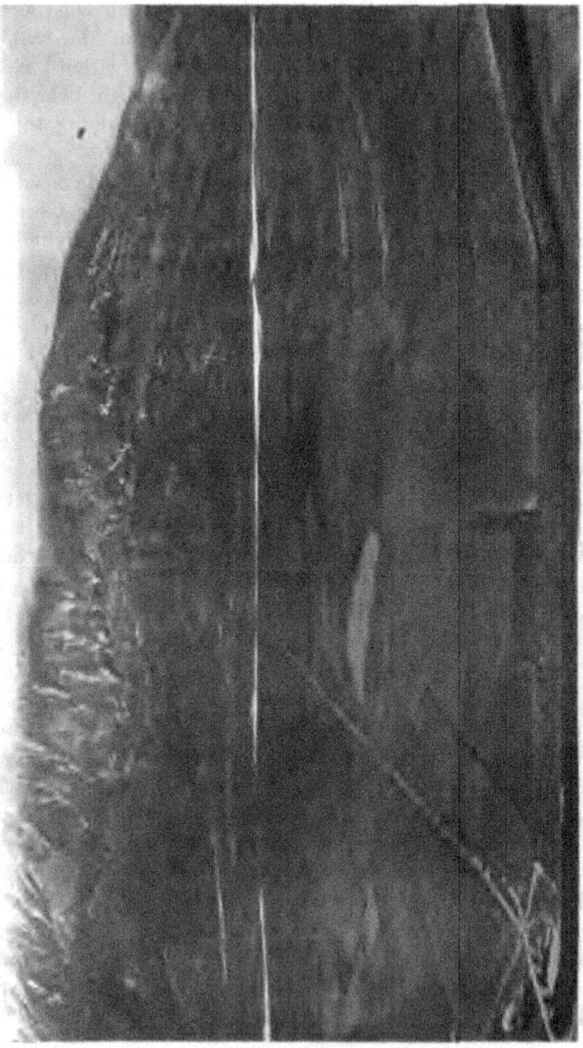

FIG. 155—The Argyrocastro basin, 50 miles southeast of Valona, and the range bounding it on the east. (Italian official photograph.)

Here was a fertile field for German diplomatic effort, and Bulgaria lent a willing ear to plans which promised immediate redress of past wrongs. Turkey was induced to return to Bulgaria the strip of land west of the lower Maritsa, thereby insuring to her a railway connection to her Mediterranean port lying wholly within her own boundaries. As a further reward for direct action against Serbia, Bulgaria should receive the coveted section of the Morava-Vardar corridor, the conquest of which would be rendered easy by Teutonic co-operation from the north. It was a bargain in valleys. In return for free use of the upper Maritsa valley and assistance in effecting the conquest of the Morava valley, Bulgaria was to receive a part of the lower Maritsa valley and a section of the Vardar valley. German diplomacy won, the geographic bargain was made, and from that moment there remained only the problem of forcibly seizing the Morava-Vardar corridor.

While conquest of the Morava valley and its continuation up the tributary Nishava was alone necessary to complete Teutonic possession of the Belgrade-Constantinople railway route, two considerations made a comprehensive campaign against the entire Morava-Vardar corridor essential. In the first place, as we have just seen, the Vardar valley had to be secured for political reasons, since its possession by Bulgaria constituted an essential part of the Teuton-Bulgar bargain. But military reasons also required its capture. It constituted the one effective line of communication leading to the Serbian armies defending the northern frontier. To cut it was to deprive those armies of reinforcements, munitions, and other supplies coming from the south. Furthermore, possession of the Morava-Maritsa corridor would never be secure so long as Serbia and her allies held the Vardar depression, for at any moment they might launch a bolt along this natural groove which would sever the Orient railway at Nish and thus undo all that had been accomplished through the new alliance with Bulgaria. For the Teuton-Bulgar forces the capture of the combined Morava and Vardar valleys thus became a single military problem.

After the establishment of the battle line across the narrow part of the Balkan Peninsula between Valona and Saloniki, this front acquired special strategic significance. It was henceforth the short barrier which on the one hand prevented a Teuton-Bulgar advance into old Greece and the utilization of the indented Greek coast for enemy submarine bases, and on the other hand prevented the Allies from sweeping in behind the Teutonic armies holding the Italian and Western fronts. When depletion of the enemy's reserves had reached the point where it was a question of his holding a short front or none at all in the Balkans, this sector was critically important—the key to swift and decisive victory. For if it could be broken, the enemy's inability to provide sufficient forces to hold the longer front farther north, made necessary by the form of the peninsula, would throw open to attack the vital bases and communications upon which the Western armies depended for their existence. The Battlefield of the Balkans became, in fact, an exposed flank of the Western front. While the line on this field was yet unbroken, although the war was in its last phase and hope of a German victory was gone, Ludendorff could say: "We still had reason to hope that we should maintain our position; in Italy and Macedonia our flanks and rear were covered." He fully realized that the defection of Bulgaria would "lay open Austria's flank, and therefore ours, to the Entente," and that "it made no difference whether our defeat came in Macedonia or in the West. . . . It was vital to do everything to secure our position in the Balkan Peninsula, in order to prevent the Entente moving into Hungary and making a flank attack on Germany and Austria." A fortnight after the short Balkan front was destroyed and the Teuton flank exposed, Von Hindenburg advised the German Chancellor that "as a result of the collapse of the Macedonian front, and of the weakening of our reserves in the West which this has necessitated, and in view of the impossibility of making good the very heavy losses of the last few days . . . the only right course is to give up the fight."[3]

[3] Ludendorff, Vol. 2, pp. 275, 353, 368, 370, 386.

Surface Features of the Balkan Peninsula

FORM OF THE MORAVA-MARITSA CORRIDOR

Let us examine for a moment the physical characteristics of the lowland corridors which have figured so prominently in the past history, both political and military, of southeastern Europe. The mouth of the Morava valley is widely open to the plains of Hungary, where the Morava River unites with the Danube some miles east of Belgrade. Southward up the river the valley narrows gradually, and the hills on either side rise to mountainous proportions; but as far up as Nish it is fairly broad, with a flat and sometimes marshy flood plain over which the river flows in a complicated meandering course, with occasional oxbow lakes and braided channels. Only at two points, where the river has probably cut through ridges of exceptionally resistant rock, does the valley narrow to a more youthful form and force the better roads to make long detours over the hills. There is usually ample room for a main road on each side of the river, while the railway crosses from one bank to the other in order to connect with the larger towns located on the valley floor (Fig. 156). The river is navigable halfway up to Nish, and throughout the entire distance the flood-plain soils yield rich harvests of maize and wheat.

From Nish the Morava-Maritsa depression leads southeastward up a branch stream called the Nishava, to a low divide within Bulgarian territory. The valley of the Nishava is more youthful than that of the Morava and is so narrow in places that the wagon road twice abandons it for a course across the mountains. The railway is able to follow it throughout, however, and in one place the valley widens into a broad basin on the floor of which lies the important town of Pirot. Here fortresses crowned the adjacent hills to guard against a Bulgar invasion of Serbia along this comparatively easy path.

After crossing the divide at Dragoman Pass, about 2,500 feet above sea level, both road and railway descend to the broad, fertile floor of the Sofia basin. As this trends northwest-southeast it is in line with the general course of the Morava-Maritsa corridor, although it drains to the northeast through a narrow outlet gorge into the Danube. At the southeastern end of the basin the low Vakarel Pass, but little higher than the Dragoman, is crossed, and road and railway easily reach the much larger basin drained by the Maritsa River and its tributaries.

The Maritsa takes a direct course toward Constantinople for more than 150 miles, then turns abruptly southward to the

FIG. 156—Wagon train passing through the flooded Morava valley, part of the great north-south corridor through the Balkan mountains. It was the cutting of this artery of communication by the Bulgars which assured the conquest of Serbia. (Copyright by Underwood and Underwood.)

Mediterranean Sea. At this sudden bend in the river stands the fortified city of Adrianople. Except for a short distance below the city, the Maritsa no longer serves as part of the great pathway to Constantinople but becomes a segment in the natural moat, consisting of the Tundja and lower Maritsa valleys, which in the past has repeatedly provided Constantinople with an admirable first line of defense against attack from the west. Above Adrianople the river is too frequently obstructed with sand bars to be of much use for navigation, but its broad basin carries the road and railway which follow the southern bank of the stream.

South of Adrianople the small Ergene River flows to the Maritsa from the east, and its valley offers a very gentle grade which the railway ascends till within a few miles of Constantinople.

FORM OF THE MORAVA-VARDAR CORRIDOR

The physical characteristics of the Morava valley as far up as Nish have already been discussed in connection with the Morava-Maritsa trench. From Nish southward to Leskovats,

FIG. 157—Austrian forces assembled along the Save River barrier, ready to attempt the invasion of Serbia. The breadth and volume of the river make the construction of temporary bridges a difficult task. (Copyright by Underwood and Underwood.)

road and railway traverse one of the open intermontane basins which so frequently occur in the midst of the Balkan ranges; but farther south the stream flows from a youthful gorge which continues up the river for ten or twenty miles before the valley again broadens out to a somewhat more mature form. Just north of Kumanovo lies the divide between the Morava and Vardar drainage, a low, inconspicuous water-parting some 1,500 feet above sea level, located in the bottom of the continuous, through-going depression and placing no serious difficulties in the way of railroad construction.

South of Kumanovo the valley broadens into a triangular basin, near the three corners of which stand Kumanovo, Skoplye, and Veles. The main Vardar River enters the basin from the west, flowing out again at the south through a narrow, winding valley which carries the railway but no good wagon road. At Demir Kapu the valley narrows to an almost impassable gorge for a distance of several miles but soon broadens again to form another basin or group of open flats, near the southern end of which lies Gyevgyeli (Fig. 159). South of this town the river once more enters a narrow gorge, the Gipsy Pass, cut through the Gyevgyeli plateau, thence to issue into a wide basin or flat-floored valley in which the river follows a braided and occasionally meandering channel to the sea. The lower course of the Vardar lies in a very broad, marshy plain terminating in the delta southwest of Saloniki. The special strategic importance of the triangular basin near Skoplye and of the Demir Kapu and Gipsy gorges will appear in subsequent pages.

While the Morava River is navigable for small boats from the mouth halfway up to Nish, the upper Vardar is too full of rapids and its lower course too full of sand bars to make river traffic practicable. The strategic value of the Morava-Vardar corridor, like that of the Morava-Maritsa, lies in the fact that, notwithstanding it occasionally narrows to gorgelike proportions, it gives an unbroken channelway clear through a rugged mountain barrier.

Let us examine next the physiographic features which serve as natural defenses of the important Morava-Maritsa and Morava-Vardar corridors.

Natural Defenses of the Two Corridors

The Morava valley is widely open to the north and is there bounded on both sides by comparatively low hills. An enemy securing a foothold in the rolling country to the east or west could enter from either of these directions as well as from the north, just as the Orient railway coming from Belgrade enters the valley from the west, 25 miles above its mouth. Hence an effective barrier against attack from the north must cover more than the actual breadth of the northern entrance to the valley. Such a barrier is provided by the natural moat of the Save and Danube Rivers, which protects the entire northern frontier of Serbia, and by the hills south of the moat which, as one progresses southward, rise into a wild, mountainous highland.

THE SAVE-DANUBE BARRIER

The Save is a large river swinging in great meanders across a broad, marshy flood plain (Fig. 157). The extensive swamp lands on either side of the river are difficult to traverse at any time, while the flood waters which spread over the lowland in spring and autumn often make the barrier quite impassable except at one or two points. South of Mitrovitsa and west of Shabats the marshy peninsula between the Drina and the Save is called the Machva (Fig. 160) and is famous for its inhospitable character. In volume the Save is of sufficient size to constitute an obstacle against invasion, but for navigation it suffers from its overlong meandering course and from frequent shifting of channels and sand bars. At no point is the stream fordable, and at Belgrade alone is it crossed by a bridge.

The Danube is a river of imposing volume, in places from one to several miles wide. Its value as a defense against invasion is very great, notwithstanding that the numerous islands which mark its braided course from Belgrade east to the Iron Gate gorge offer some advantages for a crossing by dividing the width into several parts easier to negotiate by pontoons. It is unfordable and unbridged. East of the braided section the river exchanges its open valley for a narrow, winding gorge which cuts through a mountainous upland and reaches its most imposing aspect at the Iron Gate near Orsova. The walls of the gorge, sometimes forest-clad, sometimes bare rock, are exceedingly steep; while the mighty volume of water constricted within its narrower channel gives a river which is both swift and deep. To cross such a barrier in the face of enemy fire would severely test the abilities of the best-trained soldiery.

It is not strange that so impressive a natural obstacle as the Save-Danube valley should have served for centuries as a bulwark against invasion of the Balkan Peninsula from the north, nor that it should long have been the physical barrier separating the dominions of the Sultan from Austrian lands. In combination with the difficult hill country to the south, the great natural moat furnished the Serbians with an admirable defensive screen, in attempting to pierce which the Teutonic armies suffered more than one costly defeat.

THE BALKAN AND RHODOPE MOUNTAINS

The natural defenses of the Morava-Maritsa corridor have only an academic interest in so far as the World War is concerned, since the greater part of its length lay securely within the territory of Germany's allies. It was never directly attacked, nor was it in serious danger of attack from the north at any time. From the south it was threatened only after the collapse of Bulgaria in the last days of the war. Had the Rumanian campaign been more wisely conducted and had it received proper Russian support, the great Balkan Range which defends the corridor on the north from near the Iron Gate of the Danube at the west to the Black Sea on the east, would have played an important rôle. This imposing wall is cut through by two rivers, the Timok and the Isker, each of which is followed by a minor railway, while a third railway crosses through a sag in the crest farther east. On the other hand, had the Saloniki army been able to undertake a large-scale offensive before the disintegration of the Bulgarian armies rendered them incapable of defending any terrain, the formidable mass of the Rhodope Mountains, through which the Struma River alone has cut a practicable gateway, must have bulked large in the military operations. As it was, they served to discourage any offensive on the part of forces which were already handicapped by the fear of Greek treachery in their rear.

Throughout its entire length the Morava-Vardar corridor is protected on the east by a complex of mountain ridges representing the western end of the great Rhodope massif and the westernmost members of the impressive Balkan Range. All of these mountains appear to have reached a mature stage of dissection, in which the maximum degree of ruggedness is attained. A maze of steep-sided ridges and peaks rises from one to several thousand feet above the bottoms of narrow valleys, while at the north the mountain barrier is reinforced by the gorge of the lower Timok River and a short section of the Danube valley. Much of this difficult country is forested, and no part of it could be crossed with ease by a hostile army.

The Vital Passes

There are, nevertheless, certain pathways through the eastern barrier which may be forced by a foe possessing superior numbers. Chief among these is the segment of the great Morava-Maritsa corridor carved by the Nishava River, which stream rises in what was before the war Bulgarian territory and flows directly through the barrier into the Morava-Vardar trench at the critically important junction near Nish. To stop this gap the fortifications of Pirot just inside the Serbian border were constructed. Zayechar, another fortified town farther north, guards the common entrance to branches of the upper Timok valleys, through which hostile forces might ascend to passes whence the drop into the Morava valley is readily effected. The Vlasina, Kriva, and Bregalnitsa Rivers, rising at or near the Serbo-Bulgarian boundary on the crest of the main range southeast of Nish and flowing westward to the Morava and the Vardar, lead to the corridor at Leskovats basin, at Kumanovo, and in the Veles region. Finally, the broadly open Strumitsa valley, formerly mainly in Bulgarian territory but heading close to the lower Vardar, afforded access to several passes from which it was but a few hours' march to the Vardar valley either above or below the Demir Kapu gorge.

It appears, therefore, that despite the protection afforded by difficult mountainous country east of the Morava-Vardar line, the corridor was open to attack at a number of critical points, provided the invading forces were sufficiently large to overwhelm resistance and drive their columns through the narrow valleys. This danger was the more acute because along much of the eastern frontier Bulgarian territory reached to the crest of the mountain barrier and in some places even beyond the crest to the western, or Serbian, slope. It should be noted, furthermore, that the hostile territory flanked the Morava-Vardar corridor throughout practically its entire length, usually lying not more than fifty miles distant, while near Vranye and just north of the Greek border westward protrusions of the Bulgarian frontier reduced the distance to a dozen miles or less. The largest and most vital artery carrying the life blood of Serbia lay dangerously near the surface, and a single stab of the Bulgarian knife might prove fatal.

THE WESTERN MOUNTAINS

West of the Morava-Vardar corridor the threat of danger was less imminent, and the natural protective screen more effective. Although Bosnia and Hertsegovina were in Austrian hands, the people were more or less hostile to their new rulers and favorably disposed toward the Serbs. Montenegro was Serbia's ally, while uncertain Albania was not an important factor in any event. Across the Adriatic lay Italy, one of the Allied powers. Only at the north, then, was there danger of an attack upon the Morava-Vardar line from the west; while farther south succor from friends, rather than attacks from enemies, was to be expected from the direction of the Adriatic.

The broad belt of mountains lying between the Morava-Vardar depression and the Adriatic shore is one of the most imposing topographic barriers in Europe. From the earliest times it has stood as an almost impassable wall cutting off the people of central Serbia from all effective intercourse with the inhabitants of the Italian peninsula. In the Middle Ages, Ragusa and other Slavonic cities on the Adriatic coast, although part of a Serbian province and the home of a flourishing school of Serbian literature, found communication with the interior so difficult and with Italy so easy that they came under Venetian instead of Serbian control. The same mountain wall which so long prevented extension of Serbian power westward to the sea likewise served for centuries as an effective barrier against the eastward migration of Western European civilization into the dominion of the Turks. In 1914 no railroad had crossed the barrier to unite the great valley of central Serbia south of Belgrade with the sea.

Included in the mountainous belt are ranges high enough to carry snow caps until the month of August, and the name "Albania" is believed by some to have its origin in the snowy appearance of that wild region. It is said that the "accursed mountains" of northern Albania and eastern Montenegro include some of the least explored lands of all Europe. Just as the mountains of Wales and the Highlands of Scotland preserve languages and customs which have been driven from the open country of England, so the fastnesses of the Albanian hills have kept alive a difficult language that is older than classical Greek and customs which render the rude inhabitants of the country a picturesque subject for study. The conquering arm of the Turk reduced the Bulgarian inhabitants of open plains to complete subjection within a comparatively short time; but a century and a quarter was required to secure a less firm hold upon the mountainous lands of Serbia, while the inaccessible wilds of Albania and Montenegro were never completely subjected to Turkish power. Montenegro was the last Serbian stronghold to yield to Turkish supremacy and the first to regain complete independence.

The physical characteristics of a belt of country so difficult to traverse deserve a word of further description. In the north the mountains, eroded on earth folds of the Appalachian type and trending northwest-southeast parallel to the northern Adriatic coast, rise from 5,000 to 8,000 feet above sea level in the higher ranges. Between the hard rock ridges streams have excavated parallel valleys on the weaker beds, but these valleys are of little real service to man since they lie at right angles

FIG. 158.—Flat, desert floor of the Monastir basin and the dissected mountain block bounding it on the west. The town of Monastir is seen at the base of the range to the left. Enemy positions on the exposed basin floor were outflanked by the Allied advance in 1916 along the ranges on the east (see Fig. 164) and west. (Italian official photograph.)

to the natural course of his movements between coast and interior. Farther south the rock structure is more complex, and the mountain ridges produced by erosion accordingly of more complicated pattern.

The Karst Region

Among the rocks involved in the mountain building, limestone is a conspicuous element, and its soluble nature has imposed a peculiarly forbidding aspect on the topography. Most of the rainfall passes underground through sink holes and smaller solution cavities and then finds its way through subterranean channels to a few principal rivers, lakes, or the sea. As a consequence much of the mountain country is dry and barren, springs are far apart, and the open watercourses difficult of access because deeply intrenched in rock-walled gorges. The "gaunt, naked rocks of the cruel karst country" are not only themselves of little value to mankind, but they render inaccessible and therefore comparatively useless many excellent harbors on the eastern coast of the Adriatic. Hence the vital importance of Fiume, located where the barrier is narrow and easily crossed near the head of the Adriatic, as an economic outlet for the Yugoslav people.

Because the limestones are purer and more abundant along the coastal border we find that the karst topography is there best developed. Farther inland the maze of hills is occasionally broken by one of the intermontane basins, the center of whose broad floor may be covered by marshland or lake while throughout its remaining portion the fertile soils derived from impure limestone and other rocks yield good returns to the cultivator. Among the largest of the basins (Fig. 152) are those in which Monastir and Ipek are located; the Tetovo basin, west of Skoplye (Üsküb), where a branch of the Vardar takes its rise; and the famous Kosovo Polye, or Plain of the Blackbirds, southeast of Mitrovitsa on the Ibar, where in its last great effort against the advancing Turk the Serbian army suffered defeat in 1389. It is largely to these areas that one must credit such measure of prosperity as is vouchsafed the dwellers of this western mountain barrier; but absence of connecting lowlands makes the basins of small service in expediting travel across the region.

It is true that certain rivers cut through the mountain ranges to reach the sea; but not one of these has carved a valley suitable to serve as a highway between the coast and the central Morava-Vardar corridor. For the most part the cross valleys are narrow and deep, and bounded by the steep, rocky walls characteristic of young gorges cut in limestone. Falls and rapids are frequent, and the headwaters usually end in a maze of ridges some distance west of the central depression. The valley of the Narenta carries a narrow-gauge railway through the mountains of Bosnia and Hertsegovina to a pass across which Sarayevo and the valley of the Save are accessible; but in 1914 the only branch line running east to the Serbian border terminated in the vicinity of Vishegrad, while the nearest railway terminus of the Serbian system was more than twenty miles across the mountains at Uzhitse, at the head of a branch of the Western Morava valley. Through the gorge of the middle Narenta the course of the railway is difficult, and the crossing of the pass is made possible only by using a rack-and-pinion arrangement, which indicates the unsatisfactory character of the route for commercial purposes. The next river of importance to the south is the Drin, which reaches the sea near Scutari; but it flows in a gorge so wild and deep that the poor trails of the district often leave it for a course across the barren hills. When a column of Serbian troops successfully negotiated this defile during the First Balkan War, the feat was hailed as a great military accomplishment. The Shkumbi valley offers an entrance from Durazzo to the rail end at Monastir, but traffic by this route must cross three mountain passes. A famous Roman road, the Via Egnatia, followed this valley; and the only other two important roads to cross the barrier in Roman times had their locations determined by the Narenta and the Drin, although in each case the stream gorge was abandoned in places for a more feasible course over the uplands. Of these former roads little remains today except rugged mule paths. From the standpoint of military geography the broad mountain belt west of the Morava-Vardar corridor is practically impassable.

There are within this western mountain belt three depressions which have relatively little value as parts of cross routes to the sea but which we must keep in mind if we are fully to understand certain aspects of the campaign against Serbia. First among these is the open Kolubara valley, at the head of which stands Valyevo (Fig. 160). A small railway of some military value traverses the valley and connects the town with the Save River. Directly south across the Malyen ridge, the Western Morava valley heads near Uzhitse and runs east to join the main trench. The Western Morava River is a mature stream meandering on a flat flood plain of considerable breadth and is bordered by a narrow-gauge railway connecting Uzhitse with the Orient Express line. Finally, the Kosovo Polye, already mentioned, forms part of a subsidiary lowland parallel to the main Morava depression (Fig. 151). Northwestward the basin is replaced by the long, narrow, winding gorge of the Ibar, which unites with the Western Morava but which is not followed throughout its length by so much as a good wagon road. To the southeast the basin is drained by the Lepenitsa River, which flows through a narrow outlet gorge at Kachanik, the so-called Kachanik Pass, to unite with the Vardar at Skoplye. An important railway leaves the Nish-Saloniki line at Skoplye and runs through the Kachanik gorge and Kosovo Polye to Mitrovitsa on the Ibar.

The Entrenched Camp of Saloniki

The rôle which the great Allied base at Saloniki played in the World War justifies particular attention to the remarkable terrain which offered, ready prepared by Nature, a practically impregnable position for an entrenched camp. The high defensive value of this remarkable terrain is due to an unusually pronounced development of the "range-and-basin" topography which we have discussed on earlier pages. Roughly speaking, two raised and tilted blocks of strongly folded crystalline lime-

FIG. 159—Map showing the natural defenses of the armed camp of Saloniki.

stone, schists, and granite forming mountain ranges, and two down-dropped strips forming long, comparatively narrow basins floored with alluvial deposits, all showing the prevailing northwest-southeast trend, form the key to the defensive system which defied the Central Powers for several years and prevented them from driving the Allies into the sea.

The Inner Wall

The southernmost block slopes gradually upward from the three-pronged Chalkidike Peninsula (Fig. 152) to fall off more abruptly into the basin on the north. Saloniki lies at the southwest base of this first block near its western end, where it is called the Khortach Range (Fig. 159). This partially wooded but arid upland is broken by minor fractures into subsidiary blocks and basins, which we need not consider here, and is dissected by stream erosion into a maze of hills offering good defensive positions. Just north of Saloniki there is a low gap in the range called the Dervend col, or pass, which is in line with a similar depression in the next block farther north (the Lahana saddle mentioned below) and with the general direction of the Struma River before its abrupt turn southeastward toward the Gulf of Orfano. In his admirable description of the geography of this portion of Macedonia, Captain A. G. Ogilvie has suggested that the Struma may formerly have continued southward to the Gulf of Saloniki and so have cut these notches before being deflected to its present course.[4] However that may be, the two passes determine the path of the main road from Saloniki to Seres and the only crossing of importance over the Khortach Range east of the Saloniki-Doiran railway.

North of this first mountain block is the long basin (Fig. 159). from two to six miles wide, stretching from the Vardar River eastward nearly to the Gulf of Orfano, a distance of some sixty miles. The floor of this great natural moat, usually barren of trees and well exposed to fire directed from the mountain wall to the south,

[4] A. G. Ogilvie: A Contribution to the Geography of Macedonia, *Geogr. Journ.*, Vol. 55, 1920, pp. 1–34.

is further defended by an almost continuous belt of lakes, marshes, and marshy rivers occupying the depression. Langadha and Beshik Lakes with their marshy shores are of themselves formidable obstacles, and passage across or between them can easily be prevented by fortifying the northward projecting bastion of the Ogurli plateau and the northern hills of the Khortach Range. Westward a smaller lake and several minor streams, together with their bordering marshes, present less noteworthy obstacles which help to make the lowland trench, when swept by fire from the heights to the south, an almost impassable barrier.

It was on the southern mountain block, fronted by the natural moat of the Langadha-Beshik basin, that the Allies constructed their inner trench system, adapted to defense by the small forces originally available and humorously called by the British the "bird-cage lines." The front line trenches were usually placed at the northern base of the mountain wall so as to sweep the smooth alluvial slopes, descending gently toward the center of the depression, with a grazing machine-gun fire. Barbed wire defenses were placed on the alluvial plain to retard the enemy and keep him subjected to this deadly fire as long as possible. A second system of trenches farther back and higher up on the face of the range commanded a broader sweep of the plain with excellent observation and artillery control. The third position consisted of a series of strong redoubts along the crest, with subsidiary intervening field works. The smooth glacis of the alluvial slope lent itself admirably to machine-gun defense, upon which the strongest reliance was placed. On the west the Vardar River and its broad marshes carried the defensive system southward to the Gulf of Saloniki, where the Allied fleet could hold the approaches to the "bird-cage lines" under the fire of its guns. Thanks to the natural features of the terrain the Allies possessed in this inner defensive system a well-nigh impregnable position.

The Outer Wall

The same may be said of the outer system, prepared and occupied when sufficient troops were available to hold the longer lines.

Northeast of the Langadha-Beshik basin the next block rises, gradually in its western half where it forms the Krusha Balkan, more steeply farther east in the Beshik Range, to drop off by a slope that is often precipitous into the Struma basin (Fig. 159). As in the case of the southern block, the Krusha-Beshik mass is in places broken into minor ranges and basins; but the general form of the tilted block remains fairly distinct. The traveler crossing the Krusha Balkan from the southwest is profoundly impressed by the long and gradual character of the ascent and astonished at the abruptness of the northeastern descent. Here and there along the northeastern scarp he may see a broad bench or terrace part way down the slope, apparently a slice of the broken mass which has dropped part way down the fracture face. A dense growth of brush covers this steep face of the range and, by limiting an enemy advance to the roads cut through it, simplifies the problem of defense. Streams have deeply trenched the more gentle backslope, dissecting the undulating surface into a maze of hills across which movement is difficult and confined to a very few roads supplemented by mule paths. Seen from the south these hills appear arid and barren; but the northern declivities of the higher ones near the crest are, like the steep main scarp, clothed with a dense cover of brush. The Lahana saddle between the higher Krusha and Beshik sectors of the range gives passage to the Saloniki-Seres road, the only important route which crosses it between Lake Doiran and the sea.

Like a great moat fronting the Krusha-Beshik range on the northeast, the Doiran-Struma basin stretches from Lake Doiran to the Gulf of Orfano. Seventy-five miles in length and varying in width from three to a dozen miles or more, this down-dropped strip of the earth's crust is one of the most striking topographic features in Macedonia (Fig. 159). The beautiful alluvial fans spreading out across it from the high Belashitsa block mountain (Pl. XI, A) and the ranges to the southeast have forced the Butkova and Struma Rivers over toward the southwestern side of the trench, where they form, together with the vast marshes and interlacing channels bordering them, a serious military obstacle dominated by the Krusha-Beshik heights. Lakes Ahinos and Butkova, bordered by marshes which vary in extent with the changing seasons, increase the strength of the barrier. For an enemy to traverse several miles of smooth, open plain in full view of the defenders on the southwestern wall of the great trench; to press on under accurately directed artillery fire from guns concealed in the ravines of the brush-covered scarp and in the valleys of the backslope; to force the crossings of rivers and marshes under this same accurate fire; to scale the steep wall of the Krusha-Beshik range where the brush largely restricts movement to roads and paths accurately registered for artillery fire and further controlled by rifle and machine-gun fire from numberless positions hidden in the dense growth; and to do all this without adequate artillery support, since the absence of concealed artillery positions on the open plain compelled the attacking forces to rely largely on guns hidden in the ravines dissecting the foothills and alluvail fans on the far northern side of the basin—this was indeed an almost impossible task. Once sufficient troops were available properly to man this natural fortress, the Krusha-Beshik wall and the Doiran-Struma moat were justly considered impregnable.

Gyevgyeli Plateau

Between Lake Doiran and the Gipsy Pass gorge of the Vardar River below Gyevgyeli stretches a broad upland usually referred to as a "plateau," upon the northeastern end of which, near Lake Doiran, rises a high ridge commanding both the plateau to the southwest and the lake basin to the northeast (Fig. 159). Possession of the Gyevgyeli plateau was essential to the Allies if they were to prevent an enemy advance down the Vardar and if their front along the Doiran-Struma moat was not to be outflanked. On the other hand, the enemy saw in the Gyevgyeli plateau the one sector of the natural defenses of Saloniki where a successful attack would achieve a maximum of advantage, since to pass it would be to open up the way to an advance into the areas behind the main Allied front from Lake Doiran to the Gulf of Orfano. Thus the plateau and its dominating ridge were in a sense the key to the outer defenses of Saloniki and witnessed some of the most bitter fighting on the Balkan front.

Southward from the Gipsy Pass the Vardar River with its multiple channels, crossing the open plain, or *campagna*, of its last basin, further reinforced by Lake Ardjan and its bordering marshes on the east and by the outlying hills and parallel marshy streams to the west and farther downstream by the vast, marshy delta of the Vardar, continues the defensive barrier to the sea. The strength of the complete system of natural positions is too obvious to require further emphasis. Behind such barriers so great an army as was finally assembled here could withstand the best that the Kaiser and his allies could send against them. The enemy repeatedly attacked the outer wall in vain and camped for months along the Doiran-Struma trench with his own lines securely based on the impregnable wall of the Belashitsa Range and adjacent mountains.

MALARIAL LOWLANDS

In closing this account of the natural features of the Balkan Peninsula reference should be made to one common difficulty encountered by both combatants in this range-and-basin country. The stagnant waters of the shallow lakes and marshes abounding on the basin floors, in a region where the temperature conditions are just right for the breeding of the malaria-bearing mosquito *Anopheles*, make southern Macedonia "one of the most malarious districts in the world." Captain Ogilvie reports, on the basis of data supplied by the Medical Directorate of the British War Office, that 70 per cent of the children in one village examined in the Struma basin were found to have enlarged spleens. "The effect of the disease upon a large European army hitherto untouched by it was naturally very great . . . and was exceedingly serious from a military point of view." In the years 1916–1918 the total admissions to hospital for malaria in the British contingent alone numbered 161,559 and in the last year represented 371.6 admissions for every 1,000 men engaged.[5]

[5] Ogilvie, A Contribution to the Geography of Macedonia, p. 17.

The present writer noted with interest the warning appearing on the menu cards of officers' messes, clubs, hotels, and restaurants in different parts of the Balkans, often repeated in four or five languages: "Don't forget your quinine," and found the quinine bottle classed with the salt and pepper shakers as a necessary part of the table equipment. Since the malaria is much worse in the marshy lowlands than on the hills, the British maintained on their Struma front a "summer line" on the hills, to which they withdrew when the disease was at its worst in the warm season, and a "winter line" down on the plain, which they could occupy with impunity when the malaria-bearing mosquitoes were hibernating. In summer this "winter line" in the plain would be held by relatively immune Greek troops.

CHAPTER XVI

MILITARY OPERATIONS ON THE BATTLEFIELD OF THE BALKANS

The Battle of the Frontier

The first blow in the World War was struck in the last days of July, 1914, when Austria launched an offensive along the entire Save-Danube line defending Serbia's northern frontier (Fig. 160). The Serbians destroyed the great bridge over the Save at Belgrade to make the barrier more secure and assailed with vigor every enemy column which endeavored to cross the river by boats or pontoon bridges. For nearly two weeks the Austrians made repeated attempts at seven different points to reach the south bank and at the same time attacked the western frontier along the line of the Drina near Losnitsa and Vishegrad. At Belgrade a crossing in the shelter of the ruined bridge was only temporarily successful. Farther east, at Semendria, an island served as the base for crossing on a pontoon bridge where the channel narrowed to 200 yards; but the invaders were first held in check, then thrown back in defeat. All attempts to cross at Obrenovats, southwest of Belgrade, failed. Far to the west Austrian troops succeeded in forcing a passage at Mitrovitsa and for some days held their ground in the marshes on the south side of the stream; while the Drina was crossed at Vishegrad. Even here the success was partial and temporary, for Vishegrad was retaken by the Serbs on August 7, and on the 10th the Serbian government reported the expulsion of the last Austrian from Serb territory. The first attempt to force the northern barrier had ended in failure.

The Battle of the Yadar

About the middle of August the Austrians made a second attempt to invade Serbia. Realizing the strength of the Save-

Danube barrier, the Austrian commander delivered his main attack from the west against the line of the Drina. This river cuts transversely across the northwest-southeast parallel mountain ridges of the region, and by taking these natural defenses in the flank the invaders hoped to push along the lowlands between them and turn the strong northern positions from the southwest. Two of these ridges and the intervening depression constituted the keys to military control of this extreme northwestern corner of Serbia (Fig. 160). The depression is formed by the valley of the Yadar River, which heads near the upper Kolubara and flows northwestward to the Drina; and by the upper Kolubara valley eastward to beyond Valyevo. Northeast of this trough lies the high ridge bearing Mount Tser, which not only commands the Yadar valley south of it but likewise the valley of the Danube and the Machva marshes from its northern slopes. Southwest of the trough lies a range known in its southeastern part as Malyen Ridge, which in turn continues eastward to Mount Rudnik, due south of Belgrade. This long range dominates from the south both the Yadar and upper Kolubara valleys, including the important Valyevo basin.

If the Austrians could cross the Drina barrier near Losnitsa and push up the Yadar valley, sweeping the ridges on either side as they advanced, they might continue over the cross-divide into the upper Kolubara and capture the important Serbian military base of Valyevo. This would effectively turn the line of the Save and threaten Belgrade from the south. A simultaneous advance southward from Shabats against the Mount Tser ridge would facilitate the maneuver, and the junction of the Shabats and Losnitsa columns would clear the northwestern corner of Serbia of Serb troops.

After a furious bombardment of the Save-Danube line superior Austrian forces effected crossings of the Save at Shabats and the Drina at Losnitsa. While this major operation was being carried out, strong demonstrations at other points were attempted. Columns seeking to cross at Belgrade were defeated. In the gorge of the Iron Gate at Orsova, where the swift current and steep walls made the attempt peculiarly hazardous, it is reported that three Austrian regiments were destroyed while trying to cross by a pontoon bridge. Renewed attempts to cross at Belgrade and Semendria were frustrated. Belgrade stood on the point of a peninsula projecting into Hungarian territory and was

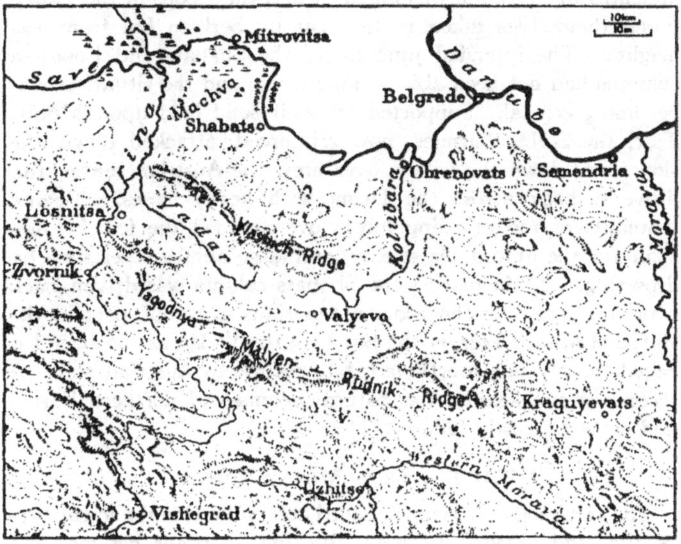

FIG. 160—Map of northwestern Serbia, showing the terrain on which were fought the Battles of the Yadar and of Mount Rudnik.

subject to attack from three sides. It was the capital of Serbia, and its capture was urgently desired for political as well as strategic reasons. That this important outpost at the very door of the enemy's country, attacked by superior numbers and bombarded by superior artillery, should have resisted capture for four months, is a striking proof of the defensive value of such barriers as the Save and Danube Rivers.

Meanwhile the main Austrian forces were pushing southeastward up the Yadar valley, while the Shabats column was slowly driving the Serbian right wing in this sector back against the slopes of the Mount Tser ridge. But the task was a heavy one, and the strong Serbian positions on Mount Tser were causing the Austrians no end of trouble. By August 20 they had advanced nearly to the head of the Yadar, but, as their progress along the difficult ridge position was much slower, the troops in the valley found themselves taken in the rear by Serbian fire from the heights. The intended junction of the Shabats and Losnitsa columns had not been able to take place, and the situation was becoming critical. Supported by their solid hold upon Mount Tser, the Serbian armies, now reinforced, attacked down the slopes and down the valley, overthrew the Austrian center and drove it back across the Drina with heavy losses. Fleeing remnants of the invading forces overcrowded the few bridges spanning the unfordable stream, and large numbers perished by drowning. By August 24 the Shabats column was driven back over the Save. Of the 200,000 Austrians who took part in the Battle of Mount Tser, or the Battle of the Yadar, as the action is called, nearly one fourth failed to return to Austrian soil. A second attack against the natural defenses of northern Serbia had proved a costly failure.

The Battle of Mount Rudnik

The victorious Serbs now pushed across the Danube to capture Semlin, and invaded Bosnia. But the Austrians had not yet given up the idea of chastising Serbia and taking from her control the vital arteries of the Balkans, and the Serbs were now compelled to fall back on their natural defenses to meet a third and still more powerful Austrian offensive. Again the line of the Drina was heavily attacked in the second week of September, while simultaneously a strong offensive developed against the Save-Danube front. The enemy was quickly halted along the line of the Save, and an Austrian army which succeeded in getting a foothold on the southern bank of the Danube east of Belgrade was overwhelmed and driven back in disorder.

On the west, however, the smaller Drina was successfully

passed on a wide front, crossings being effected at Losnitsa and at a number of other points between the junction with the Save on the north and Vishegrad far to the south. Slowly and methodically the new advance began, parallel to the grain of the country. Strong enemy forces drove the defenders back along the ridges and pushed up the Yadar and other parallel valleys. By the middle of November the Austrians had crossed the minor divide at the head of the Yadar, pushed into the upper Kolubara depression, and captured Valyevo. Four days later they had pushed eastward along Malyen Ridge well toward Mount Rudnik, and farther north had forced the Serbian armies to take refuge behind the lower Kolubara valley. The Serbian commander, unable to defend both the northern and western fronts with a small army insufficiently supplied with munitions, ordered his forces to fall back on the strong mountain mass which on the west defended the main Morava valley and which was dominated by Mount Rudnik. His right wing now rested on the Danube east of Belgrade, and his left blocked the route leading from Uzhitse eastward down the Western Morava valley to the Morava-Vardar corridor. This alignment of the Serbian forces involved the evacuation of Belgrade, which the Austrians occupied December 2. An enemy push down the Western Morava would, however, be still more serious, for it would at a single blow cut the vital railway artery of Serbia and outflank the main Serbian armies in the north, besides depriving them of the great industrial center and chief Serbian arsenal at Kraguyevats.

Then came the stroke which startled the world. Launching an offensive from the solid buttress of Mount Rudnik and sweeping down the slopes of Malyen Ridge, the Serbian veterans overwhelmed two Austrian corps with disaster. Valyevo was recaptured, Uzhitse cleared of the enemy, and the beaten Austrians once more driven across the Drina in disorder. Belgrade was reoccupied by Serbian forces December 15, and the badly chastised Austrian armies, weakened by over 100,000 casualties, executed a humiliating retreat, while their commander was officially disgraced for the crime of failure. For the third time the natural defenses of northern Serbia, heroically defended, had defied the attempted Austrian invasion.

It was now evident that Austria, with many of her troops engaged on the Russian front, could not assemble forces competent to dislodge the Serbians from their favorable defensive positions. Comparative quiet ensued on this front for many months, a quiet which might indeed have been broken earlier had not Italy, then neutral, served notice on Austria that a disturbance of the *status quo* in the Balkans before suitable "compensations" had been given to Italy could not be tolerated. By early fall of 1915 conditions had changed. Italy's entrance into the war made it unnecessary for Austria longer to consider her feelings. The great Russian retreat made it possible to withdraw Austrian troops from that front for service against Serbia. At the same time the need of controlling the Morava-Vardar trench was increasing. The Central Powers were besieged by the Allies, and an outlet to neutral lands and to the sea was a pressing necessity. The Turks needed munitions and the Central Powers needed food. A successful campaign was also required to wipe out the disgrace of past defeats at Serbian hands and to impress wavering neutrals with Teutonic military prowess. Hence was initiated the diplomatic campaign already described, which culminated in the peaceable conquest of the Maritsa valley and the accession of the Bulgarian armies to the ranks of the Central Powers. Conditions were now ripe for a combined Teuton-Bulgar campaign designed to conquer the entire Morava-Vardar trench.

The Conquest of Serbia

Early October, 1915, found some 300,000 Germans and Austrians massed on the Save-Danube line, while a larger number of Bulgars were concentrating in the mountains along the eastern border. The main Serbian army stood behind the northern defensive line to meet the Austro-German attack, smaller forces alone being detailed for operations on the east. A Bulgarian offensive was to be met by the Greek army acting in concert with an Anglo-French expeditionary force. At the last moment the whole scheme of Serbian defense was shattered by the treachery of King Constantine, who not only repudiated Greece's treaty with Serbia and refused the promised support of his army but gave secret assurances to the Bulgarian government that if its armies attacked the Serbians their flank would be secure from any interference by the Greeks. The entire length of the Morava-Vardar trench was thus thrown open to attacks from the east while the main Serb armies were trying to protect the northern entrance.

On October 6 the Austro-German assault was launched. Heavy artillery fire, which the Serbians could not match, protected the columns attempting to force a passage across the river barrier. Nevertheless, the crossing was a costly undertaking. Many of the invaders were driven back to the northern bank or caught on the southern side and annihilated, before large forces after two or three days' hard fighting securely established themselves on the southern bank. It is interesting to note that the principal crossings were effected above Belgrade, below Belgrade, at Semendria, Ram, and Gradishte—all five points close to the northern entrance of the Morava valley, all of them except the last located at the ends of Hungarian railways capable of bringing supplies directly to the points of crossing, and all of them near sand-bar islands in the river which were utilized to good advantage in several and possibly in all cases. There also was heavy cannonading at Orsova, the only other railhead on the Danube frontier; but no crossing of the difficult gorge near the Iron Gate seems to have been made until later, possibly after threat of envelopment caused withdrawal of the main body of defenders from the northeastern corner of Serbia. When the crossing was effected here it was with the aid of an island in the river below Orsova.

After the Danube barrier had been forced, the southward progress of the Teutonic armies was remarkably slow. For six weeks the average rate of advance was about one mile a day. Despite their enormous superiority in big guns, it cost the Austro-Germans much time and the loss of many men to drive the Serbs from successive defensive positions in the hills. More than two weeks elapsed before the Danube was freed from the Serbian menace, and thus rendered available for boat transport of munitions to Bulgaria and Turkey. Austrian forces crossing the Drina near Vishegrad, the only rail end on the northwestern frontier, found themselves unable to dislodge the Serbs from their mountain fastnesses and after ten days' fighting had made no progress toward the head of the Western Morava valley.

Meanwhile Bulgaria declared war on Serbia, and Bulgarian armies poured through the gaps in the eastern mountain barrier and descended tributary valleys to the Morava-Vardar trench. One column advanced down the Vlasina valley to the Leskovats basin, another reached Kumanovo and Skoplye (Üsküb) by the Kriva depression, while a third descended the Bregalnitsa to Veles (p. 591). Vranye, Kumanovo, Skoplye, and Veles, defended by inadequate Serbian forces, were captured within less than two weeks, and the vital artery of Serbia was cut in four places. Few could doubt that these wounds would prove fatal.

Farther north one Bulgarian army was attacking the fortifications of Pirot in order to open a way down the Nishava valley to Nish, while other forces had captured Zayechar and were trying to advance along the upper Timok branches to reach the Morava trench above and below Nish. Progress in this field was much slower than farther south, however, and the Serbs maintained themselves in the mountainous northeastern corner of their country until the fall of Pirot and Nish developed the danger that Bulgarians pushing north down the Morava, and Austro-Germans advancing up the valley to meet them, might close the neck of the salient northeast of the trench and capture the forces fighting there. Under pressure of this threat the Serbs withdrew to the southwest; and about November 13, or more than a month after the campaign opened, the entire Morava-Maritsa trench was in the hands of the Central Powers, and the reconstruction of the Orient railway could be prosecuted. The Morava-Vardar trench as far south as Veles was also in their control, and there remained only the problem of rendering the tenure of both trenches secure by pushing the Serbian and Franco-British forces west to the Adriatic and south to the Aegean.

The disastrous results of the Bulgarian occupation of the Morava-Vardar trench now began to be more manifest. Munitions and other supplies for the Serbian armies in the north were becoming exhausted, and the one artery along which they could freely flow had been severed. The quantities which could reach the Serbian front over rough mountain trails were utterly inadequate. Reinforcements were sadly needed; but the one railway leading north from the Anglo-French base at Saloniki followed the Morava-Vardar trench and so was in the hands of the enemy, while the rough mule paths over the western mountain barrier could bring neither troops nor supplies from Italy. Had the broad belt of mountain and karst intervened between the Morava-Vardar trench and the Bulgarian frontier, and had the more open valleys of the east but led westward to the Adriatic, the history of the Balkan campaign might have been very different.

It was supposed that when the Austro-German forces reached the higher mountainous region bordering the Western Morava valley, and it became difficult, if not impossible, to bring up their heavy guns, the rate of advance would become even slower than before. The fact that the advance was actually accelerated has been interpreted to mean that the failure of Serbian supplies weakened the defense more than the unfavorable local topography injured the plans of the offensive. The Teutons moved rapidly across the Western Morava, and the Serbian army took up a position running eastward along the mountain crests south of the valley, then southward along the ridge west of the Morava-Vardar trench, and southwestward across the Kachanik gorge. It will immediately appear that the Kachanik position was the strategic key to this entire battle front. In the rear of the Serbian armies, which were now facing north and east, runs the straight subsidiary trench formed by the Lepenitsa valley, Kosovo Polye, and the Ibar valley. The gateway to this trench is the narrow Kachanik gorge. A railway from Skoplye runs through the gorge to Mitrovitsa at the northern end of the

Kosovo Polye, thereby more than doubling the military value of the depression. If the Bulgarian forces already in possession of Skoplye should succeed in breaking through the Kachanik gorge into the plain of Kosovo, they could strike north and east against the rear of the Serbian armies and convert retreat into disaster. Little wonder, then, that the "Kachanik Pass" figured so prominently in the war despatches during this period.

But if Kachanik was the key to the Serbian position, Veles was the key to Kachanik. Should the Anglo-French troops coming up the Vardar from Saloniki capture Veles and debouch into the triangular lowland to the north, they would take in the rear the Bulgarian army trying to break through the Kachanik position. It would not be necessary for the Anglo-French force to enter the Kachanik gorge; the mere threat of enclosing the Bulgarians in the valley between the Serbs up at Kachanik and their allies down at the valley mouth would be sufficient to bring the Bulgars out of the trap in order to fight on the lowland, where, if defeated, they could retire northeastward into a region fully under their control. The threat would become imminent the moment Veles fell to the Allies. Such were the topographic relations responsible for the rather striking fact that an Anglo-French attack upon Veles relieved the pressure upon Serbian forces in the mountains far to the north.

The strategic value of Veles was fully appreciated by the Bulgarian commanders, and heavy reinforcements were evidently poured into the Vardar trench at that point. All efforts of the Allied armies failed to achieve their purpose; Veles remained in Bulgarian hands, and Bulgarian attacks on the poorly equipped Serbs defending Kachanik gorge proceeded without serious interruption. When it became apparent that the Kachanik position could not long be held, the Serbian armies at the north and east fell back toward the Ipek basin, while those farther south retired on the Monastir basin. All danger to the Teutonic occupation of the Morava-Vardar trench north of Veles was thus removed, and the remainder of the campaign consisted in squeezing the remnants of the shattered Serb forces and their Montenegrin allies westward through Albania and southward through Montenegro to the sea; and in driving the Anglo-French army and the Serbs near Monastir back into Greece. The first of these movements progressed with exceeding slowness because of the difficult character of the country; and the terrors of the Serbian retreat over rugged mule paths and through wild mountain gorges in the cold and snow of winter can scarcely be imagined. But from the standpoint of military geography the second movement alone merits special consideration.

The Serbian forces retreating southward toward Monastir made a heroic stand on the mountain range separating the Vardar valley from the Monastir basin. At Babuna Pass, a sag in the crest crossed by a wagon road, the remnants of King Peter's armies held overwhelmingly superior forces of the enemy at bay for more than a week. But the unhappy Serbs, betrayed by Constantine and denied effective aid from other quarters because of the blundering and bungling of their powerful Allies, were doomed to expulsion from even this corner of their country. The continuous pressure of enemy attacks became unsupportable, the pass was abandoned, and early in December the Bulgarians were pouring southward over the plain of the broad basin to capture Monastir.

When the French and British pushed up the Vardar valley toward Veles they seized as their base for a great armed camp the triangle of mountainous ground (Fig. 165) lying between the Vardar River and one of its tributaries known as the Cherna. The position had certain topographic advantages which enabled it to be held for a long time in the face of superior forces; but suffered from one serious disadvantage which ultimately compelled its evacuation. Both the mountain ridges and the river trenches afforded admirable natural defenses. The gorge of the Cherna is steep-sided and the stream unfordable. The most important bridge, a few miles above the river's mouth, was destroyed by the French after they had failed in an effort to move westward and join the Serbs, who were then fighting at Babuna Pass to prevent the Bulgars from getting into the Monastir basin. For defensive purposes the larger Vardar River, protecting the eastern side of the triangle, possessed much tactical value, because it is both wide and unfordable and its valley is steep-sided—in one place a veritable gorge.

But it is in the Vardar valley, however, that the chief disadvantages of the situation become apparent. All munitions and other supplies, as well as all reinforcements for the armed camp had to come from Saloniki over the single-track railway running up the Vardar trench. The railway lies close to the river all the way and for several miles is actually on its eastern bank, or outside the triangle. Its position was thus dangerously vulnerable, and its vulnerability was peculiarly aggravated by the fact that in the Demir Kapu gorge, the Iron Gate of the Vardar, the line is squeezed in between the base of high cliffs and the swiftly flowing river, crosses the river on a bridge at one point, and passes through a tunnel at another. If the Bulgarians, attacking the sides of the triangle, should destroy the bridge, tunnel, or narrow roadbed in the gorge, the forces within the triangle would be caught in a trap. Hence it was that when the dispersal of the Serb armies to the northwest had so far progressed as to free additional Austro-German and Bulgarian troops for action against the Allied armies at the south, the evacuation of the triangle was considered imperative.

It has been estimated that at this time the forces of the Central Powers in the south probably outnumbered those of the Allies in the proportion of three to one, or even four to one. That the triangle should have been held so long in the face of greatly superior numbers bears eloquent testimony to the strength of the natural topographic barriers formed by the Cherna and Vardar Rivers, as well as to the efficiency of the troops which were responsible for its defense. The triangle was a prominent salient projecting far beyond the general Allied front; it possessed a vulnerable point, the Demir Kapu gorge, on the eastern side of the salient; and it was the center of a line the two wings of which were less effectively protected by natural barriers and all of which was inadequately supplied with lines of communication.

Early in December the withdrawal from this dangerous situation began. The French and British retired from the triangle and blocked the gorge against pursuit by blowing up the tunnel and bridge. The British farther east were forced back toward the south by a series of furious Bulgarian assaults, and the Serbs on the west were likewise compelled to withdraw southward into Greek territory. During this same month General de Castelnau, whose skillful use of topographic barriers in other fields has already been detailed, was sent to inspect the Saloniki region. Presumably he reported favorably on the obvious strength of the military obstacles which we have described somewhat fully in the preceding chapter; but whether as a result of his inspection, or independently of it, the remarkable natural defenses of the parallel ranges and basins were gradually organized into an impregnable entrenched camp.

The Macedonian Campaign of 1916

A series of local combats, rising in cases to the dimensions of battles of some importance, marked the period of readjustment that followed the conquest of Serbia and preceded the almost complete calm which reigned on this front from the late fall of 1916 till the middle of September, 1918. Late in July, 1916, the Serbian army, reconstituted in exile on the island of Corfu and transported to the Macedonian front some weeks before, launched a heavy attack on the strong Nidje mountain mass, constituting the base of the famous triangle between the Vardar and Cherna Rivers (Fig. 165). But this offensive did not seriously shake the Bulgarian front, solidly based on an extremely difficult terrain. The French captured some ground in the western end of the great Doiran-Struma moat (Fig. 159), but stopped short before the formidable wall of the Belashitsa Range. A serious menace developed when the Bulgarian army, after having had the fortress guarding Rupel Pass treasonably delivered into their hands by order of the faithless Constantine, prepared to attack both wings of the Allies in an attempt to envelop their whole front. Pushing southward over the open floor of the broad Monastir basin (Fig. 158) the enemy forced the Allied left wing back on the mountain range separating this lowland from the Ostrovo basin next to the east, through which latter run the main road and railway from Florina to Saloniki. On this natural barrier, which is continued northeastward by the high Kaimakchalan massif and the Nidje Mountains (Fig. 165), the turning movement on the west was halted. On the Allied right the Bulgars occupied the Kavala region, delivered up by Greek troops which went over to Germany, and advanced against the Lake Ahinos sector of the Struma trough, where they were finally checked by British forces solidly established on the outer wall of the entrenched camp of Saloniki. Supported by strong natural positions, General Sarrail launched a series of counter-blows which effectively blocked the Teuton-Bulgar design at envelopment, although gaining little ground for the Allies.

On September 12, however, the Allies began a vigorous assault designed to restore the situation, particularly on their left flank, by recovering the important Monastir basin. The whole front moved forward to the attack, from the Ostrovo basin to the Gulf of Orfano; but the chief advance soon developed on the left. Here Serbian troops pushing northwestward from the Ostrovo region drove the enemy across minor ridges and valleys to the main ridge bordering on the southeast the great elbow of the Cherna, and then attacked the high Kaimakchalan massif buttressing this ridge and forming a solid support for the Bulgar line for some miles beyond. Meanwhile French and Russian forces on the left of the Serbs pushed across the southern end of the Monastir basin, to recapture Florina in the mouth of a canyon debouching from the steep and straight western mountain wall.

After a terrific struggle lasting many days, the Bulgarians were hurled from their mountain stronghold of Kaimakchalan (Fig. 165) on September 30 by the unconquerable Serbs. There now began a slow, bitter, and bloody advance northward through the mountain ranges bordering the Monastir basin on either side (Figs. 158 and 164), the French attacking on the west, the Serbs on the east. To advance northward through the plain was practically impossible, for while the streams crossing it flowed *on* the surface and often contained little or no water, thus offering such poor defensive positions that the Bulgars made little use of them, the exceptional flatness and barrenness of the plain deprived the attackers of any cover, and the enemy on the heights could from their dominating positions take the attacking columns in flank and rear. But if progress could be made on the mountains, then the lines in the plain, taken in the rear by fire accurately directed from the heights, must be abandoned. Step by step the immensely difficult terrain was conquered, and stage by stage the enemy in the plain fell back toward the north. Yet progress was slow, and it was not until November 19 that the Allies re-entered Monastir. By this time the power of the offensive had been so worn down by the almost superhuman task that the strong enemy lines on the hills immediately northwest of the town could not be broken. The season was, moreover, far advanced, and the campaign of 1916 drew to its close. Nearly two years later the visitor to Monastir could still stand in one of the towers rising well above the houses and watch the Allies' shells bursting along the Bulgar lines just above him, or run the gauntlet of Bulgar shells dropped in the town and on the roads leading southward over the basin plain.

Local Combats

In the first year of the war Italy had landed an expeditionary force at the port of Valona in southern Albania and had gradually converted the surrounding country into an entrenched camp of considerable strength. The topography of the region lent itself admirably to this purpose. Protected on the south and west by the high limestone ridge of Karaburun, terminating in Saseno Island (Fig. 151), from which big guns can sweep the sea approaches and guard the entrance, and on the north by small islands tied to the mainland by low beaches, the harbor itself could be made reasonably secure from attack by water. By land the only approach from the south is from the Chimara basin over a steep limestone mountain wall (Fig. 153) which was passable by difficult mule path alone until the Italian army engineers built a magnificent auto road, zigzagging dizzily up the 2,000-foot scarp. Here the terrain offered excellent opportunity to render the pass impregnable. On the east a mountain wall (Fig. 161), then the deep trench of the Sushitsa branch of the Voyussa, another mountain range, and finally the greater trench of the Voyussa itself provide a succession of excellent natural defenses. An attack from the north would first have to pass the trench and marshy plain of the lower Semeni River, overcome the Malakastra ridge (Fig. 162), then cross the broad trench of the lower Voyussa valley, after which the north-south Sushitsa trench and its bordering ranges must be taken in the flank and conquered.

After the conquest of Serbia and the rapid overwhelming of the little Montenegrin army, Austrian forces pushed southward through Albania, while the Italians evacuated Durazzo and withdrew within the strong entrenched camp of Valona. After possessing themselves of the strong Malakastra position the Austrians appear to have settled down to contain the Italians within their natural fortress, without seriously attempting the difficult task of dislodging them from the strong terrain south of the Voyussa. The southern side of the river in this part of its course is the higher (Fig. 163) and dominates the northern bank, where the Italians established bridgehead defenses. From these, machine guns could sweep the valley across which any advance from the Malakastra positions must come. Inasmuch as the surface from the river to the foot of the ridge is as flat as a floor, very broad, and for the most part devoid of any covering larger than grass, a grazing fire could sweep attacking columns with murderous effect for a time sufficiently long to assure their annihilation. Both the Malakastra ridge and the hills south of the Voyussa supporting the Italian line are composed of a partially hardened argillaceous sand which cuts like cheese, stands fairly well, and is admirably adapted to the excavation of trenches and dugouts.

For more than two years the Italians held the Voyussa line without suffering serious interference from the Austrians. Then by a brisk action early in July, 1918, General Ferrero captured the Malakastra ridge and drove the enemy well back over the level plain of the Semeni farther north. Unfortunately for the Austrians the ridge does not reach westward to the sea. In winter the passage around the western end is blocked by marshes, but in summer these become sufficiently dry and hard to permit the passage of cavalry. Taking advantage of this fact the Italians launched a frontal attack on the main ridge positions, while

FIG. 161—Italian military road leading up to the pass giving access into the Chimara basin (see Fig. 153). The mountains shown here form the natural defenses of the Valona region on the east and south.

cavalry swept around through the plain, passed west of the ridge, and captured part of the Semeni plain on the north. Finding themselves attacked in the rear, the Austrians abandoned the whole ridge and fled precipitately northward. So sudden was their flight that six Austrian airplanes, returning from a scouting expedition, landed on their home field to find themselves in the Italian lines. The Italian front was now reinforced with another strong position, but in a later attack the Austrians recovered part of the important ground they had lost in the Battle of Malakastra Ridge.

FIG. 162—The exposed floor of the broad valley traversed by the Voyussa River, long the main Italian defensive position, and the Malakastra ridge beyond, on which the Austrian front was based.

Among the local struggles on the Balkan front in 1917 and 1918 should be noted a series of combats on the important Gyevgyeli plateau sector, between Lake Doiran and the Vardar (Fig. 159). Here the Bulgars held most of the high ridge west of the lake, and so commanded the British positions on the plateau farther west. In April and May the British troops made determined efforts to drive the Bulgars from the dominating height. Some ground was gained, but from the ravines descending the flanks of the ridge the enemy met every advance with a murderous fire which brought the operation to a halt. In the following spring, and again during the great Battle of Moglenitsa in September, 1918, the British assailed the Bulgarians' natural stronghold with little success, until the final collapse of the Bulgarian center opened the way to a general advance northward.

Other local combats of greater or lesser importance interrupted only to a limited degree the general calm which characterized the Balkan theater of war for many long months before the final Allied offensive in September, 1918. The battle front had become stabilized on a line stretching across one of the narrowest parts of the Balkan Peninsula. From the Adriatic Sea it followed the strong Voyussa valley-Malakastra ridge positions, then looped transversely across the wild, rugged limestone ranges and deep gorges of south-central Albania to reach the lake-filled Okhrida and Presba basins. From the northern part of Presba Lake the front crossed the high range west of the Monastir basin, traversed the level floor of that lowland just north of the town of Monastir, and scaled the high mountain wall lying next east, in the great bend of the Cherna River (Fig. 164). East of the Cherna, at the base of the mountainous triangle between it and the Vardar, the line crossed the Nidje massif (Fig. 165), where the Bulgarian center was based on the formidable wooded heights of Mount Sokol and Mount Dobropolye, after Kaimakchalan (previously captured by the Serbs) the strongest points on this part of the enemy front.

From the Cherna-Vardar triangle the line crossed the Gipsy Pass gorge to the Gyevgyeli plateau and followed the Doiran-

Struma basin to the sea. Throughout this latter sector the opposing armies faced each other across the vast expanse of the broad moat, each combatant defied by the impregnable natural barriers which faced him. On the side of the Allies the southern bank of the Struma River, the base of the Krusha-Beshik scarp, the steep slope itself, and the crest were strongly organized with trenches, redoubts, and other field works defended by wire entanglements which for intricacy of pattern and skillful adaptation to the terrain (they were so disposed as to confuse the assailants and shepherd them unwittingly into death traps swept by machine guns), as well as for height, in places exceeded anything observed by the writer on other fronts. The absence of concealed gun positions on the plain, by keeping the artillery at a great distance, and the scarcity of big guns on both sides made it impossible for either combatant to concentrate a heavy and accurate fire on the trenches of the other. As a consequence the main defensive positions were usually on the slopes toward the enemy, not on the counter slopes as was so often the case in the quite different terrain of the French front; although from the Lahana saddle westward to Lake Ardjan the British prepared an important secondary defense line on the backslope of the Krusha Balkan. On the steep fracture face of the Belashitsa Range, north of the broad Struma moat, the first main position of the Bulgar front lay along the base of the scarp, another, easily distinguishable with field glasses, some distance up the slope, and a third, apparently less continuous, along the crest. This remarkably steep scarp, which stared the British in the face for so many dreary months, was commonly described by the British soldier, with reprehensible irreverence but commendable accuracy, as "a hell of a position." From the Adriatic to the Aegean the front was nowhere else so nearly impregnable on both sides as along the Struma moat.

The Battle of Moglenitsa

The battle front across the comparatively narrow portion of the Balkan Peninsula described in the preceding paragraphs was a key position of the utmost importance. So long as the Allies were weaker than their opponents, there was always the danger that a Teuton-Bulgar attack might break the Allied center between the entrenched camps of Saloniki and Valona, overrun Old Greece, precipitate her withdrawal from her half-hearted alliance with the Entente Powers, and open the Greek coast to German

FIG. 163—The Voyussa River north of Valona, showing the steeper southern bank, along which the Italians long had their front, and the more gently sloping northern bank, where they held bridgehead positions. Part of the inner defenses protecting the temporary bridge are shown in the right foreground.

submarines—a possible catastrophe of the first magnitude to the Allied cause. On the other hand, when the Allies waxed stronger the Teuton-Bulgar line was the only dam which prevented the Allied flood from pouring northward along the Morava-Vardar corridor and subsidiary basin routes to debouch into the Hungarian plain behind the Austrian armies facing the Italians along the Piave-Trentino front. If this dam broke, irreparable disaster must follow. As we have seen on an earlier page, the geographic form of the peninsula, broad at the north and narrowing toward the south, would impose on the Austrians

FIG. 164—Mountainous terrain in the great bend of the Cherna, bordering the Monastir basin on the east. (Italian official photograph.)

the tremendous task of building a new dam in the north several times longer than the first, in case the southern one were broken. At this stage of the struggle such a task was utterly beyond the power of Austria. Practically the whole strength of her rapidly decaying army had to be concentrated on the Italian front, and only a few battalions were left to aid the Bulgarians by holding a short sector in Albania. Germany, suffering one disaster after another on the western front, could do little toward building up a new line in the southeast; like Austria she had withdrawn her divisions from the Balkans and was herself asking Austria for help in the west. Such troops as she might spare in a last desperate effort to hold Bulgaria in line would merely advance the date of German defeat on the western front. Bulgaria faced the Allied Armies of the Orient practically alone; and it was a geographic certainty that if the Bulgarians could be eliminated from the struggle, the form of the country would make the establishment of a new southern front impossible. Austrian collapse must follow, swift and sure. And with Austria-Hungary thrown open to the Allies, the German armies would be menaced in the rear. The Bulgarian battle line was indeed a wall beyond which lay Allied victory. Soon after Hindenburg should hear of its rupture, he would write Chancellor Max that the last hope of imposing peace on the Allies had vanished.

That critical wall was becoming alarmingly weak. "Bulgaria was rotting from within. The common people had had enough of the war; they were hungry, weary, and restless. King Ferdinand was growing unpopular. German influence was decreasing in proportion as the divisions lent for the victories of 1915 and 1916 decreased. And the army itself, worn out by war, by insufficient food, and by long inaction, would probably be unable to resist an unexpected and sweeping attack. Perhaps the new government of Premier Malinoff, which in June replaced the pro-German ministry of Radoslavoff at Sofia, was quite willing, before intrusting itself to the good-will of the Entente, that such an attack should come. A defeat would justify a separate and much desired peace."[1]

[1] C. J. H. Hayes: A Brief History of the Great War, New York, 1920, p. 343.

At the same time the Allied flood which was soon to beat against that wall was waxing ever more formidable. The weak expeditionary force of heterogeneous units had grown into a large army trained by long association to operate more nearly as a single machine. The Grecian menace which hampered Sarrail had been greatly reduced, and that general had been succeeded by the resourceful Guillaumat, who proposed and planned a vast operation against the Bulgar front. Guillaumat in turn was succeeded by the energetic Franchet d'Esperey, who executed the great offensive.

It was decided to rupture the Bulgarian front by an attack against its center. This would permit the attacking forces, upon bursting through, to cut the chief communications of the Bulgarian right wing in the Monastir region, leading northward out of the basin, as well as the vital railway artery through the Morava-Vardar corridor, supplying that part of the Bulgarian left wing in the Vardar sector. Completely isolated from each other and with the enemy on their supply lines far in the rear, both wings would quickly share the disaster suffered by the center.

The Bulgarian center was, as previously noted, solidly established, along the mountainous base of the Cherna-Vardar triangle, on the crest and southern slopes of the high Nidje Mountains, which dominate the Moglenitsa basin from the northwest (Fig. 165). This crest, followed by the Serbo-Greek frontier, trends from southwest to northeast, from the Kaimakchalan massif on the southwest past the peaks of Sokol, Dobropolye, Kravitsa, Vetrenik, Koziak, and Blatets, to the Kechikaya massif on the northeast. Despite its oblique trend, the crest gives off a number of ridge spurs which extend approximately east or west from several of the peaks mentioned and which form excellent successive lines of defense against an attack from the south. Thus it happened that after the capture of the Kaimakchalan massif by the Serbs, the Bulgarian first line passed through Mounts Sokol and Dobropolye, while their second line reached the crest at Mount Koziak, which formed its principal buttress. A third line was based in part on east-west ridges north of Mount Koziak.

Another interesting feature of the terrain consists of several important ridges which extend northward from the oblique crest, to reach the river trenches bounding the famous triangle. One of these reaches north from Mount Koziak direct to the Cherna gorge near the middle of the triangle's western side. A

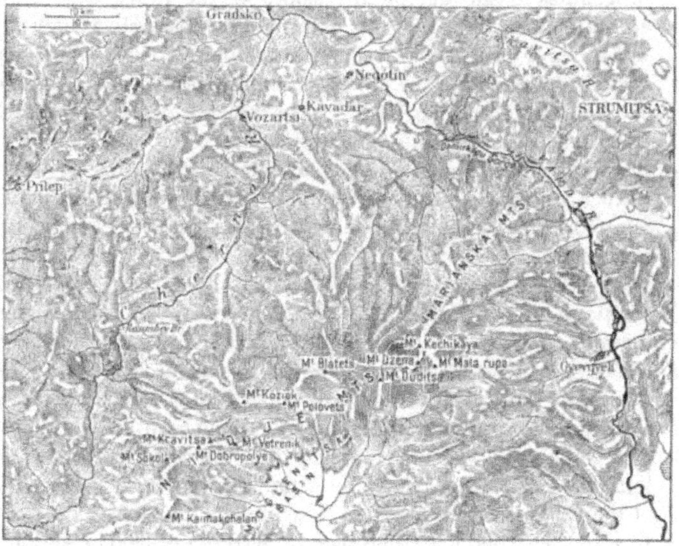

FIG. 165—Map of the Moglenitsa basin and the mountainous triangle between the Cherna and Vardar Rivers, along the Serbo-Greek frontier.

second, known as the Marianska range, runs northeast from the Kechikaya massif to overlook the Demir Kapu gorge of the Vardar, about the middle of the triangle's eastern side. Between these two another ridge leads from near Mount Blatets straight to the apex of the triangle, at the river junction. If the Bulgarian front on the main oblique crest was broken, these ridges would form highland pathways along which the victors might advance to the bounding river gorges, at the same time clearing the triangle of enemy troops by fire directed into the valleys and upon the lower ridges from advantageous positions on the heights.

The initial attack would be the most difficult. From the lofty Sokol and the rocky peak of the Dobropolye the Bulgarians dominated the whole surrounding country. Roads in the exposed Moglenitsa basin and on the slopes of the range were under accurate observation and fire, and could be used only at night. It was a heavy task, given the poor communications of the region, secretly to assemble the stores of munitions and other supplies necessary for a crushing blow. Auto-camions, ox carts, and mule trains carried on the laborious work week after week, toiling up the steep slopes along zigzag roads and bridle paths by night, resting concealed by day. When all else was prepared the troops selected for the assault were moved up, and the artillery preparation began.

On September 15, 1918, the blow fell. Against the formidable positions of the Sokol and Dobropolye Serbian and French infantry launched one assault after another in quick succession, while the British delivered a series of violent attacks on the Gyevgyeli plateau front just across the Vardar, which prevented Bulgar aid from that sector being sent to the hard-pressed center. The Dobropolye was first submerged by the rising tide of attack, later the Sokol fell. In two days practically the whole first line of the Bulgarian center was in Allied hands, but the Koziak buttress of the second line still resisted. Serb troops held in readiness to pour through the opening and overrun the triangle were still unable to begin their task. The next day Mount Koziak was conquered, the third line pierced, and Serb forces began the advance northward along the highland pathway leading to the Cherna gorge. When this barrier was reached the rapid advance was checked for a time, the Bulgarians having destroyed the bridges and *passerelles* as they retired.

Several days of fighting were required to break the enemy front on the northeastern end of the crest, from Mount Blatets to the Kechikaya massif. When the Allies finally broke through, the advance along the Marianska ridge to the Demir Kapu gorge was rapid, and the Bulgarians in the Vardar region found their railway communication cut at a critical point in their rear. Meanwhile the northward advance along the other ridges had proceeded apace, and Allied forces were astride the vital road leading from Prilep in the Monastir basin to Gradsko in the Morava-Vardar corridor. The Cherna-Vardar triangle was cleared of the enemy, the Bulgarian army cut in two, and the severed wings denied the means of orderly retreat.

Then began the flow of the Allied flood northward through the Morava-Vardar corridor, as the breach in the Bulgar wall was opened wider and wider. Nish was reached October 12 and the Orient railway cut, severing Turkey from her German overlord and condemning her to an ignominious separate peace. A few days later Allied troops stood on the Danube, and Austria-Hungary tottered at the threat of a new invasion it could not ward off. It was then that the Italians struck their blow (p. 536), and the Dual Monarchy collapsed like a house of cards.

Meanwhile the Bulgarian armies had been definitely eliminated from the war. Their defense of the strong mountain front from the Sokol northeastward, while sufficiently vigorous to cause the Allies some difficulty, was not such as to indicate that the advantages of the position were properly utilized. It has been intimated, perhaps with good reason, that the Bulgar command, trusting to the natural strength of the Sokol-Dobropolye massif, had inadequately garrisoned it. The speed with which the difficult mountains of Macedonia were overrun by the victors, and the small number of Bulgarian casualties other than prisoners, gave rise to the suspicion that the débâcle was not wholly involuntary on the part of the Bulgarians. Ludendorff supports this view at some length and believes "no other explanation exists for the rapid advance of the Entente troops over broken country, eminently suited for defense,"[2] while Colonel Bujac of the Serbian army indignantly rejects such an interpretation in his

[2] Erich von Ludendorff: Ludendorff's Own Story, August, 1914–November, 1918: The Great War from the Siege of Liège to the Signing of the Armistice As Viewed from the Grand Headquarters of the German Army, 2 vols., New York, 1919; reference in Vol. 2, pp. 365–369.

interesting account of the Battle of Moglenitsa.[3] However this may be, the Bulgarian troops certainly made comparatively poor use of the topographic advantages for defense which the mountain barriers of the Balkans offered them, once their retreat was well begun. At the Cherna, along the Vardar and the parallel trench of the Lakavitsa, as well as on the ridge between these two, stiff engagements were fought; and before the formidable mountain front along the Doiran-Struma moat the British and Greeks encountered "peculiarly stubborn resistance." But even here the Struma valley gateway to Sofia was opened by September 27. Three days later the Bulgarian forces laid down their arms. In the victory of Moglenitsa Bulgaria's sudden and disastrous attack on the exposed Serb flank in 1915 had been amply avenged.

[3] Colonel Bujac: L'offensive de septembre 1918 sur le front de Macédoine, Paris, 1919, p. 70.

INDEX

INDEX

Made in United States
North Haven, CT
04 November 2026

21766795R00368